Optimization in Computational Chemistry and Molecular Biology

Nonconvex Optimization and Its Applications

Volume 40

The titles published in this series are listed at the end of this volume.

Optimization in Computational Chemistry and Molecular Biology

Local and Global Approaches

edited by

C.A. Floudas
Princeton University

and

P.M. Pardalos
University of Florida

KLUWER ACADEMIC PUBLISHERS

DORDRECHT / BOSTON / LONDON

A C.I.P. Catalogue record for this book is available from the Library of Congress.

Chemistry Library

ISBN 0-7923-6155-5

Published by Kluwer Academic Publishers,
P.O. Box 17, 3300 AA Dordrecht, The Netherlands.

Sold and distributed in North, Central and South America
by Kluwer Academic Publishers,
101 Philip Drive, Norwell, MA 02061, U.S.A.

In all other countries, sold and distributed
by Kluwer Academic Publishers,
P.O. Box 322, 3300 AH Dordrecht, The Netherlands.

Printed on acid-free paper

Printed in the Netherlands.

Contents

Preface

A tantalizing problem that cuts across the fields of computational chemistry, biology, medicine, engineering and applied mathematics is how proteins fold. Global and local optimization provide a systematic framework of conformational searches for the prediction of three dimensional protein structures that represent the global minimum free energy, as well as low energy biomolecular conformations.

This book contains refereed invited papers submitted at the conference on *Optimization in Computational Chemistry and Molecular Biology : Local and Global Approaches* held at Princeton University, May 7-9, 1999. The conference brought together the most active researchers in computational chemistry, molecular biology, local and global optimization and allowed for the exchange of ideas across discipline boundaries of applied mathematics, computer science, engineering, computational chemistry and biology. The conference themes included advances in local and global optimization approaches for molecular dynamics and modeling, distance geometry, protein folding, molecular structure refinement, protein and drug design, and molecular and peptide docking.

We feel that this book will be a valuable scientific source of information to faculty, students, and researchers in optimization, computational chemistry and biology, engineering, computer science and applied mathematics.

We would like to take the opportunity to thank the authors of the papers, the anonymous referees, and the Department of Chemical Engineering and the School of Engineering and Applied Sciences of Princeton University for supporting this effort. Special thanks and appreciation go to John L. Klepeis for assisting us in the the preparation of the camera ready Latex form of this book. Finally, we would like to thank the Kluwer Academic Publishers for their assistance.

Christodoulos A. Floudas Panos M. Pardalos
Princeton University University of Florida

September 1999.

Optimization in Computational Chemistry and Molecular Biology, pp. 1-18
C. A. Floudas and P. M. Pardalos, Editors

Predicting Protein Tertiary Structure using a Global Optimization Algorithm with Smoothing

Aqil Azmi
Richard H. Byrd
Elizabeth Eskow
Robert B. Schnabel
Computer Science Department
University of Colorado at Boulder
Campus Box 430, Boulder, CO 80309, USA

Silvia Crivelli
Physical Biosciences and Life Sciences Divisions and
National Energy Research Supercomputer Center
Lawrence Berkeley National Laboratory
Berkeley, CA 94720, USA

Thomas M. Philip
Teresa Head-Gordon
Physical Biosciences and Life Sciences Divisions
Lawrence Berkeley National Laboratory
Berkeley, CA 94720, USA

Abstract

We present a global optimization algorithm and demonstrate its effectiveness in solving the protein structure prediction problem for a 70 amino-acid helical protein, the A-chain of uteroglobin. This is a larger protein than solved previously by our global optimization method or most other optimization-based protein structure prediction methods. Our approach combines techniques that "smooth" the potential energy surface being minimized with methods that do a global search in selected subspaces of the problem in addition to locally minimizing in the full parameter space. Neural network predictions of secondary structure are used in the formation of initial structures.

Keywords: Protein Structure prediction, global optimization, smoothing.

1   Introduction

The problem of predicting the three-dimensional structure of a protein given its primary sequence of amino acids continues to challenge the biochemistry community. Assuming the native structure exists at the global minimum of the free energy surface, the solution can be found if one is given a potential energy function that is able to model the free energy, and a global optimization algorithm capable of finding the global minimum of this function. This work attempts to solve the latter problem, while also using a perturbation of the potential energy function known as "smoothing". The problem of finding this global minimum among an enormous number of local minima within a very large parameter space is a daunting challenge.

The global optimization method presented is related to the work of [18], which incorporates sampling and local minimizations in an iterative fashion, providing some theoretical guarantees of success. In practice, the amount of work required to reach the theoretical guarantee is prohibitive, but our method combines sampling and local optimization in small-dimensional, appropriately chosen subspaces, with additional techniques in the full-dimensional problem space to provide a reasonable alternative. The method uses a two phased approach which creates initial minimizers in Phase 1 and improves upon them in Phase 2. It has proven to be successful in predicting structures of homopolymers [3], small proteins, and molecular clusters [4]. This paper will discuss recent advances in predicting the structure of the 70 amino acid A-chain of uteroglobin.

The idea behind smoothing the potential energy function is to soften the function by reducing abrupt function value changes while retaining the large-scale structure of the original function. As a result of dampening high gradient values and fine grain fluctuations in the original function, nearby minimizers merge as smoothing removes the barriers between them, and the total number of local minima is reduced [20]. We have observed that the smoothed function also produces greater variation in the structure and energy values of resulting minima when used within the global optimization algorithm.

In the first phase of the global optimization, initial structures are created and then passed to a second phase for improvement. We make use of neural network predictions of secondary structures such as alpha-helices and beta-sheets [10] to create the initial structures. Initial structures with the predicted secondary structure already formed are better candidates for improvement in the second phase of our algorithm than those without this secondary structure.

The next two sections discuss the energy functions used in this research, and the smoothed variant of the energy function. Section 4 describes the global optimization algorithm, and computational results are presented in section 5. The final section includes a summary, and directions for future research.

2   The Energy Function

The potential energy function used in this research is the AMBER molecular mechanical force field [8], and is defined as

$E_{\text{AMBER}} =$

$$\sum_{\text{bonds}} K_r(r - r_{eq})^2 + \sum_{\text{angles}} K_\theta(\theta - \theta_{eq})^2 + \sum_{\text{dihedrals}} \frac{V_n}{2}[1 + \cos(n\phi - \gamma)] \tag{1}$$

$$+ \qquad \sum_{i<j}\left(\varepsilon_{ij}\left[\left(\frac{\sigma_{ij}}{r_{ij}}\right)^{12} - 2\left(\frac{\sigma_{ij}}{r_{ij}}\right)^6\right] + C\frac{q_i q_j}{r_{ij}}\right), \tag{2}$$

where (1) represents the bonded interactions and (2) represents the pairwise nonbonded interactions.

In order to account for hydrophobic effects, an empirical solvation term has been added to E_{AMBER} during certain portions of the global optimization algorithm, which will be referred to as E_{SOLV}. This potential was formulated from simulations of methane molecule pairs in water [9], using the semiempirical Pratt-Chandler theories [16, 17]. The E_{SOLV} potential consists of a small number of terms of the form

$$\sum_{i,j \in \mathcal{A}} K_s \exp\left(\frac{-(r_{ij} - \rho)^2}{v}\right),$$

where the sum is taken over all aliphatic carbon pairs of the protein. The advantages of using this potential are that it provides a well-defined model of the hydrophobic effect of small hydrophobic groups in water, and is computationally tractable and differentiable. Future work will fine-tune this potential for better performance in determining the relative energies of a variety of protein folds.

We use predictions of secondary structure based on using a neural network trained on a large data bank of known proteins [12, 11, 22]. Given the primary sequence, this system predicts, on a per amino-acid basis, whether the secondary structure of each residue should be alpha helix, beta sheet, or coil, and provides an additional indicator of the strength of each prediction. These predictions are utilized within two biasing functions, which are added to E_{AMBER} during the local minimizations in phase one for creation of initial structures to be used in the global optimization algorithm. The technique of applying predicted structural information during energy minimization is known as the "antlion method" of Head-Gordon and Stillinger [10], whose purpose is to deform the hypersurface of the objective function such that the basin surrounding the structures with appropriately predicted secondary structure is widened and dominates. The first biasing function is

$$E_{\phi\psi} = \sum_{\text{dihedrals}} k_\phi[1 - \cos(\phi - \phi_0)] + k_\psi[1 - \cos(\psi - \psi_0)] \tag{3}$$

which biases the backbone torsional angles of a residue, where ϕ_0 and ψ_0 are the dihedral angles of a perfect α-helix or β-sheet, and k_ϕ and k_ψ are force constants related to the strength of the prediction from the neural network. The values of ϕ_0 and ψ_0 used for α-helices are $-62°$ and $-41°$ respectively. The second function encourages the predicted helical hydrogen bonds to form between the oxygen of residue i and the hydrogen of residue $i + 4$, for residues i and $i + 4$ which are predicted to be helical, and has the form

$$E_{HB} = -w_{i,i+4}/r_{i,i+4}. \tag{4}$$

In this function, $w_{i,i+4}$ is the weight output by the neural network, and provides a strong incentive for an intramolecular hydrogen bond to form when residue i is strongly predicted to be helical. Future work will include the development of a function similar to (4) for β-sheets, possibly using a matching algorithm to determine potential hydrogen bond participants. These neural net predictions are also used to determine how to focus the computational effort on different parts of the protein in Phase I of our algorithm.

An additional perturbation to the energy function, which has proven to be a tremendous aid in solving global optimization problems involving protein structure, is our analytical "smoothing" function which is described in the next section.

3 Smoothing

The idea of smoothing is to reduce abrupt function value changes without destroying the basic structure of the original function. The smoothed function value at a point is commonly formulated by taking a weighted average of the energy function in a neighborhood of the point using a distribution function centered at the point. Smoothing reduces the total number of local minima of a problem by reducing and ultimately removing the barriers between nearby minimizers, thus causing them to merge. This technique, called spatial averaging, has been studied in various ways [5] [6] [7] [13] [19] [21]. Using a Gaussian distribution function, the smoothing transformation is

$$\tilde{f}_{s,\lambda}(x) = \int H(f(\acute{x}),s) \cdot e^{-\|x-\acute{x}\|^2/\lambda^2} d\acute{x} \tag{5}$$

where λ and s are the smoothing parameters. The parameter λ determines the scale of the Gaussian distribution, while the parameter s is used with the function H to transform the original function $f(x)$ into a function with no poles. The transformation $H(f,s)$ is necessary to make the function integrable, and also further dampens the function. In the work of [13] this transformation consists of approximating $f(x)$ by a sum of Gaussian functions, while in the work of [21] the transformation consists of truncating $f(x)$ at some fixed maximum value.

We utilize a new family of smoothing functions, which is not integration based, but instead uses an algebraic method of smoothing applied to the nonbonded portion (2) of the potential energy function. The bonded interactions (1) are not considered for smoothing since the relative contribution to potential energy is small and since they are relatively nicely behaved from an optimization viewpoint. The new smoothing functions have the form

$$\tilde{E}_{\text{SMOOTH}<\gamma,P>} = (1) + \sum_{i \neq j}^{n} e_{ij} \left[\left(\frac{1+\gamma}{r_{ij}^2 + \gamma} \right)^P - 2 \left(\frac{1+\gamma}{r_{ij}^2 + \gamma} \right)^{P/2} \right] + C \frac{q_{ij}}{\sigma_{ij}} \sqrt{\frac{1+\gamma}{r_{ij}^2 + \gamma}} \tag{6}$$

where i and j are any two atoms and $r_{ij} = d_{ij}/\sigma_{ij}$, d_{ij} being the Euclidean distance, and γ and P are two smoothing parameters. The smoothing parameter $\gamma > 0$ is used to remove the pole at $d = 0$; as γ increases the value of the smoothed potential at $d = 0$ decreases. The other smoothing parameter P is used to widen the valley of the minimizer. Note that (6) reverts to the equations for E_{AMBER} (1) + (2) if we set $\gamma = 0, P = 6$; in other words, smoothing is turned off. On the other hand, with any $\gamma > 0$, the poles at $r_{ij} = 0$ disappear for both the Lennard-Jones and electrostatic components of the function. This smoothing function is easy to utilize computationally while still possessing the desirable properties of a smoothing function for this problem class. For more information on our smoothing technique, including its comparison with other smoothing approaches and its application to global optimization problems, see [20, 1, 2].

4   The Global Optimization Algorithm

Our global optimization method incorporates the basic features, sampling and local minimizations, of existing stochastic methods, but it is only able to solve large-scale problems because it also incorporates phases that focus on small dimensional subproblems. These phases constitute a significant departure from previous stochastic methods and account for a major portion of the computational effort of the method.

The basic algorithm that we have utilized previously consists of two phases. The first phase generates initial three dimensional structures for the given sequence of amino acids describing the protein. A buildup procedure is used that samples on the set of dihedral angles for each amino acid some fixed number of times, and selects the angle values that produce the best partial energy for the part of the chain built so far, before proceeding to the next amino acid. A subset of the best structures generated by this buildup procedure are then selected as start points for full dimensional local minimizations, and some number of the best minimizers generated are passed to the second phase of the algorithm. A variation of this procedure, using neural net secondary structure prediction information, will be described later in this section.

The second phase accounts for most of the computational effort and success of the method. The basic idea of this phase is to select a configuration from the list of local minimizers, and then select a small subset of its variables for improvement. The subset consists of a small number of dihedral or torsional angles of the protein. An interesting new technique for making that choice will be described below. Once the subset has been determined, a stochastic global optimization procedure similar to the one in [18] is executed to find the best new positions for the chosen dihedral angles, while holding the remaining dihedral angles fixed. The global optimization procedure samples over the entire $-180° \rightarrow 180°$ angular range of each of the chosen dihedral angles, and performs small-scale local minimizations over the subspace of selected dihedral angles from those sample point configurations with the lowest energy of all sample points within their "critical radius", as described in [18]. Instead of using a probabilistic stopping criteria, iterations of sampling plus small-scale local minimizations are performed a fixed number of times because the goal of the small-scale global optimization is not to locate *all* minima of the small-dimensional problem, but only some number of the best ones. Some of the best resulting configurations are then "polished" by applying a full dimensional local minimization using all the variables. The new full-dimensional local minimizers are then merged with those found previously and the entire phase is iterated a fixed number of times. The incorporation of smoothing functions into this phase will be discussed below.

The framework for the basic global optimization algorithm is outlined below in Algorithm 4.1.

Algorithm 4.1 – Framework of the Basic Global Optimization Algorithm for Protein Problems

1. **Phase 1: Generation of Initial Configurations**

   (a) **Protein Sample Point Buildup:** Build up sample configurations from one end of the protein to the other by sequentially generating the dihedral angles for each amino acid: randomly sample the angles for current amino acid a fixed number of times and select the angle values that give the lowest energy function value for the partial protein generated so far.

   (b) **Start Point Selection :** Select a subset of the best sample configurations from step 1a to be start points for local minimizations.

   (c) **Full-Dimensional Local Minimizations :** Perform a local minimization from each start point selected in step 1b. Collect some number of the best of these minimizers for improvement in Phase 2.

2. **Phase 2: Improvement of Local Minimizers:** For some number of iterations:

   (a) **Choose the configuration to improve and the small-scale problem parameters:** From the list of full-dimensional local minimizers, select a local minimizer to improve and a small subset of dihedral angles from that minimizer to be optimized.

   (b) **Global Optimization on a small subset of variables :** Apply a fairly exhaustive small-scale global optimization algorithm to the energy of the selected configuration using the selected small subset of the dihedral angles as variables.

   (c) **Full-Dimensional Local Minimization :** Apply a local minimization procedure, with all dihedral angles as variables, to the lowest configurations that resulted from the global optimization of step 2b.

   (d) **Merge the New Local Minimizers :** Merge the new lowest configurations into the existing list of local minimizers.

While Algorithm 4.1 was successful in finding the global minimum for small proteins (poly(ala) with 5,10,20,30 and 40 residues, and met-enkephalin with 5 residues), in order to find the global minimizer for poly(ala) with 58 residues, the algorithm was modified to use the smoothed potential energy function (5) in both Phase 1 and Phase 2. Essentially, the 2 phases were executed in the same fashion as in Algorithm 4.1, with the exception that the smoothed potential was used in all the sampling and local minimizations. Additionally, 2 steps were required to transform the solution back to the original potential energy landscape:

- At the completion of Phase 1, a local minimization was performed on each of the initial minimizers created in the smoothed space to desmooth the minima, that is, to find minimizers of the original potential surface using the smoothed minima as start points. The minima were then ranked according to the function values of the original function, and the lowest were chosen for improvement to begin the balancing portion of Phase 2. (From this point on, step 2(a) uses smoothed function values in making its selections of configurations.)

- At the completion of Phase 2, a local minimization was performed using each of the resulting smoothed minima as start points, to again desmooth the minima.

This modified algorithm using a smoothed objective was able to find lower local minimizers for large proteins in less time than Algorithm 4.1 (see [2] for results on poly(ala)-58).

When we applied Algorithm 4.1 with smoothing to uteroglobin, a 70 amino acid helical protein, Phase 1 was found to be inadequate in building up sample configurations with any secondary structure at all. (This was not our experience for any of the poly(ala) test cases, where Phase one formed the desired helical structures, and met-enkephalin has essentially no secondary structure. But the structure of uteroglobin is far more challenging.) To deal with these difficulties, the new approach uses a preprocessing step referred to as Phase 0, to insert secondary structure into an initial configuration using the neural network predictions of alpha and beta, described above. Starting from a completely extended conformer containing no secondary or tertiary structure, a local minimization is first performed using the biasing functions (3) and (4) in addition to E_{AMBER} (1 and 2). From the resulting "biased" minimizer another local minimization is performed without the biasing functions. The local minimization using the biasing function encourages the formation of α-helices in regions where the neural network predictions for α are strong. Because the network predictions may not be completely correct, the biasing terms may either force helical formations in regions where they do not belong, or discourage them from forming in regions where they should be located. The local minimization without the biasing functions allows for some correction in areas where the predictions may be wrong. The output from Phase 0 is a single configuration which contains at least partially correct secondary structure, but does not contain correct tertiary structure. Given the ability of the biochemistry community to predict secondary structure fairly well, and the great difficulty and complexity of the full protein structure prediction problem, it seems very reasonable to try to utilize secondary structure predictions in our methods.

The effect of Phase 0 is to set the angles for residues predicted to be alpha or beta at appropriate values. Phase 1 (step 1a) is modified so that these values are left fixed and only those dihedral angles which have not been predicted to be alpha or beta by the neural network are sampled, using as input the configuration generated in Phase 0. In addition, during the local minimizations of step 1(c), the same two-step minimization strategy as described for Phase 0 is used. That is, local minimizations on the sample points are performed using the biasing functions, followed by local minimizations from the "biased" minimizers without the biasing functions. In this manner, a diverse set of initial configurations is created, each containing secondary structure that was predicted by the neural network. Algorithm 4.2 gives the framework for the new global optimization method, which includes the creation of initial configurations using neural net secondary structure prediction and smoothing in the minimizer improvement phase.

An important part of Algorithm 4.2 is the heuristic used to determine which configuration is selected at each iteration of the second phase. We consider an initial configuration and any configurations generated from it (via global optimization of a small sub-space of dihedral angles followed by a full-dimensional local minimization) to be related, such that the latter is a "descendent" of the former. For some fixed number of iterations, the work in Phase 2 is balanced over each of the k sets of configurations consisting of the k initial minimizers and all of those minimizer's descendants. First, each of the k initial minimizers is chosen for improvement. Then, at each iteration for the remainder of the balancing phase, the set of configurations with the least amount of work performed on its members so far is selected, and the best configuration in this set that hasn't already been used is chosen. Configurations are rated in terms of their smoothed energy function value, and best refers to the lowest in energy value. After the fixed number of iterations of the balancing phase have been performed, the remaining iterations of the local minimizer improvement phase select the best configuration that has not already been selected, regardless of where it descended from. We have found that the combination of the breadth of search of the configuration space that the balancing phase provides with the depth of search that the non-balancing phase allows is useful to the success of our method.

In this study, we restrict our choice of dihedral angles (used to define the small-scale subproblem in step 1 of Phase II) to those angles not labelled as alpha-helical (or beta) by a call to the DSSP [12] program which identifies the secondary structures of a given protein. To choose from among the available angles we use an angle-choice heuristic that we have developed recently. The new heuristic involves dividing the protein into some number (in this case 10) contiguous regions, and determining which region is doing worst. To make this determination, for each region $\mathcal{R}$ we compute a partial energy consisting of the sum of the non-bonded interaction energies between atoms in $\mathcal{R}$ and atoms not in $\mathcal{R}$, plus the dihedral energy of the torsion angles in $\mathcal{R}$. To appraise the goodness of these partial energies, we construct a *comparison pool* consisting of about 50 of the best conformations chosen so that no two have the same general structure. The DSSP program mentioned above is used to identify structure similarities. For each region in the selected conformation we compare its partial energy to the partial energies of the corresponding region of each conformation in the comparison pool, and determining its rank in this set. The region with the worst ranking is then selected as being the worst region. To select angles likely to improve this region, we compute the partial derivatives of the worst region with respect to every dihedral angle that we are allowing to change. The 6 or so angles with partial derivatives greatest in absolute value are then chosen to define the subspace for the small-scale problem.

Algorithm 4.2– Framework of the New Global Optimization Algorithm

I **Phase 0: Generation of Initial Configuration containing predicted Secondary Structure**

1. Local minimization on an extended conformer using the biased energy function.

2. Local minimization on the output of step I.1, using the unbiased potential energy function

II **Phase 1: Generation of Initial Configurations**

1. **Protein Sample Point Buildup:** Build up sample configurations from one end of the protein to the other by sequentially generating the dihedral angles for each amino acid that was not predicted to be α or β by the neural network: for the current amino acid, randomly sample the set of dihedral angles a fixed number of times and select the dihedral angle that gives the lowest energy function value for the partial protein generated so far.

2. **Start Point Selection :** Select a subset of the best sample points from step II.1 to be start points for local minimizations.

3. **Full-Dimensional Local Minimizations :**

   (a) Perform a local minimization using the biasing functions, from each start point selected in step II.2.

   (b) Perform a local minimization using the unbiased potential energy function from each of the minimizers from step II.3.a.

   Collect some number of the best of these minimizers for improvement in Phase 2.

III **Phase 2: Improvement of Local Minimizers: For some number of iterations:**

1. **Choose the configuration to improve and the small-scale problem parameters:** From the list of full-dimensional local minimizers, select a local minimizer to improve and a small subset of dihedral angles from that minimizer to be optimized.

2. **Global Optimization on a small subset of variables :** Apply a fairly exhaustive small-scale global optimization algorithm to the energy of the selected configuration using the selected small subset of the dihedral angles as variables, and the smoothed potential energy function for sampling and local searches.

3. **Full-Dimensional Local Minimization :** Apply a local minimization procedure, with all dihedral angles as variables, to the lowest configurations that resulted from the global optimization of the step III.2, using the smoothed potential energy function.

4. **Merge the New Local Minimizers :** Merge the new lowest configurations into the existing list of local minimizers.

IV **Postprocessing phase:** Desmooth the minimizers from Phase 2 by performing local minimizations using the original potential energy function

5   Computational Results

We have tested Algorithm 4.2 on a 70 amino-acid helical protein, the A-chain of uteroglobin. Figure 1 shows the crystal structure for this protein, obtained from the protein data bank entry 2UTG.

Figure 1: Crystal structure for the A-chain of Uteroglobin

Phase 0 of our new algorithm attempts to formulate a single configuration containing secondary structure consistent with the neural net predictions for this protein. This phase starts with a structure of the target protein that is the minimum closest to the fully extended form with all backbone pairs assuming the values $\phi = 180$ degrees and $\psi = -180$ degrees. A local minimization is performed using E_{AMBER} (1 and 2) plus the biasing functions (3 and 4) with both the strength for the force constants as well as the prediction of which residues are α obtained from the neural network predictions. This minimization is followed by a local minimization using only E_{AMBER}. Each of the 2 minimizations cost roughly 6000 function evaluations. The first result of Phase 0 is shown in Figure 2.

Figure 2: Resulting minimizer from Phase 0

The protein structure in Figure 2 has reasonably well formed helices in three of the four helices found in the crystal structure, and extended formations in the regions predicted to be coil. One of the four helices found in the crystal is much more disordered in our result, however, as a consequence of weak predictions in that region of the sequence.

In order to assess the ability of Phase 0 if it were given a more accurate secondary structure prediction, another Phase 0 was executed with a modified prediction file which uses the secondary structure of the crystal to predict helix rather than coil in the region of the disordered helix. This run obtained a better formed helix in that region. This new structure is shown in Figure 3.

Figure 3: Second minimizer from Phase 0 with improved secondary structure

We decided to use the structure of Figure 3 as the starting configuration for Phase 1, as well as the modified prediction file, because we anticipate that future improvements in the network prediction algorithm will largely correct weak predictions for α-helical proteins.

In Phase 1, 60 sample configurations were generated in Step II.1, and the six configurations with the lowest energy values were selected for local minimization. The dihedral angles in each of the 70 amino acids were sampled 100 times, and the energy of the protein so far was evaluated each time. The cost for sampling each of the 60 configurations in terms of function evaluations is 100 samples times N, where N is the number of amino acids, which is 70 in the case of uteroglobin. Since these function evaluations only evaluate the portions of the energy that change with each buildup step, the total cost of this build-up was much less than the cost of 7000 evaluations. Each of the 6 local minimizations cost roughly 6000 function evaluations, using a limited-memory BFGS code [14] varying all 3375 Cartesian coordinates (1125 atoms) of the protein, on the energy surface of (E_{AMBER} + $E_{\phi\psi}$ + E_{HB}). In step II.3.b, the local minimizations performed to "unbias" the structures on (E_{AMBER} + E_{SOLV}) used a BFGS method with an internal coordinate representation which fixes the bond angles and bond lengths of the protein while allowing only the dihedral angles to vary. There are approximately 400 dihedral angle internal coordinate parameters in Uteroglobin, and each internal coordinate minimization cost, on average, 3000 function evaluations. Function evaluations on the Aspen Avalanche machines used in these computations take approximately 2 seconds each to execute.

Table 5.1 gives the energy(E_{AMBER} + E_{SOLV}) and root mean squared (r.m.s.) deviation from the crystal structure, computed by measuring distances between corresponding carbon-alpha atoms, for the minimizers found in Phase 1. These minimizers are also shown in Figures 4-6. Only one of these structures (-2760, the third best in energy value) has a

Table 5.1: Minimizers found in Phase 1

Energy ($E_{AMBER} + E_{SOLV}$) in Kcal/Mol	RMSD from Crystal structure in Angstroms
-2780	13.17
-2773	12.16
-2760	9.34
-2673	12.78
-2601	20.61
-2552	13.99

structure that even vaguely resembles that of the crystal structure, and the r.m.s. values reflect this, having values that are considered very high in the remaining cases. Also, it is evident from this data that there is not a direct correlation between energy values and the deviation in distances from the crystal structure. Further discussion on this topic follows the Phase 2 results. From Figures 4 through 6 of these configurations, it is possible to see a relationship between compactness and lower energy values of the structures.

The list of initial input configurations for Phase 2 consisted of the six minimizers from Phase 1. Eighteen iterations of phase 2 were executed, 6 balancing iterations and 12 non-balancing, as described in the previous section. Six dihedral angles were optimized in the small-scale global optimization of step III.2. Each small-scale global optimization performed an average of 5000 function evaluations for sampling and 2000 total function evaluations for the small dimensional local searches. In this experiment, each balancing iteration performed 8 full-dimensional local minimizations (step III.3), and each non-balancing iteration performed 12. The full-dimensional local mininizations over the internal coordinate parameters cost on average 2400 function evaluations each. Thus, all the minimizations in Phase 2 required almost 500,000 function evaluations, and constituted the principal computational cost of the algorithm.

The smoothing parameters used were $P = 5$, and $\gamma = .05$. Based on the results of experiments performed in [1], we conducted some additional testing on smoothing parameters for uteroglobin, and modified the parameters slightly from those found to be most effective for several smaller proteins. For a detailed discussion of the choice of smoothing parameters for these other proteins, see [1].

Table 5.2 contains results from this run of Phase 2 of Algorithm 4.2. The 15 lowest energy minimizers are presented, as well as their r.m.s. deviations from the crystal structure, and the iteration of Phase 2 in which they were (first) found. The lowest energy value found from all the resulting minimizers of Phase 2 (using internal coordinate minimization) is -2944 with an r.m.s. of 7.95 Angstroms, but the minimizer with the smallest r.m.s. deviation from the crystal structure, again measured on the carbon-alpha atoms, has an energy value of -2877, and is marked by an "*". The r.m.s. for that minimizer is 7.24 Angstroms, and furthermore, the structure has the correct folds, but the larger outer bend is more compact, and the 2 central bends are further apart from each other than in the crystal structure. All of the configurations with r.m.s. of 8 Angstroms or less can be considered moderately good fits of the crystal structure and all have similar structures.

Figure 4: Phase 1 minimizers with corresponding energy values $(E_{AMBER} + E_{SOLV})$ of -2780 and -2773.

Figure 5: Phase 1 minimizers with corresponding energy values $(E_{AMBER} + E_{SOLV})$ of -2760 and -2673.

Figure 6: Phase 1 minimizers with corresponding energy values $(E_{AMBER} + E_{SOLV})$ of -2601 and -2552.

Table 5.2: Best 15 minimizers found in Phase 2

Energy $(E_{AMBER} + E_{SOLV})$ in Kcal/Mol	RMSD from Crystal structure in Angstroms	Iteration of Phase 2 where found
-2944	7.95	15
-2929	8.02	18
-2911	7.92	11
-2904	10.81	4
-2898	7.88	18
-2894	7.78	1
-2890	7.93	18
-2889	7.86	10
-2887	7.70	15
-2882	10.30	5
-2877.8	10.32	5
-2877.7	7.87	15
-2877.4*	7.24	10
-2873	8.60	15
-2871	8.46	10

Figure 7 shows a tube diagram of the crystal structure, in order to be able to compare it to the 2 minimizers discussed above, with energy values -2877 and -2944. Figures 8 and 9 show these 2 configurations.

These results show that in the case of uteroglobin, Algorithm 4.2 is able to find configurations with tertiary structures close to that of the crystal structure. However, the fact that the lowest energy minimizers in Table 5.2 do not necessarily have the smallest r.m.s. deviations indicates that the potential energy function may not model the molecular behavior sufficiently accurately. To investigate this further we evaluated the energy of the crystal structure using $(E_{AMBER} + E_{SOLV})$. In order to do this, it was necessary to put hydrogen atoms in their appropriate locations, and do a local minimization on this new structure using Cartesian coordinates. The local minimization was done with the parameterization over all the Cartesian coordinates, in order to allow the bond lengths and bond angles to be optimized in accordance with E_{AMBER} optimal values. The r.m.s. of the new structure compared to the crystal structure is 2.40 Angstroms, which is a good fit. The energy is -3090.74, in Cartesian coordinates. The values reported in Table 5.2 are internal minimizer values, which have fewer degrees of freedom than Cartesian minimizers and thus higher energy values. When a few key minimizers from Table 5.2 are minimized in the Cartesian coordinate parameterization over $E_{AMBER} + E_{SOLV}$, the resultant energy values (shown as internal → Cartesian) are: $-2944 \rightarrow -3182$, $-2929 \rightarrow -3160$, and $-2877.4 \rightarrow -3107$.

These results indicate that the energy values of minimizers found in Phase 2 are already considerably lower than the energy value of a structure that is very close to the crystal structure. This lack of correspondence is even more pronounced when minimizing on E_{AMBER} without E_{SOLV}. So while it is unknown whether Algorithm 4.2 has found the global minimizer of $E_{AMBER} + E_{SOLV}$, this global optimization process has shown that the immediate need at this stage is for an improved energy function that more closely resembles

Figure 7: Tube diagram of the crystal structure for the A chain of Uterglobin

Figure 8: Tube diagram of Phase 2 minimizer with the lowest r.m.s. deviation from the crystal structure, with energy value ($E_{AMBER} + E_{SOLV}$) of -2877.

Figure 9: Tube diagram of Phase 2 minimizer with the lowest energy value over all minimizers found, with energy value ($E_{AMBER} + E_{SOLV}$) of -2944.

the actual energy for low energy configurations. Future work on the solvent term E_{SOLV} will hopefully alleviate this problems.

We also examined the role of smoothing in the success of our algorithm. A 6 iteration run with no smoothing (i.e. $P = 6$, and $\gamma = 0$) resulted in the lowest minimizer energy value of -2863. In comparison, the run described above with smoothing parameters $P = 5$ and $\gamma = 0.05$ found a lowest minimizer energy value of -2904 for the same computational effort. Fewer total iterations were performed for this comparison than in our final calculation, but the indication is that without smoothing the performance of the algorithm is limited. This is similar to the results observed when using the algorithm with and without smoothing on other protein problems [2], and supports the assessment that smoothing is an important and useful component of our global optimization approach for protein structure prediction.

6   Summary and Future Research

We have presented a new global optimization strategy that combines smoothing the potential energy function by an algebraic technique, with a sophisticated global optimization strategy. Initial configurations are built up with "biasing" functions that are formed from neural network predictions of secondary structure. Results for the helical protein uteroglobin show that the method has successfully predicted the correct tertiary structure folds for this protein. Uteroglobin, with 70 amino acids and 1125 atoms is a substantial sized target, but it is necessary to continue testing the method on other targets, including those with more complicated structures than those found in helical proteins.

Preliminary work has begun on a new method for building up initial configurations in Phase 1 that uses no neural network information. Using partial energy evaluations centered around one amino acid at a time, and sampling from a selection consisting of an equal percentage of alpha, beta and coil backbone dihedral angles, the method has successfully generated α-helices for Uteroglobin that are at least as well formed as those in Figure 2 of section 5 without any apriori knowledge or prediction. Future work to generalize the method to beta sheets is currently being pursued.

In order to utilize neural network predictions of beta-sheets, a new biasing function is being developed that will enable the the predicted information for beta to be incorporated into secondary structure formulations. Matching algorithms [15] may be used to assist in the construction of beta-sheets from pairs of beta-strands.

Our results point out that the development of accurate empirical potential energy functions remains crucial to the success of protein structure prediction via optimization. An improved treatment of solvation effects is a key component in creating improved empirical energy functions. The large set of configurations generated by Algorithm 4.2 for uteroglobin will be useful for providing examples of various folds that can be used to assess the ordering of energy values produced by candidate functions. It may be possible to improve the parameterization of the equations in the current solvent term, E_{SOLV}, using this information, or a new solvent term may be needed.

The extreme challenge of designing algorithms for solving the protein folding problem involves a tradeoff between using chemical knowledge where possible, yet developing somewhat general purpose optimization techniques. Our collaboration of optimizers and chemists is attempting to determine the balance of the two areas that will best aid in solving this difficult problem.

References

[1] Azmi, A. (1998), " Use of smoothing methods with stochastic perturbation for global optimization (a study in the context of molecular chemistry)", Phd thesis, University of Colorado.

[2] Azmi, A., Byrd, R.H., Eskow, E. and Schnabel, R.B. (1998), " New smoothing techniques for global optimization in solving for protein conformation".

[3] Byrd, R.H., Eskow, E., van der Hoek, A.,Schnabel, R.B, Shao, C-S., and Zou, Z.(1996), "Global optimization methods for protein folding problems", *Proceedings of the DIMACS Workshop on Global Minimization of Nonconvex Energy Functions: Molecular Conformation and Protein Folding*, P. Pardalos, D. Shalloway, and G. Xue, eds., American Mathematical Society, 29-39.

[4] Byrd, R.H., Eskow, E., and Schnabel, R.B. (1995), "A New Large-Scale Global Optimization Method and its Application to Lennard-Jones Problems", Technical Report CU-CS-630-92, Dept. of Computer Science, University of Colorado, revised 1995.

[5] Coleman, T., Shalloway, D. and Wu, Z. (1992), "Isotropic Effective Energy Simulated Annealing Searches for Low Energy Molecular Cluster States", Technical Report CTC-92-TR113, Center for Theory and Simulation in Science and Engineering, Cornell University.

[6] Coleman, T., Shalloway, D. and Wu, Z. (1993), "A Parallel Build-up Algorithm for Global Energy Minimizations of Molecular Clusters Using Effective Energy Simulated Annealing", Technical Report CTC-93-TR130, Center for Theory and Simulation in Science and Engineering, Cornell University.

[7] Coleman, T., and Wu, Z. (1994), "Parallel Continuation-Based Global Optimization for Molecular Conformation and Protein Folding", Technical Report CTC-94-TR175, Center for Theory and Simulation in Science and Engineering, Cornell University.

[8] Cornell, W.D., Cieplak, P, Bayly, C.L., Gould, L.R., Merz, K.M., Ferguson, D.M., Spellmeyer, D.C.,Fox, T., Caldwell, J.W., and Kollman, P.A. (1995), "A second generation force field for the simulation of proteins, nucleic acids, and organic molecules", *J. Am. Chem. Soc., 117*, 5179-5197.

[9] Head-Gordon, T. (1995), " A new solvent model for hydrophobic association in water, I. Thermodynamics", *J. Am. Chem. Soc., 117*, 501-507.

[10] Head-Gordon, T., and Stillinger, F.H. (1993), "Predicting polypeptide and protein structures from amino acid sequence;antlion method applied to melittin", *Biopolymers, 33*, 293-303.

[11] Head-Gordon, T., and Stillinger, F.H. (1993), "Optimal neural networks for protein structure prediction", *Phys Rev. E., 48*, 1502-1515.

[12] Kabsch, W. and Sander, C. (1983), "Dictionary of protein secondary structures: pattern recognition of hydrogen-bonded and geometrical features", *Biopolymers, 22*, 2577-2637.

[13] Kostrowicki, J., Piela, L., Cherayil, B.J., and Scheraga, A. (1991), " Performance of the Diffusion Equation Method in Searches for Optimum Structures of Clusters of Lennard-Jones Atoms", *J. Phys. Chem., 95*, 4113-4119.

[14] Liu, D.C and Nocedal, J. (1989), "On the limited memory BFGS method for large scale optimization methods", *Mathematical Programming, 45*, 503-528.

[15] Lovasz, L. and Plummer, M.D. (1986), *Matching Theory*, Elsevier, Amsterdam.

[16] D Pratt, L. R. and Chandler, D. (1977), "Theory of the hydrophobic effect", *J. Chem. Phys. 67*, 3683-3704.

[17] Pratt, L. R. and Chandler, D. (1980), "Effects of solute-solvent attractive forces on hydrophobic correlations", *J. Chem. Phys., 73*, 3434-3441.

[18] Rinnooy-Kan, A.H.G. and Timmer, G. (1984), "A stochastic approach to global optimization", *Numerical Optimization*, Boggs, P., Byrd, R. and Schnabel, R.B. eds. SIAM, Philadelphia, 245-262.

[19] Shalloway, D. (1991), "Packet annealing: a deterministic method for global minimization, with application to molecular conformation." Preprint, Section of Biochemistry, Molecular and Cell Biology, *Global Optimization*, C. Floudas and P. Pardalos, eds., Princeton University Press.

[20] Shao, C.-S., Byrd, R.H., Eskow, E., and Schnabel, R.B., " Global Optimization for Molecular Clusters Using a New Smoothing Approach", submitted to *Journal of Global Optimization.*

[21] Wu, Z. (1993), "The Effective Energy Transformation Scheme as a Special Continuation Approach to Global Optimization with Application to Molecular Conformation", Technical Report CTC-93-TR143, Center for Theory and Simulation in Science and Engineering, Cornell University.

[22] Yu, R.C. and Head-Gordon, T. (1995), "Neural network design applied to protein secondary structure prediction", *Phys Rev. E., 51*, 3619-3627.

Optimization in Computational Chemistry and Molecular Biology, pp. 19-46
C. A. Floudas and P. M. Pardalos, Editors

Methodology for Elucidating the Folding Dynamics of Peptides : Met-enkephalin Case Study

J. L. Klepeis
Department of Chemical Engineering
Princeton University
Princeton, NJ 08544
john@titan.princeton.edu

C. A. Floudas
Department of Chemical Engineering
Princeton University
Princeton, NJ 08544
floudas@titan.princeton.edu

Abstract

The ability to characterize the energy surface and reaction pathways of peptides is an important step in understanding the protein folding process. However, this problem may be intractable for realistically modeled systems, such as those represented by all atom force–fields. In this work, a method for mapping the connectivity of relatively low energy regions through the identification of first order transition states is introduced. The methodology relies on the use of a deterministic global optimization approach for identifying low free energy ensembles of conformers [20]. The technique is applied to both unsolvated and solvated forms of the oligopeptide met-enkephalin.

Keywords: Protein folding, transition states, global optimization, enkephalin.

1   Introduction

A fundamental problem in the area of computational chemistry and molecular biology is how a protein folds to its correct conformation in a reasonable time scale. The complexity of this problem is due to a lack of information regarding the folding pathways on the protein's intricate energy surface.

To better understand protein folding it is necessary to examine a protein's energy hypersurface. Such examination often begins with the identification of the most stable energy structure – the global minimum energy conformation. The prediction of a protein's tertiary structure from only sequence information, is a hallmark of computational chemistry. A number of reviews on methods for solving this multiple-minima problem are available [15, 31].

The characterization of the energy surface must also include the identification of other stable and metastable configurations. Mathematically, these structures correspond to stationary points of the energy function. In particular, local minima represent stable conformations, while (first order or higher order) saddle points constitute transition states that connect two stable structures. A folding pathway defines the connection between two stable conformations (local minima) through a series of transition states (saddle points).

Since the folding pathway may include a number of intermediates, a rigorous description of the energy surface would require the identification of all local minima and saddle points of the energy function. This precept has been used to study a flexible, helical forming tetrapeptide [12, 37, 38]. For large molecular systems, such an approach faces computational challenges. In this work, a new methodology is outlined which maps out low lying energy regions by first locating low (free) energy minima through a deterministic global optimization technique [20]. These minima are then used to initiate searches for first-order transition states. Connectivity between two minima is established by retracing the paths back down from each transition state.

Folding pathways between any two minima can be identified through the combination of minimum-transition-minimum triples. Once the connectivity of the energy surface has been established, transition rates between minima can also be calculated using Rice-Ramsperger-Kassel-Marcus (RRKM) theory [22, 23]. Finally, following the ideas given in [37], a "rate connectivity graph" can be constructed. This graphical representation of the energy surface is based on the transition rates between minima. Such a graph is similar in nature to the connectivity tree and energy disconnectivity graph concept, which uses finite energy (or temperature) partitioning to define basins of attraction [9, 13].

In this work, the approach for finding low (free) energy minima, as proposed in [20], is outlined in Section 3. A detailed description of the methodology for determining transition states and energy surface connectivity is given in Section 4. Finally, in Section 5, the results for the oligopeptide system of met-enkephalin are given. Both unsolvated and solvated forms of the molecule are studied.

2   Modeling

2.1   Potential energy

To provide a detailed description of the protein system, a semi-empirical all atom force–field is employed. There exist many parameterizations for molecular potential functions, including AMBER [35, 36], CHARMM [10], ECEPP [27, 28, 29], and MM3 [5]. In this work, the ECEPP/3 (Empirical Conformational Energy Program for Peptides) model, which is the latest installment of the ECEPP force–field, is utilized [30]. For this force field covalent bond lengths and bond angles are fixed at their equilibrium values, which implies that all residues of the same type have essentially the same geometry in various proteins. Based on these approximations, the conformation is only a function of the dihedral angles. That is, ECEPP/3 accounts for energy interaction terms which can be expressed solely in terms of the independent torsional angles.

The total conformational energy is calculated as the sum of the electrostatic, nonbonded, hydrogen bonded, and torsional contributions, as shown in the following equation :

$$
\begin{aligned}
E_{\text{ECEPP/3}} \;=\;\; & \sum_{(i,j)\in ES} \frac{q_i q_j}{r_{ij}} \;+\; \sum_{(i,j)\in NB} F_{ij}\frac{A_{ij}}{r_{ij}^{12}} - \frac{C_{ij}}{r_{ij}^{6}} \\
+\;\; & \sum_{(i,j)\in HB} \frac{A'_{ij}}{r_{ij}^{12}} - \frac{B_{ij}}{r_{ij}^{10}} \;+\; \sum_{k\in TOR} \frac{E_k}{2}\left(1 + c_k \cos\left(n_k \theta_k\right)\right)
\end{aligned}
\tag{1}
$$

Here r_{ij} is the distance between atoms i and j. The sets ES, NB, and HB correspond to those pairs of atoms (i,j) which contribute electrostatic, nonbonded and hydrogen bonded energies, respectively. For the electrostatic term, the dielectric constant and conversion factor have been incorporated into the parameters q_i and q_j. A_{ij}, C_{ij}, A'_{ij} and B_{ij} are parameters that define the well depth and width for a given nonbonded or hydrogen-bonded interaction. F_{ij} is equal to 0.5 for 1–4 interactions and 1.0 for higher order interactions. A torsional term is also included for those dihedral angles in the set TOR. θ_k represents any dihedral angle, while c_k takes a value from [-1,1], and n_k reflects the symmetry type for the particular dihedral angle. Additional contributions are calculated for special structural features, such as proline rings and disulfide bridges. The main energy contributions are computed as the sum of terms for each atom pair whose interatomic distance is a function of at least one dihedral angle.

2.2 Solvation energy

Solvation energy can be modeled either explicitly or implicitly. To make the computations tractable, a continuum model based on empirical correlations with solvent accessible volumes is employed [8], which is represented by the following equation :

$$
E_{\text{HYD}} \;=\; \sum_{i}(\text{VHS}_i)(\delta_i)
\tag{2}
$$

Here VHS_i corresponds to the volume of the hydration shell, while δ_i is the free energy density parameter for the atom or functional group. The main assumption of this model, and any geometrically parameterized solvation model, is that hydration free energies can be calculated from an average free energy of interaction of each atom (or functional group) with a layer of solvent known as the hydration shell. The total free energy of solvation is taken to be the sum of the free energies for each of the atoms (or functional groups) in the peptide.

In general, the analytical form for calculating the volume of a hydration shell (VHS_i) is not suitable for force field models using pairwise intramolecular potential, such as ECEPP/3. Furthermore, direct truncation at the pairwise double–overlap term would lead to large errors. In this work, the RRIGS (Reduced Radius Independent Gaussian Sphere) approximation is used to efficiently calculate the exposed volume of the hydration shell [8]. This method uses a truncated form (double overlap) for VHS_i, while also artificially reducing the van der Waals radii of all atoms other than atom i when calculating VHS_i. These reductions effectively decrease the contribution of the double overlap terms, leading to a cancellation of the error that results from neglecting the triple and higher overlap terms. In addition, the characteristic density of being inside the overlap volume of two intersecting spheres is not represented as a step function, but as a Gaussian function; this provides continuous derivatives of the hydration potential. Therefore, the solvation energy contributions can easily be added at every step of local minimizations since the RRIGS approximation has the same set of interactions as the ECEPP/3 potential. The δ_i are RRIGS specific, and were determined by a least square fitting of experimental free energy of solvation data for 140 small organic molecules [8].

2.3   Free energy

A strict interpretation of Anfinsen's thermodynamic hypothesis requires the global minimization of the conformational free energy to predict the native structure of a protein. In practice, however, most protein models include only potential and solvation effects. One reason for this neglect for including entropic effects is that a rigorous free energy model requires infinite sampling to associate accurate statistical weights with each microstate.

Other approximate calculations exist for estimating these statistical weights (and thus entropic effects). The most simplistic model would rely on only the Boltzmann weight associated with each microstate. A more sophisticated approximation, known as the harmonic approximation, utilizes second derivative information to characterize the basin of attraction. More complex schemes try to mimic the anharmonic trajectory along the energy surface. These quasi-harmonic approximations generally require the use of MC/MD simulations.

In this work, entropic effects are included via the harmonic approximation [14, 17, 18]. The development of this model can be understood physically by first considering the partition function for the system :

$$ Z \;=\; \exp^{-\frac{(E-TS)}{k_B T}} \;=\; \exp^{-\frac{E}{k_B T}} \exp^{\frac{S}{k_B}} \tag{3} $$

In Equation (3) the partition function is the product of the Boltzmann factor ($\exp[-E/k_B T]$) and the number of states available to the system ($\exp[S/k_B]$). At a given stationary point, the harmonic approximation is equivalent to :

$$ E(\theta) \;=\; E(\theta_\gamma) + \frac{1}{2}\left(\theta - \theta_\gamma\right) H(\theta_\gamma)\left(\theta - \theta_\gamma\right) \tag{4} $$

Here γ identifies the stationary point, and the stationarity condition ($\nabla E(\theta_\gamma) = 0$) is used to eliminate the gradient term. In this way, each basin of attraction is characterized by properties of its corresponding minima, which include the local minimum energy value, $E(\theta_\gamma)$, and the convexity (Hessian) information around the local minimum, $H(\theta_\gamma)$. An analogous representation of this system is N_θ independent harmonic oscillators, each with its own characteristic vibrational frequency. The minimum can then be characterized by the occupation of each normal mode.

To develop an expression for the entropic effect, Equation (4) can be substituted into Equation (3). By summing over all energy states, the partition function becomes :

$$Z_\gamma^{\text{har}} = \exp^{-\frac{E(\theta_\gamma)}{k_B T}} f(T) \prod_i^{N_\theta} \frac{1}{\lambda_i} \tag{5}$$

In Equation (5), $f(T)$ is a function dependent only on temperature, while λ_i represent the eigenvalues of $H(\theta_\gamma)$. Comparison of Equations (5) and (3) implies that :

$$\exp^{\frac{S}{k_B}} \propto \prod_i^{N_\theta} \frac{1}{\lambda_i} \tag{6}$$

Equation (6) can be rewritten in terms of the harmonic entropic contribution, S_γ^{har} :

$$S_\gamma^{\text{har}} \propto -k_B \ln\left[\text{Det}\left(H(\theta_\gamma)\right)\right] \tag{7}$$

A more rigorous derivation of the harmonic approximation leads to the following expression for the harmonic entropy [14, 17, 18] :

$$S_\gamma^{\text{har}} = -\frac{k_B}{2} \ln\left[\text{Det}\left(H(\theta_\gamma)\right)\right] \tag{8}$$

This can be used to calculate relative free energies via the following equation :

$$F_\gamma^{\text{har}} = E(\theta_\gamma) + \frac{k_B T}{2} \ln\left[\text{Det}\left(H(\theta_\gamma)\right)\right] \tag{9}$$

Finally, each microstate can be assigned a statistical weight (p_γ^{har}) by considering the ratio of the partition function for that microstate (Z_γ^{har}) to the total partition function :

$$p_\gamma^{\text{har}} = \frac{\left[\frac{1}{[\text{Det}(H(\theta_\gamma))]}\right]^{1/2} \exp(-\frac{E(\theta_\gamma)}{k_B T})}{\sum\limits_{i=1}^{N_\gamma} \left[\frac{1}{[\text{Det}(H(\theta_i))]}\right]^{1/2} \exp(-\frac{E(\theta_i)}{k_B T})} \tag{10}$$

To develop a meaningful comparison of relative free energies, the total partition function (denominator of Equation (10)) must include an adequate ensemble of low-energy local minima, as well as the global minimum energy conformation. Therefore, an efficient method for identifying low energy ensembles must be employed. It should also be noted that the harmonic approximation does not require the explicit inclusion of a contribution based on the density of states because each local minimizer is accounted for only once (in contrast to counting methods).

3 Locating Low (Free) Energy Minima

3.1 Traditional formulation

In its static form, the protein folding problem in dihedral angle space is posed as an unconstrained nonconvex global minimization problem with periodic variable bounds. This problem is represented by the following formulation :

$$\min_{\theta} \quad E(\theta) \tag{11}$$

$$\text{subject to} \quad -\pi \ \leq \ \theta_i \ \leq \ \pi, \quad i = 1, \ldots, N_\theta$$

Here θ represents the vector of independent torsion angles. The global minimization of E, which may include both potential and solvation energies, requires efficient global optimization methods due to the extremely rugged nature of the energy hypersurface. Along these lines, the αBB deterministic global optimization approach has been found to be particularly efficient in locating the global minimum energy conformations of isolated and solvated peptides [7, 19, 21]. This branch and bound method addresses a broad class of problems and provides theoretical guarantees of convergence to the global minimum of nonlinear optimization problems with twice-differentiable functions [1, 2, 3, 4, 6].

The αBB algorithm can be summarized in the following manner :

1 The global minimum is bracketed by developing converging sequences of lower and upper bounds.

2 The lower and upper bounds are refined by iteratively partitioning the initial domain.

3 Upper bounds on the global minimum are obtain by local minimizations of the original energy function, E.

4 Lower bounds belong to the set of solutions of the convex lower bounding functions, L, which are constructed (in each subdomain) by augmenting E with the addition of separable quadratic terms.

The development of the convex lower bounding functions, L, ensures the deterministic nature of the algorithm and guarantees convergence to the global minimum. These functions have the following generic form :

$$L \ = \ E + \sum_{i}^{N_\theta} \alpha_i \left(\theta_i^L - \theta_i \right) \left(\theta_i^U - \theta_i \right) \tag{12}$$

In Equation (12), θ_i^L and θ_i^U refer to the current lower and upper bounds on the variable θ_i. For each separable quadratic term, the α_i correspond to non-negative parameters which must be greater or equal to the negative one-half of the minimum eigenvalue of the Hessian of E over the current domain, as defined by $\left[\theta^L, \theta^U \right]$. The properties and proofs that endow the αBB approach with deterministic guarantees are detailed elsewhere [25].

3.2   Ensembles of low energy conformers

To effectively include entropic contributions via the harmonic approximation, the general protein folding formulation, as given by Equation (11), must be modified. Specifically, in addition to locating the global minimum energy conformer, an ensemble of low energy local minima must be generated. Once this ensemble has been compiled, a free energy ranking can be performed (at a particular temperature) to locate the the relative free energy global minimum.

The nature of the αBB algorithm allows for the rigorous treatment of enumerating all local minima. These approaches are discussed in detail elsewhere [20]. In particular, one rigorous formulation involves the solution of a system of nonlinear equations that correspond to the stationarity condition, $\nabla E(\theta) = 0$. Through a reformulation, this problem becomes one of identifying all multiple global solutions. This method has been implemented using the αBB technique to locate all stationary points (local minima and saddle points) of triatomic molecules and small peptides [37, 38].

Given the astronomical number of local minima for even relatively small oligopeptides, this work addresses the problem of identifying low energy, rather than all, local minima. In particular, the generation of low energy ensembles of conformers is accomplished by algorithmic modifications of the general αBB procedure, rather than reformulation of the original problem. Rigorous implementation of the global optimization algorithm requires the minimization of a *convex* lower bounding function, L, in each domain. Each unique solution can then be used as starting points for the minimization of the original energy function in the current subdomain. By storing these minima a list of low energy conformers can be constructed.

A simplistic method for enhancing the distribution of local minima produced in each subdomain would involve the use of multiple random starting points when minimizing E. This method is better than just using random starting points (over the entire domain), since the subdomains become localized in regions of low energy as the separation between the lower and upper bounding sequences decreases.

However, this approach does not capitalize on all the available information provided by the lower bounding functions. In a strict implementation, each lower bounding function exhibits a single minimum in each subdomain. Since the values of α directly define the convexity of the lower bounding functions, these parameters can be modified to provide nonconvex underestimators. The modified lower bounding function, $\hat{L}$, will possess more than one minimum and can be minimized several times in each subdomain. Because the lower bounding functions smooth the upper bounding function, the locations of these minima also provide important information on the locations of low energy minima on the original energy surface. Therefore, the location of the minima of $\hat{L}$ are used as starting points for local minimizations of E, which results in an improved set of low energy conformers. In addition, these conformations are also localized in regions with low energy as the size of the subdomains decrease. This approach, which will be referred to as the EDA (Energy Directed Approach), is illustrated in Figure 1.

A second approach enhances the search for free energy local minima by accounting for entropic contributions during the course of the branch and bound algorithm. This is accomplished by calculating and including the relative entropic values at each minima of the upper and lower bounding functions, as shown in the following equation :

$$F^{\mathrm{har}} \;=\; U(\theta_{\mathrm{min}}) + \frac{k_B T}{2} \ln \left[\mathrm{Det} \left(H(\theta_{\mathrm{min}}) \right) \right] \tag{13}$$

The expression is similar to Equation (9) except that $U(\theta_{\mathrm{min}})$ now represents either the local minima of E or L. The thermodynamic temperature used in Equation (13) must also be specified as an additional input parameter. A similar modification for Monte Carlo minimization has also been attempted [26].

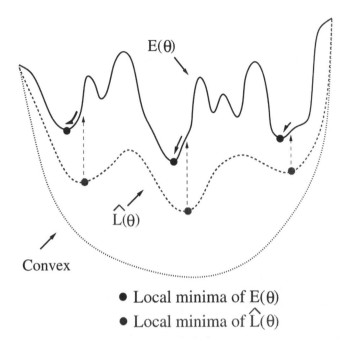

Figure 1: Using multiple minima of the lower bounding function ($\hat{L}$) to find low energy conformers of the upper bounding function (E).

This approach, which will be referred to as the FEDA (Free Energy Directed Approach), can be used to deterministically identify the global minimum free energy conformer at a given temperature. However, since the goal is to identify an ensemble of low (free) energy conformations, the FEDA is implemented using the modified lower bounding functions, $\hat{L}$.

3.3 Algorithmic Description

The αBB methodology for locating ensembles of low energy conformers requires interfaces between αBB and several programs. These include the potential and solvation energy modules, ECEPP/3 [30] and RRIGS [8], respectively; a local nonlinear optimization solver, NPSOL [16], which is used to locally solve both the upper and lower bounding problems; and UBC, an upper bound check module that is used to verify the quality of the upper bound solutions.

The algorithmic steps are summarized in the following :

1 Initialize the best upper bound to $+\infty$.

2 Initialize α values for all global variables (variables that will be partitioned). All other variables are treated locally within the periodic interval.

3 If FEDA is used, specify a thermodynamic temperature.

4 Partition the domain along one of the global variables.

5 Construct a lower bounding function ($\hat{L}$) in each subdomain. Perform three local minimizations according to the following procedure :

    A Generate 50 random points and perform function evaluations of $\hat{L}$.

    B Use the point with the minimum value as a starting point for local minimization of $\hat{L}$.

    C If FEDA is used, calculate and add $-TS^{\text{har}}$.

    D Store unique solutions.

If the minimum valued solution (of $\hat{L}$ in this subdomain) is greater than the current best upper bound, the region is fathomed (discarded from list of lower bounds).

6 The unique local minima of $\hat{L}$ are used as starting points for the minimization of the upper bounding function, E, in each subdomain. Perform two additional local minimizations according to the following procedure :

    A Generate 50 random points and perform function evaluations of E.

    B Use the point with the minimum value as a starting point for local minimization of E.

    C Check gradient and Hessian conditions (to verify point is a local minimum).

    D If FEDA is used, calculate and add $-TS^{\text{har}}$.

Update the current best upper bound as the minimum of all upper bounds, including those generated in the current subdomain.

7 Select the subdomain with the lowest lower bound value for further partitioning.

8 Update α values.

9 If the best upper and lower bounds are within the specified tolerance the program will terminate; otherwise it will return to step 4.

4   Identifying Folding Pathways

4.1   Illustrative example

To rigorously characterize the energy surface, and thus the folding pathways, of a given protein, the full set of minima and transition states must be identified. A methodology that builds on the concepts outlined in the previous section has been developed.

In particular, the ability for the EDA and FEDA to identify ensembles of low energy minima provides a propitious starting point for further examination of the energy hypersurface. These low energy and free energy minima are used to initiate local searches for first-order transition states. These transition states can then be used to identify minimum-transition-minimum triplets. At the very least, this procedure will provide a robust mapping of low energy regions, which can be enhanced by increasing the number of low energy minima used to initiate transition state searches.

The usefulness of this approach is illustrated by a small test case. In Figure 2 a two-dimensional contour plot for unsolvated alanine (NH_2–CH–CH_3–C=O–OH) is given. Both the χ and ω dihedral angles are fixed at optimal values. The energy surface exhibits 7 local minima and 12 transition states. When starting a transition state search (and triplet search) with the 3 lowest energy minima, all 7 local minima and 9 of the 12 transition states are identified. The number of found transition states can be increased to 11 (of 12) by initiating the search with 4 low energy minima.

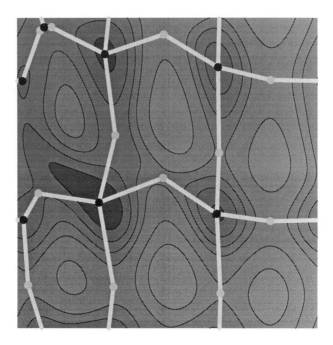

Figure 2: Energy contour plot for unsolvated alanine. The black circles denote local minima, while the grey circles denote transition states.

4.2 Eigenmode following

To implement the transition state search, an efficient method for locating saddle points is essential. In particular, eigenmode techniques have been shown to be powerful methods for locating both minima and saddle points [11, 32, 33, 34].

The eigenmode class of local solvers utilize second derivative information and are basically modified implementations of the Newton-Raphson method for locating stationary points. This is best illustrated by first examining the Newton-Raphson step :

$$\Delta\theta \;=\; -H^{-1}g \tag{14}$$

Here g and H are the gradient vector and Hessian matrix at a point on the energy surface. Equation (14) can also be written in terms of eigenvectors and eigenvalues :

$$\Delta\theta \;=\; -\sum_{i}^{N_\theta} \frac{g_i e_i}{\lambda_i} \tag{15}$$

Here g_i is the gradient vector along the e_i eigenmode, and λ_i is the corresponding eigenvalue. In actuality, the Newton-Raphson method is only guaranteed to converge to a local minimum when the point is already in the vicinity of a local minimum; that is, all eigenmodes have positive eigenvalues. This is because eigenmodes with positive eigenvalues lead to lower energy, while those with negative eigenvalues will tend toward higher energy. For this reason, the Newton-Raphson search can not be directed towards a particular type of stationary point.

In contrast, the eigenmode methods attempt to influence the step by shifting the eigenvalue. A simple eigenmode implementation which guarantees convergence to a minimum is to use the absolute value of the eigenvalues in Equation (15). The general form for an eigenmode step is given by the following :

$$\Delta\theta \;=\; -\sum_{i}^{N_\theta} \frac{g_i e_i}{\lambda_i - L_i} \tag{16}$$

The eigenvalues, λ_i, are now shifted by the parameters L_i. By controlling the signs of the $\lambda_i - L_i$ expressions, the eigenmode algorithm can be made to converge to stationary points that match the sign of $\lambda_i - L_i$. For example, if one $\lambda_i - L_i$ is enforced to be negative, the algorithm will converge to a first order saddle point. In contrast to Newton-Raphson, eigenmode methods can be used to locate specific classes of stationary points, regardless of the eigenvalues at the starting point. In this particular implementation, these methods can be used to step away from local minima and converge to first order transition states.

A number of formulations have been proposed to determine the shift parameters. In this work, the method proposed in [34] is adopted. The rules for choosing the shift parameters depend on the role of the particular eigenmode. If the eigenmode is selected for minimization (decrease in energy), the following rules apply :

$$L_i \;=\; \begin{cases} 0 & \text{if } \lambda_i > 0 \\ 2\lambda_i & \text{if } \lambda_i < 0 \end{cases} \tag{17}$$

These conditions enforce $\lambda_i - L_I > 0$, as required for minimization. In contrast, when the eigenmode is selected for maximization the following rules are used :

$$
L_i = \begin{cases} \frac{1}{2}(\lambda_i + \sqrt{\lambda_i^2 + 4g_i^2}) & \text{if } \lambda_i > 0 \\ 0 & \text{if } \lambda_i < 0 \end{cases} \tag{18}
$$

When $\lambda_i > 0$, the direction of movement is reversed using an empirical shift parameter.

In this work, the eigenmode method is used to perform $2N_\theta$ first order transition states searches from each local minimum. These searches correspond to an initial step away from the minimum in both the positive and negative directions for each eigenmode of the system (low and high). Once the initial step is taken, the particular eigenmode (for maximization) is chosen by checking the largest overlap with the previously followed eigenmode. This procedure is called the eigenmode following method. Another approach (not used here) would be to select the eigenmode for maximization based on the lowest eigenvalue (least eigenvalue method). Each first order transition state can be linked to the two local minima it connects by perturbing the system away from the saddle point along the eigenmode corresponding to the negative eigenvalue (positive and negative step). This provides the minimum-transition-minimum triples that define the connectivity of the energy surface.

4.3   Transition rates

Once the triples are identified, rates of transitions can be calculated using Rice–Ramsperger–Kassel–Marcus (RRKM) theory. If the system is initially at minimum A, the probability that a transition to minimum B will occur (through transition state TS) is proportional to the ratio of the partition functions at TS and A. The rate for this transition is equal to :

$$
W_{A \to TS \to B} = \frac{kT}{h} \frac{Z_{TS}}{Z_A} \tag{19}
$$

$\frac{kT}{h}$ is a temperature dependent factor that sets the average rate for transitions (at a given temperature). The ratio of partition functions Z_{TS} and Z_A represents the probability that the transition state can be attained. This implies that once the transition state is encountered via thermal fluctuations from minimum A, the transition to minimum B is essentially complete. If the harmonic approximation is again employed (as in Equation (5)), Equation (19) becomes :

$$
W_{A \to TS \to B} = \frac{\prod_i^{N_\theta} f_{i,A}}{\prod_i^{N_\theta - 1} f_{i,TS}} \exp\left[-\frac{(E_{TS} - E_A)}{kT} \right] \tag{20}
$$

Here E_{TS} and E_A are the energies of the system at the transition state and minimum A, respectively. The f_i represent the vibrational frequencies of the protein, which can be obtained through the solution of the following problem :

$$
\left(H - (2\pi f)^2 I \right) \theta = 0 \tag{21}
$$

As before, H is the second derivative matrix of the energy function. I now represents the true inertia tensor of the molecular system.

It is important to note that the product of frequencies for the transition state (in the denominator of Equation (20)) involves one less mode than for a local minimum. In particular, this mode is the eigenmode corresponding to the negative eigenvalue and represents the reaction coordinate for the system. For this reason, the factors $\frac{kT}{h}$ (one from Equation (19) and one from the extra vibrational frequency for minimum A) negate each other.

A similar relationship can be written for the reverse transition. That is, the rate of going from minimum B through transition state TS to minimum A is equal to the following :

$$W_{\text{B}\rightarrow\text{TS}\rightarrow\text{A}} \quad = \quad \frac{\prod\limits_{i}^{N_\theta} f_{i,\text{B}}}{\prod\limits_{i}^{N_\theta-1} f_{i,\text{TS}}} \exp\left[-\frac{(E_{\text{TS}}-E_{\text{B}})}{kT}\right] \tag{22}$$

In addition, the transition rate between any two minima can be calculated by summing the rates for all transitions along the pathway(s) connecting these two minima.

4.4   Rate connectivity

To better analyze the connectivity of the minima a "rate disconnectivity graph", as proposed in [37], can also be constructed. This graphical description of the energy surface stems from the concepts originally given in [12], which introduce the ideas of an energy tree as derived from energy barriers between minima. The ideas were later refined to include a generalization for graphically representing the clustering of minima for finite energy or temperature scales [9]. This "disconnectivity graph" shows how disconnected minima merge as energy or temperature increases.

In [37] this concept was taken a step further by defining a "rate disconnectivity graph"; a graphical representation based on transition rates instead of energy barriers. Here the vertical axis represents the transition rate, which begins at the transition rate cutoff. If all minima are connected, there will be a frequency low enough so that all minima are grouped under a single node. That is, transitions must exist which connect all minima in the graph. The transitions with the lowest frequency are branched near the top of the graph. As the cutoff frequency is increased, certain transitions are eliminated and the graph begins to bifurcate.

Once bifurcation occurs, the graph can be used to check the connectivity between two minima. Minima within the same group are connected by a transition pathway, while minima from different groups are not connected. The maximum point (vertically) along the pathway between two minima identifies the slowest transition. As the rate cutoff continues to increase, fewer transitions remain connected and a finer partitioning will appear. Eventually, at a frequency above the highest transition rate, all minima will define their own group.

5   Computational Studies : Met-Enkephalin

Met-enkephalin (H–Tyr–Gly–Gly–Phe–Met–OH) is an endogenous opioid molecule found in the human brain, pituitary, and peripheral tissues and is involved in a variety of physiological processes. The peptide consists of 24 independent torsional angles and a total of 75 atoms, and has played the role of a benchmark molecular conformation problem. All 24 dihedral angles are considered variable, with the 10 dihedral angles of the backbone residues acting as global variables (variables on which branching occurs). Both approaches (EDA and FEDA) for finding low energy conformers were applied to the isolated form of met-enkephalin using the ECEPP/3 force–field. The EDA was also applied to the RRIGS + ECEPP/3 solvated form of this pentapeptide.

A detailed presentation of the results, including the derivation of thermodynamic quantities and identification of a folding transition, can be found in [20]. In this work the unsolvated FEDA results (with 10 independent runs and initial temperature values varying from 50 to 500 K in 50 K increments) and the solvated EDA results are utilized. The ensemble of local minima generated by these runs are used to identify a set of distinct conformers by checking for repeated and symmetric conformations. An additional criterion, which requires that at least one dihedral angle varies by more the $50°$ when comparing each pair of structures, is imposed. The unique conformations are then used to generate ranked distributions according to energy and free energy values. The density of unique conformers is determined by separating the energy ranking into discrete 0.5 kcal/mol energy bins relative to the global minimum energy structure. After performing this analysis, the density of states was found to follow a Boltzmann-like distribution.

In the case of isolated met-enkephalin, the 10 FEDA runs generated a total of 87974 distinct local minima. The potential energy ground state conformation for met-enkephalin consists of a type II' β–bend along the N–C' peptidic bond of Gly^3 and Phe^4, with a total energy of -11.707 kcal/mol. The 10 solvated EDA runs produced 72784 distinct minima. In this case, the ground state exhibits a more extended backbone than the unsolvated form, although the aromatic rings are still proximate. When considering the harmonic free energy, the prediction of the free energy global minimum (FEGM) can be calculated over a range of temperatures. Table 1 compares the FEGM structures at 300 K to the ground state structures for both the unsolvated and solvated systems.

The unsolvated results indicate that the FEGM structure possesses a potential energy contribution 1.808 kcal/mol higher than the ground state structure. The change in dihedral angle values, especially those for the ϕ and ψ angles, also point to an important structural variation between these conformations. In fact, the FEGM conformation exhibits central residues within the α helical region of the Ramachandran plot; which is a significant change from the type II' β turn of the ground state structure. For the solvated form, the relative energy difference between the FEGM and ground state structures is even larger than for the unsolvated case. In addition, the central residues for the FEGM conformation are essentially fully extended – even more so than the ground state structure. These results emphasize the importance that entropic effects can have in defining the free energy of the system.

The qualitative effect of adding entropic contributions can be determined by comparing the distribution of distinct minima for the original energy and the free energy (300 K) of the system. Table 2 provides the number of distinct minima in 0.5 kcal/mol (free) energy

Table 1: Dihedral angle values for the ground state and free energy global minimum (FEGM at 300 K) structures of met-enkephalin. Columns 3 and 4 present the results for the unsolvated form, while columns 5 and 6 are for solvated met-enkephalin. The last two rows indicate the energy and harmonic free energy values (kcal/mol).

Residue	DA	Unsolvated		Solvated	
		Ground	FEGM (300 K)	Ground	FEGM (300 K)
Tyr[1]	ϕ	-83.4	179.8	-168.2	-168.4
	ψ	155.8	-18.2	-30.9	-34.3
	ω	-177.1	-178.1	178.6	-178.9
	χ_1	-173.2	178.2	-173.5	178.7
	χ_2	79.3	81.3	-100.9	-100.8
	χ_3	-166.3	177.3	19.3	179.0
Gly[2]	ϕ	-154.3	-59.8	78.5	177.8
	ψ	85.8	-37.6	-86.5	-179.9
	ω	168.5	-178.8	-177.3	180.0
Gly[3]	ϕ	83.0	-67.0	162.4	-180.0
	ψ	-75.0	-40.1	92.2	179.9
	ω	-170.0	179.7	172.6	179.7
Phe[4]	ϕ	-136.9	-70.9	-150.3	-155.3
	ψ	19.1	-39.5	159.8	147.2
	ω	-174.1	-179.8	-178.1	-176.8
	χ_1	58.9	173.9	65.8	-179.5
	χ_2	94.5	-102.6	-87.4	-111.7
Met[5]	ϕ	-163.5	-161.0	-75.0	-78.7
	ψ	160.9	122.1	113.9	-51.1
	ω	-179.8	-178.0	-178.4	179.7
	χ_1	52.9	-174.7	-172.3	-67.2
	χ_2	175.3	174.0	176.1	-178.8
	χ_3	-179.9	179.0	-180.0	-179.9
	χ_4	-178.6	-60.1	60.0	-180.0
F^{har}		-11.707	14.175	-50.060	-28.604
E		-11.707	-9.899	-50.060	-46.030

Table 2: Number of distinct minima in ranked (free) energy bins. Each bin represents a 0.5 kcal/mol range above the previous bin. The 0 K results do not include entropic effects.

Bin	Unsolvated		Solvated	
	0 K	300 K	0 K	300 K
1	2	5	10	19
2	3	16	14	98
3	12	42	34	378
4	46	97	117	885
5	47	208	326	1730
6	87	403	717	2812
7	161	846	1440	4451
8	297	1524	2611	5390
9	543	2597	3891	6301
10	828	4032	5567	6736
11	1066	5726	6677	6675
12	1527	7499	7624	6295
13	2244	9315	7650	5756
14	2818	10862	7047	5113
15	3657	12004	6375	4361
16	4472	12167	5534	3437

bins within 8.0 kcal/mol of the FEGM. As previously mentioned, the distributions are Boltzmann-like, with the distributions becoming more dense at 300 K. For the solvated case this increased density is especially dramatic. These results indicate that, as temperature increases, the relative stability of entropic contributions offsets substantial differences in energy. At large bin number the number of minima begins to decrease, especially for the solvated case. This is most likely due to inadequate sampling of high energy minima, which would most likely fill these free energy bins.

It is also useful to calculate relative free energies for clusters of low energy conformers because it is impossible to determine the population of a given structure class based on a pointwise approximation of entropic effects for individual structures. By clustering conformers into classes, the error associated with the harmonic approximation should also be reduced. In this work, structures are grouped according to the Zimmerman codes for the central residues of the peptide [39]. In particular, two conformers belong to the same cluster if the central three residues possess the same Zimmerman code. The relative free energy of a cluster is defined as :

$$F_{\text{cluster}} = -k_B T \ln \left[\sum_{i \in C} p_i^{\text{har}} \right] \tag{23}$$

Here, the individual p_i^{har} are calculated via Equation (10), and the set C defines conformations with the same structural classification. A reference free energy is also used to normalize the probabilities. The results for unsolvated and solvated met-enkephalin at 300

Table 3: Clustered relative free energies for met-enkephalin. The information provided in this table includes : Zimmerman code*, number of individual conformers in cluster, total probability ($\sum p_i^{\text{har}}$) and free energy of cluster (F_{cluster}).

Unsolvated				Solvated			
Code	Number	$\sum p_i^{\text{har}}$	F_{cluster}	Code	Number	$\sum p_i^{\text{har}}$	F_{cluster}
CD*A	2128	0.263	0.796	E*EE	148	0.0474	1.818
C*DE	1360	0.125	1.239	EE*E	152	0.0445	1.856
AAA	327	0.111	1.309	D*E*E	149	0.0273	2.147

* S. S. Zimmerman, M. S. Pottle, G. Némethy, and H. A. Scheraga, *Macromolecules*, 10, 1-9 (1977).

K are shown in Table 3.

In general, the structural characteristics of the cluster with the lowest free energy are not necessarily the same as that of the FEGM. The results for unsolvated met-enkephalin indicate that the CD*A class of structures is dominant at 300 K, in part due to the large number of low energy structures within this class. This group of structures is similar to the ground state conformation, except that the type II' β-bend has shifted from the Gly^3-Phe^4 to the Gly^2-Gly^3 backbone region. Figure 3 compares the ground state structure to the lowest free energy structure within the CD*A cluster. For the case of solvated met-enkephalin, the FEGM structure actually belongs to the dominant cluster type (E*EE). Figure 4 illustrates the differences between the extended FEGM structure class and the ground state conformer.

To meaningfully characterize the energy surface, an obvious choice is to focus the analysis on connectivity to the ground state conformation. In addition, transitions between the ground state structure and representative structures of the dominant free energy cluster at 300 K were selected for folding pathway determination. For these reasons, the set of low energy conformers used to initialize the transition state searches are based on the union of the 1000 lowest energy minima and the 1000 lowest free energy minima at 300 K. In this way, the connectivity between low free energy structures and the ground state structure could be explored, while maintaining a tractable number of starting structures.

For each unique local minima the transition state search entailed 48 eigenmode following runs; that is, 2 initial step directions for all 24 unique eigenmodes. These searches were implemented in parallel using MPI on NPACI's HP Exemplar. Following these runs, the set of unique first order transition states was identified and ranked. For each transition state, the next step involved the determination of the minimum-transition-minimum connectivity, which required 2 additional eigenmode searches corresponding to the positive and negative step along the reaction coordinate. For unsolvated met-enkephalin, this procedure resulted in the identification of 51272 total stationary points, with 22775 local minima and 28497 first order transition states. The corresponding results for solvated met-enkephalin were 76828 total stationary points, 34722 minima and 42106 first order transition states.

To quantify the connectivity to the ground state, the transition rates were calculated for each minimum-transition-minimum triple. This information can be used to compile the

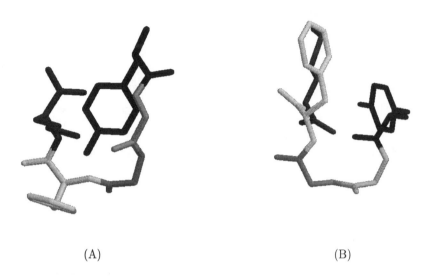

(A)   (B)

Figure 3: Ground state structure for unsolvated met-enkephalin (A). Lowest energy unsolvated structure for dominant free energy cluster at 300 K (B). Only heavy atoms are represented.

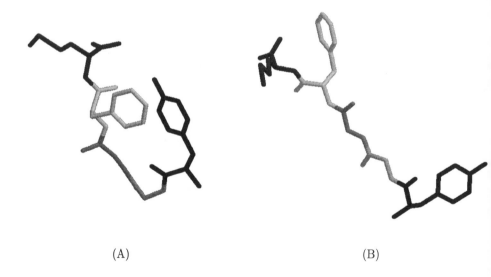

(A)   (B)

Figure 4: Ground state structure for solvated met-enkephalin (A). Lowest energy solvated structure for dominant free energy cluster at 300 K (B). Only heavy atoms are represented.

Table 4: Number of local minima connected to the ground state conformation for transition rate cutoffs of 10^7, 10^9 and 10^{11} Hz.

Transition Rate Cutoff (Hz)	Unsolvated # Connected Minima	Solvated # Connected Minima
10^7	15753	25148
10^9	1685	20660
10^{11}	131	1040

rate disconnectivity graph, and thus identify the number of connected minima for various transition rate cutoffs. Table 4 shows the number of minima connected to the ground state conformation for various transition rate cutoffs. Initially, at a 10^7 Hz rate cutoff, a large fraction of the minima ($\approx$ 70 %) are still connected to the unsolvated ground state conformation. However, as higher rate cutoffs are applied, the node containing the ground state becomes quite small, with only 131 minima at a 10^{11} Hz cutoff. In contrast, the solvated ground state is connected to a larger fraction of minima at all cutoff frequencies. This is particularly evident for the 10^9 Hz rate cutoff. These observations imply that for the solvated system there exist faster transitions which link the ground state conformation to larger regions of the energy surface.

These results can be visualized by plotting the rate disconnectivity graphs for the various rate cutoffs. The corresponding graphs are depicted in Figures 5 and 6. For unsolvated met-enkephalin these graphs reveal that the branch connecting the ground state conformation to the majority of other minima involves relatively slow transitions; that is, between 10^7 and 10^8 Hz. For this reason, the ground state cluster for 10^9 Hz cutoff comprises only a small portion of the complete graph. In contrast, Figure 6 shows that the solvated ground state conformer is part of the main graph because several connected branches exhibit transitions higher than 10^{10} Hz. Even for the 10^{11} Hz cutoff the node containing the ground state conformation is quite dense when compared to the corresponding unsolvated disconnectivity graph. These graphs illustrate the liquid-like behavior of the solvated system.

The examination of folding pathways was accomplished by characterizing the transition between the ground state and the lowest free energy minimum of the dominant cluster. In the case of unsolvated met-enkephalin, this transition characterizes the shifting of the type II' β-bend from the Gly^2–Gly^3 to the Gly^3–Phe^4 backbone region. The representative structure for the dominant free energy cluster (denoted as CD^*A_1) at 300 K exhibits an energy of -9.402 kcal/mol, which is 2.305 kcal/mol higher in energy than the ground state structure. For solvated met-enkephalin, the fully extended to ground state pathway was explored. In this case, the dominant free energy cluster (E^*EE) is represented by the FEGM at 300 K. This structure (denoted as E^*EE_1) has an energy of -46.030 kcal/mol, which is about 4 kcal/mol higher in energy than the ground state conformer. In addition, the second lowest free energy minimum of the E^*EE cluster (denoted as E^*EE_2) was included in the pathway analysis. This minimum exhibits a slightly higher energy of -45.712 kcal/mol.

One way to characterize the folding transition is to examine the pathways connecting two

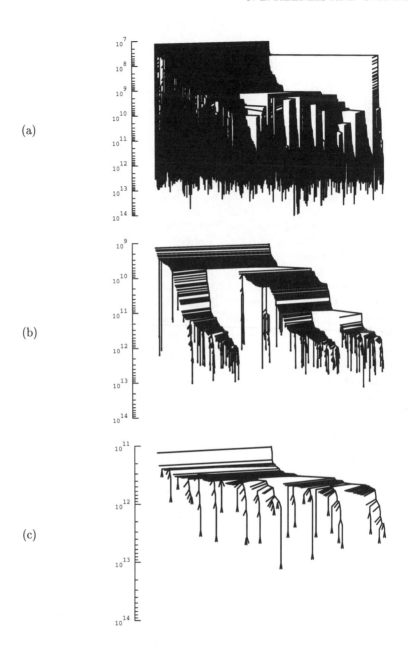

Figure 5: Rate disconnectivity graphs of unsolvated met-enkephalin for transition rate cutoffs of 10^7 (a), 10^9 (b) and 10^{11} (c) Hz. Only the node containing the ground state conformation is shown (at the top of the graph). The ground state conformer always corresponds to the rightmost branch.

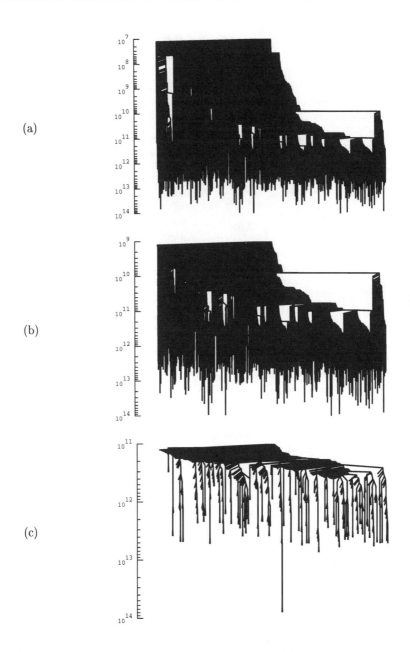

Figure 6: Rate disconnectivity graphs of solvated met-enkephalin for transition rate cutoffs of 10^7 (a), 10^9 (b) and 10^{11} (c) Hz. Only the node containing the ground state conformation is shown (at the top of the graph). The ground state conformer always corresponds to the rightmost branch.

Table 5: Number of unique pathways between two minima. The first column denotes the length of the pathway in terms of the number of transitions. For unsolvated met-enkephalin, the overall transition is between the ground state conformer and the lowest free energy conformer for the dominant free energy cluster (CD^*A_1). For solvated met-enkephalin two transitions with the ground state conformer are considered : the first (E^*EE_1) and second (E^*EE_2) lowest free energy conformers for the dominant free energy cluster (E^*EE).

# of	Unsolvated	Solvated	
Transitions	$CD^*A_1 \leftrightarrow$ Ground	$E^*EE_1 \leftrightarrow$ Ground	$E^*EE_2 \leftrightarrow$ Ground
5	–	–	2
6	–	–	–
7	4	13	17
8	18	108	16
9	148	1082	135
10	1228	8583	436
11	10024	62867	2595
12	78936	102124	14154

minima. Through graph theory techniques an exhaustive search can be used to enumerate the full set of pathways that connect these two minima. However, such a search will quickly become impractical when the total number of connected minima is large. To avoid this problem, the enumeration was conducted for pathways comprising less than 12 transitions states. The number of unique pathways for different numbers of transition states (for the minima defined in the preceding paragraph) are given in Table 5. These results indicate a stronger connectivity among the solvated minima, which corroborate the results from the disconnectivity graphs. In particular, when considering the transition between the ground state and the representative structure from the dominant free energy cluster, the solvated case exhibits an order of magnitude increase in the number of pathways with an equal number of transitions. In addition, when considering the second solvated transition, a number of relatively short pathways can be identified.

Since the length of the pathway is not necessarily proportional to the overall transition rate between the two minima, a more detailed analysis of these folding pathways must be considered. A visual method for accomplishing this goal is to trace a pathway between the two minima on the rate disconnectivity graph. The overall transition time is related to the location of the highest branch (slowest transition) in the connectivity tree. Figure 7 illustrates the tracing of folding pathways for unsolvated and solvated met-enkephalin. For unsolvated met-enkephalin the traced pathway includes 7 transition states. To display the entire pathway, the disconnectivity graph with a 10^7 transition rate cutoff was constructed. This implies that the lowest transition rate occurs in the range of 10^7 to 10^8 Hz. In the case of solvated met-enkephalin, the shortest pathway connecting E^*EE_2 to the ground state includes only 5 transition states. In addition, the transition rates between these minima are relatively fast, and a high cutoff of 10^{11} Hz is sufficient for displaying the entire traced pathway.

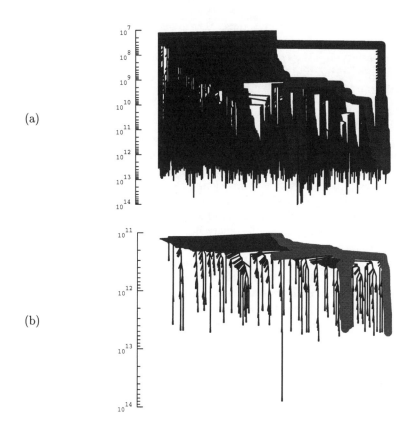

Figure 7: Tracing pathways on rate disconnectivity graphs. (a) shows a 7 transition state pathway (for unsolvated met-enkephalin) connecting CD*A$_1$ to the ground state on a disconnectivity graph with a 10^7 rate cutoff. (b) shows a 5 transition state pathway (for solvated met-enkephalin) connecting E*EE$_2$ to the ground state on a disconnectivity graph with a 10^{11} rate cutoff. The traced pathways are indicated by thick grey lines which extend to the root node. The highest branch constitutes the slowest transition. The ground state always corresponds to the rightmost branch.

Another method for visualizing the folding pathways is given in Figure 8. In this figure, $\phi - \psi$ conformational plots are used to illustrate the structural changes which occur along the pathway. The same unsolvated and solvated pathways shown in Figure 7 are graphed. In each plot, five lines are drawn, with each line corresponding to one of five residues for met-enkephalin. Black dots pinpoint the $\phi - \psi$ values for the transition states along the pathway, while minima are represented by grey dots. The plots imply that along the unsolvated pathway, large structural variations correspond to changes in ϕ values. In contrast, many residues in the solvated pathway undergo only small changes, while two residues display significant variation in both ϕ and ψ variables. Similar plots can be constructed for all dihedral angles, and used to identify important variables along the folding pathway.

6 Conclusions

An important step in better understanding the protein folding process is the characterization of energy surfaces. Rigorously, this characterization requires the identification of all stationary points of the energy function. Along these lines, deterministic methods have been applied to small peptide systems [37]. However, the jump to larger peptide systems (with astronomically large numbers of minima) is currently intractable.

In this work, a new methodology for analyzing energy surfaces of peptides was introduced. The approach is based on the identification of low (free) energy local minima using a modification of the αBB deterministic global optimization technique [20]. Once these minima have been identified, the low energy regions are explored by transition state searches and the determination of minimum-transition-minimum triples.

The methodology was applied to both unsolvated and solvated forms of met-enkephalin. On the order of 80000 local minima were identified for both forms. In addition to typical force–field components, entropic contributions were accounted for through a harmonic approximation. The results were used to determine the free energy global minima at 300 K, as well as the dominant free energy cluster at this temperature.

In order to map out the low energy regions, the union of 1000 lowest energy and 1000 lowest free energy minima was compiled and used to initiate transition state searches. First order transition states, as well as minimum-transition-minimum triples, were identified through a parallel implementation of eigenmode following methods. The final set of stationary points included 22775 minima and 28497 first order transition states for unsolvated met-enkephalin, and 34722 minima and 42106 first order transition states for solvated met-enkephalin.

Transition rates and connectivities were then calculated for these structures. To explore the connectivity to the ground state conformations, rate disconnectivity graphs were constructed for different rate cutoffs. This analysis indicated higher transition rate connectivity for the solvated form of met-enkephalin.

In addition, the transitions between specific minima were examined. These transitions were first characterized by the number of pathways (for a given number of transition states). Visual interpretation of these results included pathway tracing on the disconnectivity graph and conformational plots of $\phi - \psi$ variable transformations along the pathway.

In conclusion, the combination of a deterministic global optimization based search for low (free) energy ensembles and eigenmode searches for transition states and minimum-

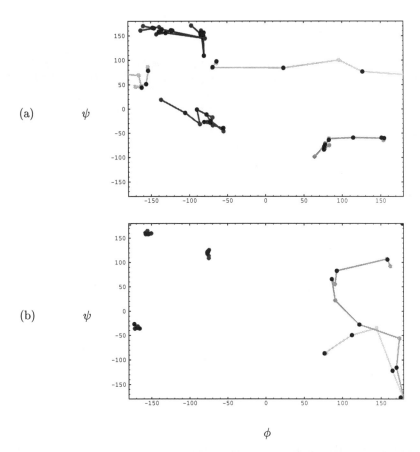

Figure 8: Tracing pathways on $\phi - \psi$ conformational plots. Each line corresponds to one of five residues for met-enkephalin. The black dots represent transition states, while grey dots are minima. The same pathways given in Figure 7 are plotted. (a) shows a 7 transition state pathway (for unsolvated met-enkephalin) connecting CD*A$_1$ to the ground state. (b) shows a 5 transition state pathway (for solvated met-enkephalin) connecting E*EE$_2$ to the ground state.

transition-minimum triples can be used to effectively characterize energy surfaces. The techniques were applied to the unsolvated and solvated forms of met-enkephalin and used to perform a comparative analysis of these systems. A number of visualization techniques were also shown to be useful for further elucidating the folding process for this peptide.

Acknowledgments

The authors gratefully acknowledge financial support from the National Science Foundation, Air Force Office of Scientific Research, the National Institutes of Health (R01 GM52032) and NPACI (for parallel computing resources).

References

[1] Adjiman C. S., Androulakis I. P., Maranas C. D. and Floudas C. A. (1996), "A global optimization method, αBB, for process design," *Comp. Chem. Eng. Vol. 20*, S419–S424.

[2] Adjiman C. S., Androulakis I. P. and Floudas C. A. (1997), "Global optimization of MINLP problems in process synthesis and design," *Comp. Chem. Eng. Vol. 21*, S445–S450.

[3] Adjiman C. S., Dallwig S., Floudas C. A. and Neumaier A. (1998), "A global optimization method, αBB, for general twice-differentiable constrained NLPs – I. Theoretical advances," *Comp. Chem. Eng. Vol. 22*, 1137–1158.

[4] Adjiman C. S., Androulakis I. P. and Floudas C. A. (1998), "A global optimization method, αBB, for general twice-differentiable constrained NLPs – II. Implementation and computational results," *Comp. Chem. Eng. Vol. 22*, 1159–1179.

[5] Allinger N. L., Yuh Y. H. and Lii J. H. (1989), "Molecular mechanics. The MM3 force field for hydrocarbons," *J. Am. Chem. Soc. Vol. 111*, 8551–8565.

[6] Androulakis I. P., Maranas C. D. and Floudas C. A. (1995), "αBB : A global optimization method for general constrained nonconvex problems," *J. Global Opt. Vol. 7*, 337–363.

[7] Androulakis I. P., Maranas C. D. and Floudas C. A. (1997), "Global minimum potential energy conformations of oligopeptides," *J. Global Opt. Vol. 11*, 1–34.

[8] Augspurger J. D. and Scheraga H. A. (1996), "An efficient, differentiable hydration potential for peptides and proteins," *J. Comp. Chem. Vol. 17*, 1549–1558.

[9] Becker O. M. and Karplus, M. (1997), "The topology of multidimensional potential energy surfaces: Theory and application to peptide structure and kinetics," *J. Chemical Physics, Vol. 106*, 1495–1517.

[10] Brooks B., Bruccoleri R., Olafson B. and States D. (1983), "CHARMM: A program for macromolecular energy minimization and dynamics calculations," *J. Comp. Chem.* Vol. *4*, 187–217.

[11] Cerjan C. J. and Miller W. H. (1981), "On finding transition states," *J. Chemical Physics* Vol. *75*, 2800–2806.

[12] Czerminski R. and Elber R. (1989), "Reaction path study of conformational transitions and helix formation in a tetrapeptide," *Proc. Natl. Acad. Sci. USA*, Vol. *86*, 6963–6967.

[13] Czerminski R. and Elber R. (1990), "Reaction path study of conformational transitions in flexible systems: Application to peptides," *J. Chemical Physics*, Vol *92.*, 5580–5601.

[14] Flory, P. J. (1974), "Foundations of rotational isomeric state theory and general methods for generating configurational averages," *Macromolecules* Vol. *7*, 381–392.

[15] Floudas C. A., Klepeis J. L. and Pardalos P. M. (1999), "Global Optimization Approaches in Protein Folding and Peptide Docking," *In DIMACS Series in Discrete Mathematics and Theoretical Computer Science, Vol. 47*, 141–171, American Mathematical Society, Providence, Rhode Island.

[16] Gill P. E., Murray W., Saunders M. A. and Wright M. H. (1986), "NPSOL 4.0 User's Guide," Systems Optimization Laboratory, Dept. of Operations Research, Stanford University, California.

[17] Go N. and Scheraga H. A. (1969), "Analysis of the contribution of internal vibrations to the statistical weights of equilibrium conformations of macromolecules," *J. Chemical Physics* Vol. *51*, 4751–4767.

[18] Go N. and Scheraga H. A. (1976), "On the use of classical statistical mechanics in the treatment of polymer chain conformations," *Macromolecules* Vol. *9*, 535–542.

[19] Klepeis J. L., Androulakis I. P., Ierapetritou M. G. and Floudas C. A. (1998), "Predicting solvated peptide conformations via global minimization of energetic atom-to-atom interactions," *Comp. Chem. Eng.* Vol. *22*, 765–788.

[20] Klepeis J. L. and Floudas C. A. (1999), "Free Energy Calculations for Peptides via Deterministic Global Optimization," *J. Chemical Physics*, Vol. *110*, 7491–7512.

[21] Klepeis J. L. and Floudas C. A. (1999), "A comparative study of global minimum energy conformations of hydrated peptides," *J. Comp. Chem.* Vol. *20*, 636–654.

[22] Kunz R. E. and Berry R. S. (1995), "Statistical interpretation of topographies and dynamics of multidimensional potentials," *J. Chemical Physics*, Vol. *103*, 1904–1912.

[23] Laidler K. J. (1989), "Chemical Kinetics," Harper Collins, New York.

[24] Li Z. and Scheraga H. A. (1988), "Structure and free energy of complex thermodynamic systems," *J. Molecular Structure (Theochem)* Vol. *179*, 333–352.

[25] Maranas C. D. and Floudas C. A. (1994), "Global minimum potential energy conformations of small molecules," *J. Global Opt. Vol. 4*, 135–170.

[26] Meirovitch H. and Meirovitch E. (1997), "Efficiency of Monte Carlo minimization procedures and their use in analysis of NMR data obtained from flexible peptides," *J. Comp. Chem. Vol. 18*, 240–253.

[27] Momany F. A., Carruthers L. M., McGuire R. F. and Scheraga H. A. (1974), "Intermolecular potential from crystal data. III.," *J. Physical Chemistry Vol. 78*, 1595–1620.

[28] Momany F. A., Carruthers L. M. and Scheraga H. A. (1974), "Intermolecular potential from crystal data. IV.," *J. Physical Chemistry Vol. 78*, 1621–1630.

[29] Momany F. A., McGuire R. F., Burgess A. W. and Scheraga H. A. (1975), "Energy parameters in polypeptides. VII.," *J. Physical Chemistry Vol. 79*, 2361–2381.

[30] Némethy G., Gibson K. D., Palmer K. A., Yoon C. N., Paterlini G., Zagari A., Rumsey S. and Scheraga H. A. (1992), "Energy parameters in polypeptides. 10.," *J. Physical Chemistry Vol. 96*, 6472–6484.

[31] Neumaier A. (1997), "Molecular Modeling of Proteins and Mathematical Prediction of Protein Structure," *SIAM Rev., Vol. 39*, 407–460.

[32] Nichols J., Taylor H., Schmidt P. and Simons J. (1990), "Walking on potential energy surfaces," *J. Chemical Physics Vol. 92*, 340–346.

[33] Simons J., Jorgensen P., Taylor H. and Ozment J. (1983), "Walking on potential energy surfaces," *J. Physical Chemistry Vol. 87*, 2745–2753.

[34] Tsai C. J. and Jordan K. D. (1993), "Use of an eigenmode method to locate the stationary points on the potential energy surfaces of selected argon and water clusters," *J. Physical Chemistry Vol. 97*, 11227–11237.

[35] Weiner S., Kollman P., Case D. A., Singh U. C., Ghio C., Alagona G., Profeta S. and Weiner P. (1984), "A new force field for molecular mechanical simulation of nucleic acids and proteins," *J. Am. Chem. Soc. Vol. 106*, 765–784.

[36] Weiner S., Kollman P., Nguyen D. and Case D. (1986), "An all atom force field for simulations of proteins and nucleic acids," *J. Comp. Chem. Vol. 7*, 230–252.

[37] Westerberg K. M. and Floudas C. A. (1999), "Locating All Transition States and Studying the Reaction Pathways of Potential Energy Surfaces," *J. Chemical Physics, Vol. 110*, 9259–9295.

[38] Westerberg K. M. and Floudas C. A., "Dynamics of Peptide Folding: Transition States and Reaction Pathways of Solvated and Unsolvated Tetra-Alanine," *J. Global Optimization*, accepted for publication.

[39] Zimmerman S. S., Pottle M. S., Némethy G. and Scheraga H. A. (1977), "Conformational analysis of the 20 naturally occurring amino acid residues using ECEPP," *Macromolecules Vol. 10*, 1–9.

Optimization in Computational Chemistry and Molecular Biology, pp. 47-55
C. A. Floudas and P. M. Pardalos, Editors

Energy Landscape Projections of Molecular Potential Functions

Andrew T. Phillips
Department of Computer Science
University of Wisconsin - Eau Claire
Eau Claire, WI 54702 USA
phillips@cs.uwec.edu

J. Ben Rosen
Computer Science and Engineering Department
University of California, San Diego
San Diego, CA 92093 USA
jbrosen@cs.ucsd.edu

Ken A. Dill
Department of Pharmaceutical Chemistry
University of California, San Francisco
San Francisco, CA 94143 USA
dill@maxwell.ucsf.edu

Abstract

Key problems in computational biology, including protein and RNA folding and drug docking, involve conformational searching over multidimensional potential surfaces with very large numbers of local minima. This paper shows how statistics provided by the CGU global optimization algorithm can be used to characterize and interpret these topographies using a 2-dimensional landscape projection.

Keywords: Energy landscapes, protein folding.

1   Introduction

The CGU (Convex Global Underestimator) global optimization algorithm [1], is very different than Molecular Dynamics (MD), Monte Carlo (MC), Simulated Annealing (SA), or

Genetic Algorithms (GA). The CGU method does not search the tops of energy landscapes, does not get caught in kinetic traps, and its speed does not depend on the shapes of energy landscapes (amino acid sequences), but only on the sizes of such landscapes. And in addition to the global minimum, the CGU method also computes a large number of low energy local minima. This is most useful for learning the shapes of energy landscapes and the nature of energy gaps, which is helpful in computational studies of folding kinetics. In fact, we have even devised a novel way to display this n-dimensional landscape, where n is the number of degrees of freedom, in a simple 2-dimensional graph that shows the distribution of all the local minima.

To obtain such an energy landscape, the CGU method searches for the globally optimal lowest energy conformation Φ_G by constructing a convex function which underestimates all known local minima and does so by the least possible amount. Based on the premise that protein folding energy landscapes are funnels with bumps ([2, 3, 4, 5]), there is information about where to find the native structure distributed everywhere throughout the landscape. To find the bottom of a funnel, even a bumpy one, you need to head generally downhill. In general, we know that most well-understood proteins must have such landscapes. The idea of a funnel is nothing more than the statement that proteins fold much faster than the "Levinthal time", the exhaustive search time, and always to the same unique state.

The CGU simply makes use of the overall funnel-like shape to guide and localize the search to regions that are estimated to be near the native structure. Since the lateral area of an energy landscape at a given depth represents the number of conformations having the same internal free energy, the funnel idea is simply that as folding progresses toward lower energies, the chain's conformational options become increasingly narrowed, ultimately resulting in the one native structure. This is fundamentally a consequence of the fact that proteins are heteropolymers.

Given a primary sequence of amino acids with n degrees of freedom $\Phi \in \Re^n$, and a potential energy function $F(\Phi)$, the CGU strategy for searching for the global energy minimum $F(\Phi_G)$ of $F(\Phi)$ involves an iterative process of three phases during each iteration: (I) sampling the landscape, (II) forming the convex global underestimator surface $U(\Phi)$, a parabolic surface under the lowest minima found so far, and (III) finding the minimum on this underestimator surface which is then used in the next iteration to localize the search region further. Figure 1 provides an $n = 2$ dimensional example of a potential function characterized by a rugged energy landscape with numerous kinetic traps, energy barriers, and narrow pathways to the native state.

In fact, this potential function, while artifical, shares many of the characteristics believed to be present in actual protein folding energy landscapes [2, 3, 6, 7, 8]. By way of comparison, Figure 2 shows the corresponding CGU underestimating surface $U(\Phi)$ for this same potential function. Landscapes such as Figure 1 are ideal for the CGU method since the convex underestimator closely approximates the funnel, and ignores the bumps, as the algorithm narrows its search region.

While the full details of the CGU method are described in [1, 9], a brief description is presented here. In phase I, $k \geq 2n + 1$ local energy minimum conformations $\Phi^{(j)}$ are generated in the search region of interest (a minimum of $2n + 1$ conformations are required for construction of the convex underestimator in n dimensions and $8n + 4$ are used in the actual implementation). These conformations are sampled from a uniform random

Figure 1: An example potential function $F(\Phi)$ with $n = 2$

distribution (over the desired search region) and then relaxed to a local energy minimum state by a Sequential Quadratic Programming (SQP) continuous minimization technique.

In phase II, the CGU function $U(\Phi)$ is then constructed as a more global surface to "fit" these k local minima (and any other known local minima in the desired search region) by underestimating all of them in the least possible amount (i.e. the L1 norm) by solving the optimization problem:

$$\min \sum_{j=1}^{k} \delta_j \tag{1}$$

where $\delta_j = F(\Phi^{(j)}) - U(\Phi^{(j)}) \geq 0$ is required for all conformations $j = 1, \ldots, k$. For $U(\Phi)$ we use a separable quadratic function of the form

$$U(\Phi) = c_0 + \sum_{i=1}^{n} (c_i \Phi_i + \frac{1}{2} d_i \Phi_i^2). \tag{2}$$

This choice is not essential but has many important benefits. First, convexity of $U(\Phi)$ is easily guaranteed by simply requiring $d_i \geq 0$ in Eq 2. Second, since c_i and d_i appear only linearly in the constraints of Eq 2, the solution to Eq 1 can be computed by a simple linear programming technique, the complete details of which are given in [1]. Third, the minimum energy conformation of $U(\Phi)$, denoted Φ_P is very easily computed by $(\Phi_P)_i = -c_i/d_i$. This conformation then serves as a *prediction* for Φ_G. In this way the CGU searches under the landscape of $F(\Phi)$ and provides a prediction Φ_P which can then be used in phase III.

Given the predicted structure Φ_P and the best known local minimum structure computed so far, denoted Φ_L, in phase III the search region is localized around Φ_P while also including Φ_L. Phases I-III are repeated over the new search regions until $\Phi_P = \Phi_L$. That is, when the CGU predicts $\Phi_P = \Phi_L$, then the method terminates, and Φ_L is declared the global minimum energy conformation.

Figure 2: The CGU underestimating function $U(\Phi)$ for $F(\Phi)$

We have conducted extensive computational testing of the CGU algorithm on a variety of molecular models [1, 9, 10, 11, 12]. These include n-chain homopolymers [1] and the Sun simplified heteropolymer model [13], with up to 36 residues (70 degrees of freedom). The computation time is determined essentially by the number and speed of local minimizations. For an n-dimensional problem, at least $2n+1$ local minima are required in order to determine the convex quadratic underestimator, but we typically use $8n + 4$. The time dependence on n of the local minimizer is therefore crucial to the computational efficiency. In the original version of the CGU implementation, the local minimizer required approximately $O(n^3)$ time per minimization, so that the total time was approximately $O(n^4)$. We are now using a local minimizer based on sequential quadratic programming, which is approximately $O(n^2)$, because far fewer function calls are needed. Therefore the total time needed to find the global minimum now appears to increase as only $O(n^3)$.

The success of the CGU algorithm in finding the global minimum depends on the energy landscape being funnel-shaped. This shape enables the convex underestimator to correctly localize the search to the region containing the global minimum. The molecular models tested in our previously described work [1, 9, 10, 11, 12] all appear to have this funnel property. As a result, the rate of success in finding the global minimum was high. Except for very small molecules, the global minimum of the energy function is not known, so success must be measured in terms of the consistency of the results obtained.

In order to determine the performance and robustness of the CGU algorithm, we have recently carried out extensive computational tests using the simplified heteropolymer model [13]. These tests included models with the number of degrees of freedom n ranging from $n = 4$ to $n = 70$. Since for this model there are 2 backbone angles (ϕ/ψ) per residue, $n = 2r - 2$ where r is the number of residues. The largest model included in these tests therefore consisted of 36 residues. A total of 20 different values of n were tested, as shown in Table 1. For each value of n, the CGU algorithm was applied 20 times, where each time a different set of $8n+4$ randomly chosen initial points (in the n-dimensional space) were used. Thus there were 20 independent solutions computed for each value of n. The lowest value of

$F(\Phi)$ obtained was taken to be the global minimum $F_G < 0$, for the corresponding value of n. The second column in Table 1 gives the CGU success rate in finding F_G. The success rate is defined as the percent of the 20 tests (with a fixed n) for which $0.99 \leq F^{(j)}/F_G \leq 1.0$, where $F^{(j)} < 0$, $j = 1, \ldots, 20$, is the minimum value obtained by the CGU on the jth test. The 3rd and 4th columns of Table 1 give the average and maximum values of $(F^{(j)} - F_G)/|F_G|$. Obviously, if the success rate is 100%, then the corresponding relative error $= 0$. The average and maximum number of CGU iterations needed are given in columns 5 and 6. Based on these results, and many similar tests, the success rate of the CGU algorithm is seen to depend primarily on how well the landscape approximates a funnel, and is essentially independent of the number of degrees of freedom.

Table 1
CGU Performance Summary

n	% success	relative error		# iterations	
		average	maximum	average	maximum
4	100	0	0	4.7	6
6	100	0	0	3	3
8	75	.037	.226	6.9	9
10	100	0	0	3	3
12	100	0	0	3	3
14	100	0	0	3	3
16	45	.041	.152	3.8	7
18	25	.281	.693	4.4	7
20	100	0	0	3	3
22	20	.081	.122	3	3
24	5	.093	.111	3	3
26	100	0	0	3	3
28	100	0	0	3	3
30	20	.065	.084	3	3
32	30	.356	.509	3	3
34	100	0	0	3	3
36	100	0	0	3	3
38	100	0	0	3	3
50	100	0	0	3	3
70	100	0	0	3	3
averages	76%	.049		3.4	

2 CGU Energy Landscape Information

The success of the CGU method described above and in [9] is based, in large part, on the choice of the underestimating function $U(\Phi)$ as a separable quadratic. While many choices for the convex underestimating function are possible, our approach is to use Eq 2 with the added restriction that the sum, over all local minimum conformations $\Phi^{(j)}$ found so far, of $F(\Phi^{(j)}) - U(\Phi^{(j)}) \geq 0$ be a minimum (see Eq 1). This choice provides important insight into the form and features of the energy landscape. This property appears to be unique to our approach.

For each degree of freedom Φ_i, the CGU associates the coefficient $d_i > 0$. We have shown in [10] that, based on the Boltzmann distribution law and the form of the CGU given in Eq 2, we can interpret $(\Phi_G)_i$ as the mean value of Φ_i and $k_B T/d_i$ as the variance σ_i^2. Hence, large d_i indicate a small variance in Φ_i from its global minimum/mean value $(\Phi_G)_i$.

Also, since the true energy landscape of $F(\Phi)$ can be thought of as a surface above an n-dimensional horizontal hyperplane, with each point in the hyperplane representing a conformation Φ, the distribution of local minima, provided by repeated iterations of the CGU method, in effect represents the energy surface $F(\Phi)$, and we have a simple way to visualize this high dimension landscape. Upon completion of the CGU method, we have available a large set of local minimum conformations $\Phi^{(j)}, j = 1, \ldots, k$ (isomers having been removed during each iteration), among which Φ_G is energetically best. We also have a "landscape" CGU which underestimates, in the same minimum sense as before, this entire set of local conformations in such a way that Φ_G remains the global minimum.

As shown in [9], this landscape CGU depends only on Φ_G and on the set of "landscape coefficients" d_i. By defining

$$\overline{\Delta\Phi} = \sqrt{\sum_{i=1}^{n} d_i(\Phi_i - (\Phi_G)_i)^2}, \qquad (3)$$

then this root mean square *weighted* deviation (RMSWD) $\overline{\Delta\Phi}$ provides a simple and convenient means for plotting the energy difference $U(\Phi) - F_G$ for any conformation Φ. Figure 3 shows this two dimensional visualization of the energy landscape for the case of an $n = 20$ dimensional counterpart to the example potential energy function shown in Figure 1. This figure plots the normalized energy gap $(F(\Phi^{(j)}) - F_G)/(F_{max} - F_G)$ for each of the local minima $\Phi^{(j)}, j = 1, \ldots, k$, and it shows their relationship to the landscape CGU energy surface. Here we denote $F_{max} = \max F(\Phi^{(j)})$ for $j = 1, \ldots, k$ (but only those minima with $\overline{\Delta\Phi} \leq 85$ are shown).

It should be noted that $F_G = -20$ and that the next lowest energy local minimum Φ', shown in Figure 3 with $\overline{\Delta\Phi} \approx 15$, has the value $F(\Phi') \approx 117$. Using the Boltzmann distribution probability, we see that the probability of observing this next lower local minimum conformation is [10]

$$P(\Phi') = \frac{\exp^{-(F(\Phi')-F_G)/k_B T}}{\sum_{i=1}^{k} \exp^{-(F(\Phi^{(i)})-F_G)/k_B T}} < 10^{-4}. \qquad (4)$$

Of course, Φ' is the next lowest energy local minimum *that we know about*, and since the CGU method is not guaranteed, nor even intended, to find all the local minima, Eq 4 is not a reliable predictor of the status of any other local minima.

The CGU method, and the corresponding energy landscape prediction, have been applied to a number of small protein models using the simplified Sun energy functions [13]. This potential explicitly treats backbone atoms, but represents sidechains as either hydrophobic or hydrophilic spheres whose radius is a function of the amino acid. Simple hydrophobic and hydrogen bonding terms are present, along with a penalty for ϕ/ψ pairs that lie outside permitted regions of the Ramachandran plot. This potential appears to be

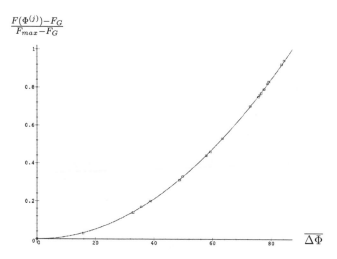

Figure 3: RMSWD Energy Landscape Projection

funnel-like [10], although with a large number of local minima. The virtue of this potential function is that it is very simple and protein-like. The disadvantages are that it is not an accurate folding potential (none is known yet), and its global optima are not known with certainty. The examples tested here include a 9-residue oxytocin, a 30-residue zinc-finger, and a 36-residue avian pancreatic polypeptide (PPT). The global minimum energy conformation obtained by the CGU method for PPT is compared with its known native conformation in Figure 4. Many local minima were obtained during this global minimization computation. Those closest to the global minimum Φ_G are shown by the energy landscape projection in Figure 5. This also shows the CGU surface obtained and how it interpolates the set of local minima which define it. It is also seen that the lowest energy local minimum (after the global minimum) has an energy difference $F^{(j)} - F_G \approx 700$, so the probability of observing any non-global minimum energy conformation is very low.

These examples show clearly the value of the CGU method and the energy landscape projection in helping to understand the relationship between the global minimum energy and the native state conformation. Many other global optimization strategies are available for searching for low energy conformations, and many can provide some form of landscape statistics as well ([14, 15, 16]). But none of these other conformational search strategies provide the CGU's level of energy landscape information for realistic 3D models.

3 Acknowledgment

The authors wish to thank the San Diego Supercomputer Center and NSF grant DBI-9996165 for providing the resources to conduct this research.

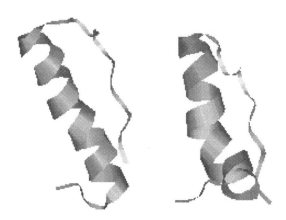

Figure 4: Native Φ_N (left) vs Model Native Φ_G (right) Structures for PPT

References

[1] Phillips, A.T., Rosen, J.B., & Walke, V.H. (1995), "Molecular structure determination by global optimization," *Dimacs Series in Discrete Mathematics and Theoretical Computer Science* 23:181–198.

[2] Dill, K.A. & Chan, H.S. (1997), "From Levinthal to pathways to funnels," *Nature Structural Biology* 4(1):10–19.

[3] Leopold, P.E., Montal, M., & Onuchic, J.N. (1992), "Protein folding funnels: A kinetic approach to the sequence structure relationship," *Proc. Natl. Acad. Sci. USA* 89:8721–8725.

[4] Socci, N.D., & Onuchic, J.M. (1994), "Folding kinetics of protein-like heteropolymers," *J Chem Phys* 100:1519–1528.

[5] Wolynes, P.G., Onuchic, J.N., & Thirumalai, D. (1995), "Navigating the folding routes," *Science* 267:1619–1620.

[6] Bryngelson, J.D. & Wolynes, P.G. (1987), "Spin-glass and the statistical mechanics of protein folding," *Proc. Natl. Acad. Sci. USA* 84:7524–7528.

[7] Bryngelson, J.D. & Wolynes, P.G. (1989), "Intermediates and barrier crossing in a random energy model (with applications to protein folding)," *J. Phys. Chem.* 93:6902–6915.

[8] Dill, K.A. (1987), "The stabilities of globular proteins," In *Protein Engineering* (eds Oxender, D.L. & Fox, C.F.) 187–192 (Alan R. Liss, Inc., New York, 1987).

[9] Phillips, A.T., Rosen, J.B. & Dill, K.A. (1999), "Convex global underestimation for molecular structure prediction," in *From Local to Global Optimization* (eds P.M. Pardalos et al.) in press (Kluwer, Dordrecht, 1999).

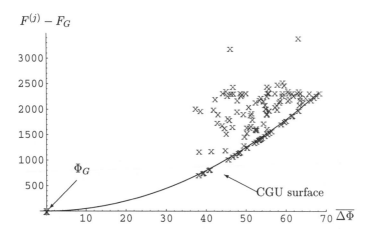

Figure 5: Energy Landscape Projection for PPT

[10] Dill, K.A., Phillips, A.T. & Rosen, J.B. (1997), "Protein structure and energy land-scape dependence on sequence using a continuous energy function," *Journal of Computational Biology* 4(3):227–239.

[11] Dill, K.A., Phillips, A.T. & Rosen, J.B. (1997), "Molecular structure prediction by global optimization," in *Developments in Global Optimization* (eds I.M. Bomze et al.) 217–234 (Kluwer, Dordrecht, 1997).

[12] Dill, K.A., Phillips, A.T. & Rosen, J.B. (1997), "CGU: An algorithm for molecular structure prediction," in *IMA Volumes in Mathematics and its Applications: Large-Scale Optimization with Applications III: Molecular Structure and Optimization* (eds L.T. Biegler et al.) 1–22.

[13] Sun, S., Thomas, P.D. & Dill, K.A. (1995), "A simple protein folding algorithm using binary code and secondary structure constraints," *Protein Engineering* 8(8):769–778.

[14] Maiorov, V.N. & Crippen, G.M. (1994), "Learning about protein folding via potential functions," *Proteins Struct. Funct. Genet.* 20:167–173.

[15] Koretke, K.K., Luthey-Schulten, Z. & Wolynes, P.G. (1998), "Self-consistency optimized energy functions for protein structure prediction by molecular dynamics," *Proc. Natl. Acad. Sci. USA* 95:2932–2937.

[16] Hao, M.H. & Scheraga, H.A. (1996), "How optimization of potential functions affects protein folding," *Proc. Natl. Acad. Sci. USA* 93:4984–4989.

Optimization in Computational Chemistry and Molecular Biology, pp. 57-71
C. A. Floudas and P. M. Pardalos, Editors
©2000 Kluwer Academic Publishers

Global Optimization and Sampling in the Context of Tertiary Structure Prediction: A Comparison of Two Algorithms

Volker A. Eyrich
Department of Chemistry and
Center for Biomolecular Simulation
Columbia University
New York, NY 10027
volker@chem.columbia.edu

Daron M. Standley
Schrödinger Inc.
1 Exchange Place, Suite 604
Jersey City, NJ 07302
standley@schrodinger.com

Richard A. Friesner
Department of Chemistry and
Center for Biomolecular Simulation
Columbia University
New York, NY 10027
rich@chem.columbia.edu

Abstract

Two global optimization algorithms, one using the branch and bound (BB) framework, and another using Monte Carlo plus minimization (MCM) are compared in terms of energy minimization, diversity of the resulting low-energy structures, and energetic rank of native-like structures. Four proteins (PDB codes 1GB1, 1CC5, 1MBD and 1GDM) are studied using each program, with the same target function, allowing a rigorous comparison of performance. In terms of the criteria used in this study, the MCM algorithm performs better than BB. The possible sources of these differences in performance are briefly discussed.

Keywords: protein folding, reduced model, global optimization, simulated annealing, Metropolis criterion, branch and bound.

1   Introduction

Global optimization algorithms are central to many protein structure prediction methods. In the idealized case where the free energy as a function of the spatial coordinates is known, and the native structure corresponds to the free energy global minimum (the thermodynamic hypothesis), the tertiary structure prediction problem is purely one of global minimization. In our experience, tertiary structure prediction is more complicated than pure global minimization for several reasons. First and foremost, is the fact that the actual free energy is not known. For this reason, structure predictions must be based on approximate potential energy functions, experimental database statistics, experimentally-derived constraints or combinations thereof. Unless these target functions can be designed so that the native is a guaranteed global minimum (which appears to be possible only in cases where a significant number of experimentally-derived constraints is used [10, 11]), the relevance of a global minimum structure is not clear. Moreover, due to the complexity of even small proteins, and the corresponding complexity of the resulting target functions, there is no objective means of determining whether a global minimum has, in fact, been reached. As a result, we must assume some uncertainty in the accuracy of the target function as well as in our ability to minimize this function.

For the above reasons, criteria used to judge the performance of a global optimization algorithm in the context of protein tertiary structure prediction may differ from criteria used in other contexts. Clearly, the ability to reach the approximate global minimum in energy is desirable, but it may not be sufficient if the uncertainties in the target function are on the order of the energy differences between competitive regions of phase space. Moreover, from the standpoint of low-resolution structure prediction, it is more important to sample as many low-energy regions as possible than it is to refine any one region. For these reasons, the following criteria form the basis for comparing the two algorithms considered in the present work:

- Energy of the lowest-energy structure;

- Diversity of low-energy structures;

- Energetic rank of native-like structures relative to the lowest energy structure.

In this paper we directly compare two global minimization algorithms that, given the above criteria, have performed well in earlier studies. The first is a branch and bound (BB) algorithm [6, 11] that is based on the αBB algorithm developed by Floudas and coworkers [5, 2]. The second is a Monte Carlo plus minimization (MCM) algorithm that follows the general framework of Li and Scheraga [8]. Both algorithms have been used extensively in our group and have been modified as needed for tertiary structure prediction using a reduced protein model with fixed secondary structure. Conclusions drawn from the present work only reflect our implementation and adaptation of these algorithms and not the original algorithms.

2   Methods

As the focus of this paper is on the results of the global minimization methods, we will only briefly outline the model, potential function, and minimization algorithms. More detailed descriptions of all aspects of the calculations can be found in our earlier publications [6, 7].

2.1   Protein Model and Target Function

The protein model represents all of the backbone atoms explicitly (using ideal geometries), but models the side chain as a single center, located at the β-carbon position. The bond lengths and bond angles of the model are constrained at ideal values, so the variables in the simulation are the backbone torsional angles. Backbone torsions in the (fixed) secondary structure elements are constrained at values fitted to reproduce the actual native secondary structure. N and C-terminal loops or random coil regions are omitted.

The potential function is composed of a sum of terms:

$$E = E_{hyd} + E_{vdw} + E_{ovlp} + E_{background} \qquad (1)$$

All four of these terms are expressed in the form of residue-residue pair interactions. E_{hyd} is based on the statistical potential of Sippl and coworkers [4] and is a linear function of the $C_\beta - C_\beta$ distance. E_{vdw} is a sum of four individual pair terms, involving both the α and β carbons on the two interacting residues. The term is parameterized by evaluating molecular mechanics van der Waals energies based on protein structures in the PDB [1, 3]. E_{ovlp} models excluded volume constraints, as determined by a database survey of minimum $C_\alpha - C_\alpha$, $C_\alpha - C_\beta$, and $C_\beta - C_\beta$ distances. $E_{background}$ is an attractive function proposed as part of the Sippl model, which is necessary to model the observed compactness of globular proteins.

2.2   Global Minimization

2.2.1   Overview

The two algorithms considered in this study employ very different search strategies. In the case of BB, the target function is augmented such that local minima are reduced (or removed, in the case of αBB), allowing the global minimum to be located. In our implementation, we are not guaranteed to find the global minimum, as we have chosen parameters such that complex problems remain tractable. In the interest of increasing diversity, the converged minima are "filled in" so higher-energy minima can be located. MCM, on the other hand, samples local minima of the original target function, so diversity is achieved by a stochastic sampling scheme.

2.2.2   The Branch and Bound Algorithm

The BB algorithm [6, 11],proceeds by bisecting regions of conformational space in the order prescribed by a lower bounding function L associated with each sub-region. The lower bound is determined by locally minimizing L. L is represented by a linear combination of the original target function T and a quadratic function S:

$$L = T + S, \tag{2}$$

where

$$S = \sum_{i=1}^{N} \alpha_i (A_i - x_i)(B_i - x_i), \tag{3}$$

In the above equation, x_i represents a spatial coordinate, A_i and B_i are the upper and lower bounds, respectively, on x_i, within a given sub-region, and α_i is the weight of S with respect to T. As long as the α_i are large enough, L is guaranteed to be convex in the sub-region; the associated minima of L thus constitute a lower bound to T within the sub-region. An upper bound U on T is similarly given by a local minimum of T in the sub-region. As the bounds are subdivided, the size of S becomes smaller, and L approaches the original function T. Based on this property, the lower bound can be shown to converge on the upper bound at the global minimum [2]. In the present work, the bounds are subdivided symmetrically, and the initial bound on each loop dihedral angle is $[-\pi, \pi]$.

Convergence is achieved when the difference between L and U is less than a cutoff value ($\epsilon = 25$ energy units, in all cases). The bounds containing the converged structure are discarded at this point, and the search for a new converged structure continues, as before, within the remaining bounds. In our implementation, we have added a diversification function E_{div}, which penalizes structures similar to structures that have already converged. This function is simply a Gaussian-shaped penalty function. For a given structure j, E_{div} is given by

$$E_{div}^{j} = H \sum_{i=1}^{N_{conv}} e^{-W R_{ij}^2} \tag{4}$$

Where H and W are the height and width of the Gaussian (in the results show here $H=400$ and $W=0.5\text{Å}^{-2}$); N_{conv} is the number of converged solutions, R_i is the RMSD (Root Mean Square Deviation) between all $C_\alpha - C_\alpha$ distances in the structures i and j.

$$R_{ij} = \sqrt{\frac{1}{N_d} \sum_{k=1}^{N_d} (d_i^{\alpha\alpha,k} - d_j^{\alpha\alpha,k})^2} \tag{5}$$

where

$$N_d = \frac{1}{2} N_{res}(N_{res} - 1), \tag{6}$$

and N_{res} is the number of residues in the protein.

We have relaxed the requirement that L be convex. In practice, this requirement would involve making the parameter α_i so large that many subspaces would be selected for subdivision and local minimization purely on the basis of size. If this were to occur, the algorithm would not converge in a reasonable time due to the size of the initial conformational space and the subsequent need for numerous local minimizations.

The BB algorithm, as described above, would continue to search for new structures as long as some subset of the original domain space remained. In practice we terminated the search when no new structures with low energies were produced since the BB algorithm tends to locate low energy structures first.

2.2.3   The Monte Carlo Plus Minimization Algorithm

The MCM algorithm represents a modified version of that proposed by Li and Scheraga [8] over a decade ago. The original implementation can be described as a method that attempts to sample the space of all discrete minima available to the system. This is accomplished by following an MC move with local minimization. Our method, on the other hand, relies on an estimation of the minimum energies of conformations by a progressive minimization scheme [7]. In this scheme the number of local minimization steps is slowly annealed in with the temperature. Partial minimization increases the possibility of making an intelligent, and hence accepted, MCM move. This is because even incomplete local minimizations sample a continuum of geometries rather than a single point in phase space as in traditional MC. Moreover, the time spent on MCM moves early in the simulation is much less than in later moves, thereby allowing a vast amount of phase space to be searched in a reasonable amount of time. To maintain structural diversity, in our opinion a crucial requirement of any tertiary structure prediction algorithm, a reasonably sized ensemble of structures is refined simultaneously.

The actual steps in the MCM algorithm are as follows:

1. Initialization: Read in the control parameters for the simulation. Specifically, the maximum number of iterations, the initial temperature (final temperatures are always zero) and minimum and maximum number of steps taken in the conjugate gradient based minimization are used to calculate increments and decrements for the two main control parameters, temperature and number of minimization steps. In addition, ensemble size, diversity parameters and communication-related parameters are initialized. All simulations start from extended structures in which the backbone dihedral angles of loops are set to zero.

2. From the current ensemble, pick a structure at random. In the parallel implementation every node maintains its own ensemble of structures but communicates with the other nodes on a regular basis.

3. Pick a loop angle of the current structure at random and change it to a random value from the interval $[-\pi, \pi]$.

4. Locally minimize the new structure by conjugate gradient based minimization with the backbone dihedral angles of loop residues as optimization variables. We do not attempt to locate a real minimum in this step (at least not in the initial stages of the simulation) and therefore limit the number of minimization steps depending on the current iteration number.

5. Compute the energy difference of the structures before the random change in the selected dihedral angles and after the minimization. Structures with lower final energies are always accepted, whereas structures with higher final energies are accepted or rejected based on the outcome of a standard Boltzmann transition probability test. Energies and dihedral angles of accepted structures are recorded and the current ensemble is updated.

6. After a given number of MCM cycles a processing node communicates with all other nodes. The current ensemble is sorted by energy and the top m structures are sent to all other nodes in a non-blocking fashion. The receiving nodes accept structures if both

Table 1: Protein Test Set

	Residues	degrees of freedom	α-helices	β-strands
CC5	76	70	4	0
GB1	54	38	1	4
GDM	149	36	8	0
MBD	147	52	8	0

their energies are lower than the highest energy in the current local ensemble and their RMS deviation from all other structures is larger than a predetermined cutoff (the diversity parameter). The highest-energy structures in the ensemble are replaced by the accepted structures.

7. Increment the maximum number of minimization steps and decrement the artificial simulation 'temperature' using the values calculated in step 1.

8. Return to step 2 unless the maximum number of iterations has been reached in which case termination is initiated

2.3   Clustering

In order to assess the diversity of the simulations, the resulting structures were clustered by structural similarity [9]. The definition of a cluster is a group of structures for which the average backbone RMSD between all members, is below a cutoff value. In all cases, a cutoff value of 5Å was used.

3   Results

In this comparative study we focus on only four systems, the proteins GB1, CC5, MBD and GDM (Table I). While both algorithms have been tested on larger and more diverse data sets, the proteins under study here should present an overview of the results to be expected from both approaches. All calculations make use of exactly the same potential energy function parameters and key software components (i.e. the same local minimization code), which allows us to rigorously compare the performance of the two approaches.

Figures 1-4 show plots of backbone Cartesian coordinate RMSD from the crystal structure versus energy for the low energy clusters obtained from BB and MCM calculations for the four systems (Clusters are represented by their lowest-energy member). The energy scales are identical and can be compared directly. Branch and Bound results represent a concatenation of several serial runs with different α-parameters and subdivision schemes. All MCM calculations were carried out on 4 to 16 workstation nodes and used identical Monte Carlo control parameters.

Tables II-IV summarize the performance of the two algorithms. Table II lists the lowest energy found by each method for each of the four proteins. Table III shows the energy below which 50 clusters were found. Finally, a summary of the ranks of native like structures relative to the lowest energy cluster is presented in Table IV. Ranks are given both for

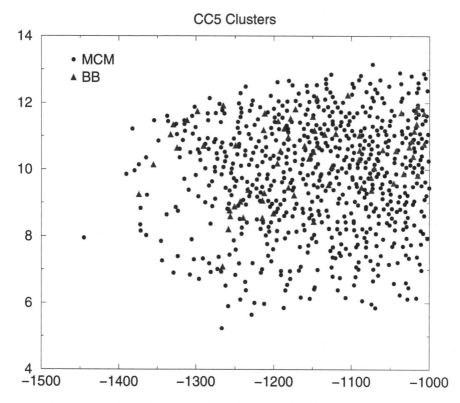

Figure 1: Low-energy CC5 clusters resulting from BB (red) and MCM (black) are represented by their energy and RMSD from the native.

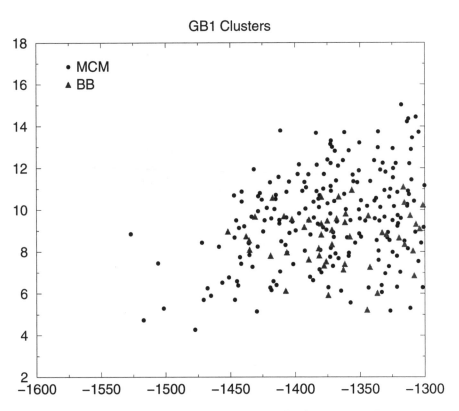

Figure 2: Low-energy GB1 clusters resulting from BB (red) and MCM (black) are represented by their energy and RMSD from the native.

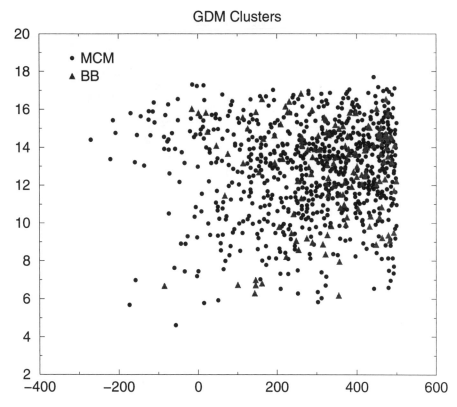

Figure 3: Low-energy GDM clusters resulting from BB (red) and MCM (black) are represented by their energy and RMSD from the native.

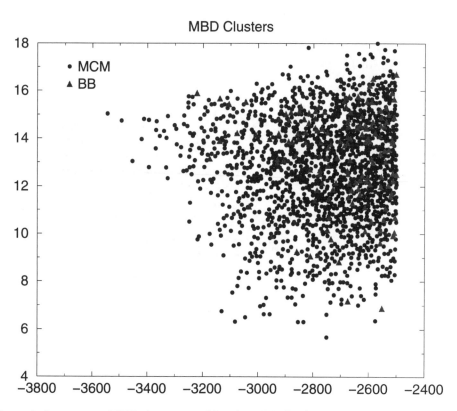

Figure 4: Low-energy MBD clusters resulting from BB (red) and MCM (black) are represented by their energy and RMSD from the native.

Table 2: Lowest Energy Found

	BB	MCM
CC5	-1374	-1445
GB1	-1452	-1527
GDM	-85	-271
MBD	-3222	-3545

Table 3: Diversity of Low-Energy Clusters: Diversity is measured by the energy below which 50 clusters (defined by a 5 Åcutoff in RMSD) were found

	BB	MCM
CC5	-1087	-1288
GB1	-1298	-1412
GDM	304	-1
MBD	-2750	-3204

clusters with an RMSD below 6Å and for clusters with an RMSD below 7Å.

While the actual CPU time required for each run is not the focus of this paper, it might be important to note that in general all BB runs were run until no more low energy structures were produced–no arbitrary CPU time limit was employed. The duration of the MCM runs on the other hand is predetermined by the MC control parameters. In this case 10,000 MC steps per node were used - a number that appears to provide reliable results across a wide range of systems. In general, the BB algorithm used 2-5 times more CPU time than the MCM algorithm. The CPU times correlate with the number of function and gradient evaluations. We have monitered the number of function and gradient evaluations for calculations carried out on the protein GB1. BB requires approximately 3.7×10^7 function and 3.6×10^7 gradient evaluations to produce a structure at -1452 energy units,

Table 4: Rank of Native-Like Clusters: Energetic Ranks (relative to the lowest-energy structure found) are shown for clusters with an RMSD from the native of less than 6 Åand less than 7 Å, respectively. RMSD values refer to to the lowest energy structure of each cluster. If no cluster was found within the indicated RMSD, a double dash is shown.

	BB		MCM	
	< 6 Å	< 7 Å	< 6 Å	< 7 Å
CC5	–	–	78	78
GB1	23	9	2	2
GDM	–	1	5	5
MBD	–	123	756	92

whereas MCM locates a structure at -1527 energy units after 1.7×10^7 function and 1.6×10^7 gradient evaluations.

4 Discussion

The proteins studied here range from 54 to 149 residues and possess 36 to 70 degrees of freedom. Previous studies have shown that the potential energy function can more or less accurately describe the energetics in these systems [7]. GDM and MBD belong to the same structural class but seem to present a qualitatively different challenge despite their similarity. We deliberately excluded larger all-beta proteins due to the fact that we are still working on additional potential energy terms that will hopefully render this structural class tractable.

Both algorithms demonstrate their ability to generate low energy structures for the four systems studied here (in fact several hundred energy units below the locally minimized native structure). These absolute lowest energies produced in the simulation represent an unambiguous criterion for performance comparison of the two approaches. In all cases MCM locates structures with energies well below the BB results. Due to the fact that we use the same local minimization code no artifacts stemming from numerical instabilities have to be considered. Converged structures from BB runs are locally minimized without the diversification function to totally relax the structure whereas MCM structures are taken directly from the simulations.

As mentioned above, our experience with the current target function indicates that a single-minded search for the absolute lowest energy structure for a given system is bound to fail. Deficiencies of the potential energy function require us to search the low energy regions as thoroughly as possible. In the cases of GB1 and GDM, for example, native like clusters are comparable in energy to non-native clusters. CC5 and MBD exhibit a radically different energy profile. Even though low-energy, native-like clusters exist, they are not competitive in energy. In those cases strategies that emphasize structural diversity during the simulation might stand a better chance of reliably locating even the energetically unfavorable but native-like regions.

Taking the energy cutoff below which 50 distinct clusters were found, MCM seems to outperform BB in this respect too. While it is not straightforward to define a criterion that measures the diversity of a search, the number of clusters located appears to be a reasonable measure, especially when one considers the further use of the results obtained from the simulations. In our approach low-energy clusters are further refined by addition of sidechains followed by atomic-level simulations with high accuracy potential energy functions. While this work is in too early a stage to be considered a workable and reliable approach, we assume that the number of non-redundant, structurally distinct alternatives presented to the atomic level simulations can only improve the possibility of locating the most native-like structure, albeit at an increased computational cost.

Since the energy functions were the same in each case, it is clear that BB either discarded bounds containing some of the structures located by MCM, or failed to identify these bounds as likely targets, and thus didn't explore them. We do not know the degree to which these two factors contribute to the lack of diversity. We do know, however, That the reduced number of clusters is not due to the clustering algorithm. Most of the converged structures in BB are dissimilar, due to the diversity term; it is just that a large number of these structures are not competitive in energy.

The ranks of the native like clusters are the most important descriptors of the simulation results. They determine the minimum number of structures required for further atomic-level detail calculations and therefore computational cost of this approach. With the exception of GDM Monte Carlo plus Minimization performs best in this respect. Improved structural diversity can obviously work against a method, if the number of non-native like clusters increases due to enhanced sampling of low energy, non-native regions.

The protein CC5, due to its 17-residue loop, presents a case where the low energy region might possibly be very ill-defined. BB seems to be unable to adequately sample the native-like region around -1250 energy units, whereas MCM locates this region despite the fact that the lowest-energy structure is about another 200 units lower. The case of MBD is similar but not quite as dramatic. BB locates more or less native-like structures with reasonable ranks although the final results cannot quite compete with the MCM results. GB1 and GDM are handled well by both methods, and ignoring the absolute energy difference between the results would most likely yield similar results in atomic-level simulation.

The results in this study are consistent with differences in the BB and MCM algorithms. While BB focuses on finding the global minimum, and must be in a sense tricked into searching higher-energy regions of phase space through the use of the diversification function, the MCM trajectories naturally pass through many high-energy minima. From this point of view, we might expect that if E_{div} were removed in BB, the algorithm would be able to locate lower energies, but at the expense of finding fewer low-energy clusters. However, when E_{div} was removed in the cases of CC5 and GDM, the minimum energies were not significantly different than those shown in Table II. The diversity of low-energy clusters was, as expected, significantly compromised. This suggests that other modifications to the original αBB algorithm, such as reducing the size of the α_i parameters, is responsible for the higher energy minima. Presently, however, reducing the α_i is the most straight-forward way of reducing the CPU time for convergence.

5 Conclusion

We have presented, for the first time, a side-by-side comparison of our implementation of BB and MCM, two algorithms that have proven useful in our earlier studies of tertiary structure prediction. Although we have not explored every possible elaboration within the BB or MCM framework, we have attempted to optimize each program for the task at hand. For example, we have also implemented BB in distance space [6, 11] while this version appears to work well in cases where distance restraints are included in the target function, it did not out-perform BB in angle space with the target function considered here (i.e., without distance restraints). The overall conclusion is that our implementation of MCM is superior to that of BB in terms of locating diverse, low-energy structures (including ones that are native-like). As expected, when the diversification term E_{div} is removed in the BB target function, the sampling is much narrower. However, we can not attribute the higher minimum energies in the case of BB to E_{div} since removing E_{div}, did not result in significantly different minimum energies.

We have also attempted to implement some of ideas inherent in MCM within the BB framework. For example, the use of multiple partial local minimizations rather than a single minimization was used in order to estimate the lower bound within each subspace. Following in the spirit of MCM, the number of minimization steps was increased as the bounds were subdivided. Although these modification improved the sampling in the initial stages of the simulation, they also slowed the time to convergence. In our implementation, we did not observe improvement in the resulting converged structures with respect to the lowest energy found, diversity of low-energy structures, or the rank of the native-like structure.

The magnitude of α needed to render L convex (and thus to guarantee a global minimum) is orders of magnitude larger than what we were able to use in simulations, due to CPU time limitations. It is perhaps surprising that we did not observe systematic improvement in the minimum BB energies by increasing the α parameters within acceptable ranges (as deemed by the CPU time). The question of whether other modifications in BB, such as using a different lower bound function, will improve performance has yet to be answered.

References

[1] Abola, E. E., Bernstein, F.C., Bryant, S.H., Koetzle, T.F., Weng, J. (1963), *Protein Data Bank. Crystallographic Databases-Information Content, Software Systems, Scientific Applications.*, Bonn/Cambridge/Chester, Data Commission of the Int'l Union of Crystallography: 107-132.

[2] Androulakis, I.P., Maranas, C.D., Floudas, C.A. (1995), "αBB:A global optimization method for general constrained nonconvex problems." *Journal of Global Optimization Vol. 7, No. 4*, 337-363.

[3] Berstein, F. C., Koetzle, T.F., Williams, G.J.B., Meyer, E.F. jr., Brice, M.D., Rodjers, J.R., Kennard, O., Shimanouchi, T., Tasumi, M. (1977), "The Protein Data Bank: A computer-Based archival file for macromolecular structures." *J.Mol.Biol. Vol. 112*, 535-542.

[4] Casari, G., Sippl, M.J. (1992), "Structure-derived hydrophobic potential." *J. Mol. Biol. Vol. 224, No. 3*, 725-732.

[5] Maranas C.D., Floudas, C.A. (1994), "A deterministic global optimization approach for molecular structure determination." *J. Chem. Phys. Vol. 100, No. 2*, 1247-1261.

[6] Eyrich, V. A., D. M. Standley, R.A. Friesner (1998), "Protein Tertiary Structure Prediction using a Branch and Bound Algorithm." *Proteins-Structure Function and Genetics (in press).*

[7] Eyrich, V. A., D. M. Standley, R.A. Friesner (1998), "Prediction of Protein Tertiary Structure to Low Resolution: Performance for a Large and Structurally Diverse Test Set." *J. Mol. Biol. (in press).*

[8] Li, Z. and H. A. Scheraga (1987), "Monte Carlo-minimization approach to the multiple-minima problem in protein folding." *Proceedings of the National Academy of Science of the United States Vol. 84*, 6611-6615.

[9] Romesburg, H. C. (1984), Cluster Analysis for Researchers. Belmont, California, Lifetime Learning Publications.

[10] Skolnick, J., A. Kolinski, A.R. Ortiz (1997), "MONSSTER: A method for folding globular proteins with a small number of distance restraints." *Journal of Molecular Biology Vol. 265, No. 2*, 217-241.

[11] Standley, D. M., Eyrich, V.A.,Felts, A.K.,Friesner, R.A.,McDermott, A.E. (1999). "A Branch and bound algorithm for protein structure refinement from sparse NMR data sets." *J. Mol. Biol. Vol. 285, No. 4*, 1689-1708.

Optimization in Computational Chemistry and Molecular Biology, pp. 73-90
C. A. Floudas and P. M. Pardalos, Editors
©2000 Kluwer Academic Publishers

Protein Folding Simulations by Monte Carlo Simulated Annealing and Multicanonical Algorithm

Yuko Okamoto
Department of Theoretical Studies
Institute for Molecular Science
and
Department of Functional Molecular Science
The Graduate University for Advanced Studies
Okazaki, Aichi 444-8585, Japan
okamotoy@ims.ac.jp

Abstract

We have performed Monte Carlo simulations based on simulated annealing and multicanonical algorithm to predict the three-dimensional structures of two oligopeptides, C-peptide of ribonuclease A and the fragment BPTI(16-36) of bovine pancreatic trypsin inhibitor. The lowest-energy conformations obtained have α-helix structure and β-sheet structure for C-peptide and BPTI(16-36), respectively, in accord with experimental implications.

Keywords: Simulated annealing, multicanonical algorithm, generalized-ensemble algorithm, α-helix, β-sheet.

1 Introduction

Proteins are the most complicated molecules that exist in nature. There exist astronomically large number of states of energy local minima in the protein systems [1]. Protein folding problem is thus one of the most challenging optimization problems in theoretical molecular science. Many efforts have been devoted to it without complete success since Anfinsen's experiments, which showed that the three-dimensional structure (tertiary structure) of a protein is determined solely by its amino-acid sequence information [2]. Simulations by conventional methods such as Monte Carlo or molecular dynamics algorithms in canonical ensemble will necessarily get trapped in one of many local-minimum states in the energy function.

Main chains of proteins have two major structural elements (secondary structure), α-helix and β-sheet. On the average about 35 %, 15 %, 25 %, and 25 % of the residues are in α-helix, β-sheet, reverse turn, and coil state, respectively. Since α-helix and β-sheet among the above four have solid, stable structures, it is of great importance to predict which residues are in these structures. Protein database from X-ray and NMR experiments have often been used to predict the secondary structures of proteins [3], but the probability of success in prediction is about 70 %.

In this article we discuss the results of our secondary and tertiary structure predictions of oligopeptide systems based on Monte Carlo simulated annealing [4] and multicanonical algorithm [5]. C-peptide of ribonuclease A and the peptide fragment of bovine pancreatic trypsin inhibitor, BPTI(16-36), are studied. By experiments, the former is known to form α-helix structure and the latter β-sheet structure, which are the two basic building elements of protein structures. Starting simulations from randomly-generated initial conformations, we show that α-helix and β-sheet structures (two of the basic structural elements of proteins) can be obtained as the global-minimum states for C-peptide and BPTI(16-36), respectively, in agreement with experiments.

The outline of the article is as follows. In section 2 we summarize the energy functions of protein systems that we used in our simulations. In section 3 we briefly review our simulation methods. In section 4 we present the results of our protein folding simulations. Section 5 is devoted to conclusions.

2   Energy Functions of Protein Systems

The energy function for the protein systems is given by the sum of two terms: the conformational energy E_P for the protein molecule itself and the solvation free energy E_S for the interaction of protein with the surrounding solvent. The conformational energy function E_P (in kcal/mol) for the protein molecule that we used is one of the standard ones. Namely, it is given by the sum of the electrostatic term E_C, 12-6 Lennard-Jones term E_{LJ}, and hydrogen-bond term E_{HB} for all pairs of atoms in the molecule together with the torsion term E_{tor} for all torsion angles:

$$
\begin{aligned}
E_P &= E_C + E_{LJ} + E_{HB} + E_{tor} , \\
E_C &= \sum_{(i,j)} \frac{332\, q_i q_j}{\epsilon\, r_{ij}} , \\
E_{LJ} &= \sum_{(i,j)} \left(\frac{A_{ij}}{r_{ij}^{12}} - \frac{B_{ij}}{r_{ij}^{6}} \right) , \\
E_{HB} &= \sum_{(i,j)} \left(\frac{C_{ij}}{r_{ij}^{12}} - \frac{D_{ij}}{r_{ij}^{10}} \right) , \\
E_{tor} &= \sum_{i} U_i \left(1 \pm \cos(n_i \chi^i) \right) .
\end{aligned}
\tag{1}
$$

Here, r_{ij} is the distance (in Å) between atoms i and j, ϵ is the dielectric constant, and χ^i is the torsion angle for the chemical bond i. Each atom is expressed by a point at its center of mass, and the partial charge q_i (in units of electronic charges) is assumed to be concentrated at that point. The factor 332 in E_C is a constant to express energy in units of kcal/mol. These parameters in the energy function as well as the molecular geometry were adopted from ECEPP/2 [6]. The computer code KONF90 [7] was used for all the Monte Carlo simulations. For gas phase simulations, we set the dielectric constant ϵ equal to 2. The peptide-bond dihedral angles ω were fixed at their usual experimental value 180° for simplicity. So, the remaining dihedral angles ϕ and ψ in the main chain and χ in the side chains constitute the variables to be updated in the simulations. One Monte Carlo (MC) sweep consists of updating all these angles once with Metropolis evaluation [8] for each update.

One of the simplest ways to represent solvent effects is by the sigmoidal, distance-dependent dielectric function [9]. The explicit form of the function we used is given by [10]

$$\epsilon(r) = D - \frac{D-2}{2}\left[(sr)^2 + 2sr + 2\right]e^{-sr} , \tag{2}$$

which is a slight modification of the one used in Ref. [11]. Here, we use $s = 0.3$ and $D = 78$. It approaches 2 (the value inside a protein) in the limit the distance r going to zero and 78 (the value for bulk water) in the limit r going to infinity. The distance-dependent dielectric function is simple and also computationally only slightly more demanding than the gas-phase case.

Another commonly used term that represents solvent contributions more accurately than the distance-dependent dielectric function is the term proportional to the solvent-accessible surface area of protein molecule. The solvation free energy E_S in this approximation is given by

$$E_S = \sum_i \sigma_i A_i , \tag{3}$$

where A_i is the solvent-accessible surface area of i-th functional group, and σ_i is the proportionality constant. There are several versions of the set of the proportionality constants and functional groups. Five parameter sets were compared for the systems of peptides and a small protein, and we found that the parameter sets of Refs. [12, 13] are valid ones [14].

3 Simulation Methods

Once the appropriate energy function of the protein system is given, we have to employ a simulation method that does not get trapped in states of energy local minima. We have been advocating the uses of Monte Carlo simulated annealing [4] and multicanonical algorithm [5] (for reviews, see Refs. [15, 16]).

3.1 Simulated annealing

In the regular canonical ensemble with a given inverse temperature $\beta \equiv 1/k_B T$, the probability weight of each state with energy E is given by the Boltzmann factor:

$$W_B(E) = \exp(-\beta E) . \tag{4}$$

The probability distribution in energy is then given by

$$P_B(T, E) \propto n(E)W_B(E) , \tag{5}$$

where $n(E)$ is the number of states with energy E. Since the number of states $n(E)$ is a rapidly increasing function of E and the Boltzmann factor $W_B(E)$ decreases exponentially with E, the probability distribution $P_B(T, E)$ has a bell-like shape in general. When the temperature is high, β is small, and $W_B(E)$ decreases slowly with E. So, $P_B(T, E)$ has a wide bell-shape. On the other hand, at low temperature β is large, and $W_B(E)$ decreases rapidly with E. So, $P_B(T, E)$ has a narrow bell-shape (and in the limit $T \to 0$ K, $P_B(T, E) \propto \delta(E - E_{GS})$, where E_{GS} is the global-minimum energy). However, it is very difficult to obtain canonical distributions at low temperatures with conventional simulation methods. This is because the thermal fluctuations at low temperatures are small and the simulation will certainly get trapped in states of energy local minima.

Simulated annealing [4] is based on the process of crystal making. Namely, by starting a simulation at a sufficiently high temperature (much above the melting temperatue), one lowers the temperature gradually during the simulation until it reaches the global-minimum-energy state (crystal). If the rate of temperature decrease is sufficiently slow so that thermal equilibrium may be maintained throughout the simulation, only the state with the global energy minimum is obtained (when the final temperature is 0 K). However, if the temperature decrease is rapid (quenching), the simulation will get trapped in a state of energy local minimum in the vicinity of the initial state.

Simulated annealing was first successfully used to predict the global-minimum-energy conformations of polypeptides and proteins [17]-[19] and to refine protein structures from X-ray and NMR data [20, 21] almost a decade ago. Since then this method has been extensively used in the protein folding and structure refinement problems (for reviews, see Ref. [22, 15]). Our group has been testing the effectiveness of the method mainly in oligopeptide systems. The procedure of our approach is as follows. While the initial conformations in the protein simulations are usually taken from the structures inferred by the experiments, our initial conformations are *randomly generated*. Each Monte Carlo sweep updates every dihedral angle (in both the main chain and side chains) once. Our annealing schedule is as follows: The temperature is lowered exponentially from $T_I = 1000$ K to $T_F = 250$ K [7]. The temperature for the n-th MC sweep is given by [7]

$$T_n = T_I \, \gamma^{n-1} , \tag{6}$$

where γ is a constant which is determined by T_I, T_F, and the total number of MC sweeps of the run. For the results presented below, each run consisted of 10^5 MC sweeps, and we made 20 runs from randomly-generated initial conformations.

3.2   Multicanonical algorithm

While a regular Monte Carlo method generates states according to the canonical distribution, generalized-ensemble algorithms [23] generate states so that a one-dimensional random walk in a pre-chosen physical quantity (for instance, the energy) is realized. Hence, any energy barrier can be overcome, and one can avoid getting trapped in states of energy local minima.

Multicanonical algorithm [5] is one of the most well-known such methods. In the "multicanonical ensemble" the probability distribution of energy is *defined* as follows:

$$P_{mu}(E) \propto n(E)W_{mu}(E) \equiv \text{constant} .\qquad (7)$$

The multicanonical weight factor then satisfies

$$W_{mu}(E) \propto n^{-1}(E) .\qquad (8)$$

Since this weight factor is not *a priori* known, one has to determine it for each system by a few iterations of trial Monte Carlo simulations. See Refs. [24] and [25] for details of the method to determine the multicanonical weight factor $W_{mu}(E)$. Once this weight factor is obtained, one performs a long production simulation run. The advantage of multicanonical algorithms lies in the fact that from this single productin run, one can obtain not only the global-minimum-energy state but also the canonical distribution $P_B(T, E) = n(E)e^{-\beta E}$ for wide range of temperatures $T = 1/k_B\beta$. The latter is accomplished by the use of the reweighting techniques [26]. Namely, $P_B(T, E)$ can be expressed in terms of the predetermined weight $W_{mu}(E)$ and the obtained distribution $P_{mu}(E)$ as follows:

$$P_B(T, E) = \frac{P_{mu}(E) \ W_{mu}^{-1}(E) \ e^{-\beta E}}{\int dE' \ P_{mu}(E') \ W_{mu}^{-1}(E') \ e^{-\beta E'}} .\qquad (9)$$

The expectation value of a physical quantity $\mathcal{A}$ at temperature T is then given by

$$< \mathcal{A} >_T = \int dE \ \mathcal{A}(E)P_B(T, E) .\qquad (10)$$

The application of multicanonical algorithm and its variants to the prediction of protein tertiary structures was proposed several years ago.[27, 28] Since then there have been various applications of the method in the protein folding problem (for reviews, see Ref. [15, 16]). A formulation of multicanonical algorithm for the molecular dynamics method was also developed [29]-[31].

For the results of multicanonical simulations presented below, we first determined the multicanonical weight factors by iterations of short preliminary runs. We then made one long production run of 1,000,000 MC sweeps from a random initial conformation for each system.

4 Results

We now present the results of our simulations based on Monte Carlo simulated annealing and multicanonical algorithm. *All the simulations were started from randomly-generated conformations.* For the results presented below, CPU time spent for one MC sweep was roughly 0.6 sec for C-peptide with the distance-dependent dielectric function and 15 sec for BPTI(16-36) with the solvent-accessible surface area term on SGI Origin 200.

The first example is the C-peptide, residues 1–13 of ribonuclease A. It is known from the X-ray diffraction data of the whole enzyme that the segment from Ala-4 to Gln-11 exhibits a nearly 3-turn α-helix [32]. It was also found by CD [33] and NMR [34] experiments that the isolated C-peptide also has significant α-helix formation in aqueous solution at temperatures near 0 °C. Furthermore, the CD experiments of the isolated C-peptide suggested that the side-chain charges of residues Glu-2$^-$ and His-12$^+$ play an important role in the stability of the α-helix, while the removal of the side-chain charge of Glu-9$^-$ enhances helix formation [33].

The NMR experiment [34] of the isolated C-peptide further observed the formation of the characteristic salt bridge between Glu-2$^-$ and Arg-10$^+$ that exists in the native structure determined by the X-ray experiments of the whole protein [32].

In order to test whether our simulations can reproduce these experimental results, we first made 20 Monte Carlo simulated annealing runs of 10,000 MC sweeps with five C-peptide analogues [7]. The amino-acid sequences of four of the analogues are listed in Table 1. The simulations were performed in gas phase ($\epsilon = 2$). The temperature was decreased exponentially from 1000 K to 250 K for each run.

Table 1: Amino-acid sequences of the C-peptide analogues studied[a].

Peptide	I	II	III	IV
Sequence				
1	Lys$^+$			
2	Glu$^-$			Glu
3	Thr			
4	Ala			
5	Ala			
6	Ala			
7	Lys$^+$			
8	Phe			
9	Glu$^-$	Glu	Leu	
10	Arg$^+$			
11	Gln			
12	His$^+$			His
13	Met			

[a] Entries for Peptides II–IV indicate that the corresponding residues in Peptide I are substituted by those with neutral side chain (and empty entries imply that no change from Peptide I is made for the corresponding residue).

In Table 2 we summarize the helix formation of all the runs [7]. Here, the numbers of conformations with segments of helix length $\ell \geq 3$ are given. ¿From this table one sees that α-helix was hardly formed for Peptide IV where Glu-2 and His-12 are neutral, while many helical conformations were obtained for the other peptides. This is in accord with the experimental results that the charges of Glu-2$^-$ and His-12$^+$ are necessary for the α-helix stability [33]. Peptides II and III had conformations with the longest α-helix ($\ell = 7$). These conformations turned out to have the lowest energy in 20 simulation runs for each peptide. They both exhibit an α-helix from Ala-5 to Gln-11, while the structure from the X-ray data has an α-helix from Ala-4 to Gln-11 [7].

Table 2: α-Helix formation in C-peptide analogues from 20 Monte Carlo simulated annealing runs.

Peptide	I	II	III	IV
ℓ				
3	4	2	3	1
4	3	2	3	0
5	1	1	0	0
6	0	1	0	0
7	0	1	1	0
Total	8/20	7/20	7/20	1/20

The agreement of the backbone structures is conspicuous, but the side-chain structures are not quite similar. In particular, while the X-ray [32] and NMR [34] experiments imply the formation of the salt bridge between the side chains of Glu-2$^-$ and Arg-10$^+$, the lowest-energy conformations of Peptides II and III obtained from the simulations do not have this salt bridge.

The disagreement is presumably caused by the lack of solvent in our simulations. We have therefore made multicanonical simulations of 1,000,000 MC sweeps for Peptide II with the inclusion of solvent effects by the distance-dependent dielectric function (see Eq. (2)) [35, 36].

As emphasized above, the results from a single simulation run in multicanonical ensemble can be used to calculate various thermodynamic quantities as functions of temperature for a wide range of temperatures (see Eqs. (9) and (10)). In Figure 1 we plot the average total potential energy and each component (in Eq. (1)) as a function of temperature. Among the component terms both electrostatic and Lennard-Jones terms vary most with the temperature. This is contrasted with our previous works on peptides with only electrically neutral side chains (Met-enkephalin [25] and homo-oligomers [37]), where the changes of the Lennard-Jones term dominate that of the total potential energy. Hence, we understand that when some of the side chains are charged in the peptide, the contributions from the electrostatic interactions become a key factor in studying the peptide conformations (together with the Lennard-Jones term that is common in any peptide).

We now examine how much α-helix formation was observed in the simulations. In Figure 2 we display the average helicity $< n >_T / N$ ($N = 13$) (where n is the total number of helical residues in the conformation) as a function of temperature for the three peptides. We observe the formation of α-helices at low temperatures and the helix-coil transitions around $T = 500 - 600$ K.

The lowest-energy conformation obtained has an α-helix from Ala-4 to Gln-11 and does have the characteristic salt bridge between Glu-2$^-$ and Arg-10$^+$. This conformation and the corresponding X-ray structure are compared in Figure 3 (the lowest-energy conformation obtained in gas phase is also shown for completeness). The figures were created with Molscript [38] and Raster3D [39]. The positions of the α-helix are identical among the

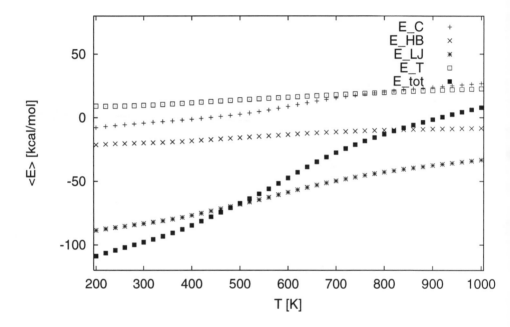

Figure 1: Average potential energies $< E >_T$ (kcal/mol) of C-peptide (Peptide II) as a function of temperature T (K). The results were obtained from a single multicanonical simulation of 1,000,000 MC sweeps.

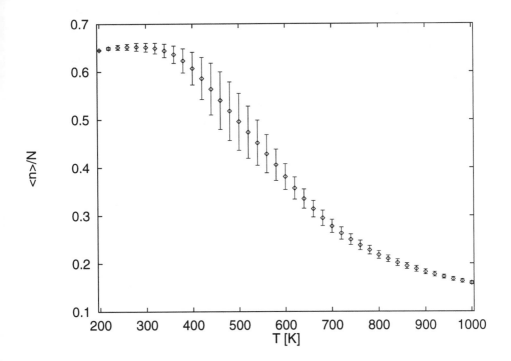

Figure 2: Average helicity $< n >_T /N$ of C-peptide (Peptite II) as a function of temperature T (K). The results were obtained from a single multicanonical simulation of 1,000,000 MC sweeps.

three structures but side-chain structures vary. The similarity between the X-ray structure and the lowest-energy conformation in aqueous solution is particularly remarkable. The root-mean-square deviations between the two are 1.4 Å and 2.7 Å for non-hydrogen atoms in the backbone and in the whole molecule, respectively.

We have also studied β-sheet formations by Monte Carlo simulated annealing [40]-[42]. The peptide that we studied is the fragment corresponding to residues 16–36 of bovine pancreatic trypsin inhibitor (BPTI) and has the amino-acid sequence: Ala^{16}-Arg$^+$-Ile-Ile-Arg$^+$-Tyr-Phe-Tyr-Asn-Ala-Lys$^+$-Ala-Gly-Leu-Cys-Gln-Thr-Phe-Val-Tyr-Gly^{36}. An antiparallel β-sheet structure in residues 18–35 is observed in X-ray crystallographic data of the whole protein [43].

We first performed 20 Monte Carlo simulated annealing runs of 10,000 MC sweeps in gas phase ($\epsilon = 2$) with the same protocol as in the previous simulations [40]. Namely, the temperature was decreased exponentially from 1000 K to 250 K for each run. The most notable feature of the obtained results is that α-helices, which were the dominant motif in the previous simulations of C-peptide, are absent in the present simulations. Most of the conformations obtained consist of stretched strands and a 'turn' which connects them. The lowest-energy structure indeed exhibts an antiparallel β-sheet [40].

We next made 10 Monte Carlo simulated annealing runs of 100,000 MC sweeps for BPTI(16-36) with two dielectric functions: $\epsilon = 2$ and the sigmoidal, distance-dependent dielectric function of Eq. (2) [41]. The results with $\epsilon = 2$ reproduced our previous results: Most of the obtained conformations have β-strand structures and no extended α-helix is observed. Those with the sigmoidal dielectric function, on the other hand, indicated formation of α-helices. One of the low-energy conformations, for instance, exhibited about a four-turn α-helix from Ala-16 to Gly-28 [41]. This presents an example in which a peptide with the same amino-acid sequence can form both α-helix and β-sheet structures, depending on its electrostatic environment.

NMR expriments suggest that this peptide actually forms a β-sheet structure [44]. The representation of solvent by the sigmoidal dielectric function, which gave α-helices instead, is therefore not sufficient. Hence, the same peptide fragment, BPTI(16-36), was further studied in aqueous solution that is represented by solvent-accessible surface area of Eq. (3) by Monte Carlo simulated annealing [42]. The parameters of Ref. [12] were used for the solvent-accessible surface area term. Twenty simulation runs of 100,000 MC sweeps were made.

As shown in Figure 4, the lowest-energy conformation obtained in solvent (structure S) involves a small but distinctive β-hairpin structure with a β-turn and three intrachain hydrogen bonds connecting two short β-strands, whereas that obtained in gas phase (structure V) has a less conspicuous β-sheet structure with only one intrachain hydrogen bond and a turn (but not a β-turn). The characteristic intrachain hydrogen bonds for these β-sheet structures together with those for the native β-sheet structure are summarized in Table 3. In view of the number of intrachain hydrogen bonds, the solvation seems to stabilize the β-sheet structure (structure S versus structure V). More detailed analysis on this point is given below when we examine the side-chain structures of these structures.

While the existence of β-turn in structure S is seen in Figure 4, the type of the β-turn can be determined by the values of the dihedral angles at the turn. The dihedral angles (ϕ,ψ) of Gly-28 and Leu-29 in structure S are $(+76.6°, -126.1°)$ and $(-95.1°, -4.3°)$, respectively.

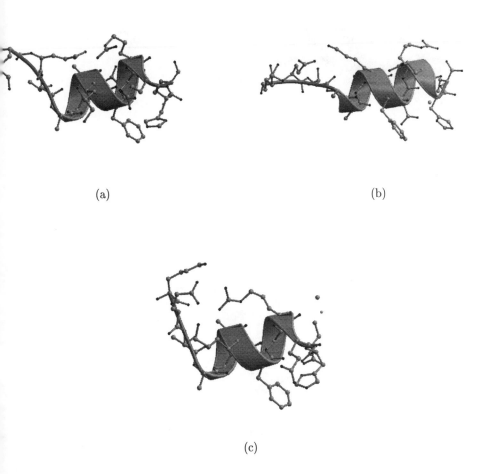

(a)   (b)

(c)

Figure 3: X-ray structure [32] of C-peptide (a) and the lowest-energy conformations of C-peptide (Peptide II) obtained from a multicanonical Monte Carlo run of 1,000,000 MC sweeps in gas phase ($\epsilon = 2$) (b) and in aqueous solution represented by the distance-dependent dielectric function (c). The figures were created with Molscript [38] and Raster3D [39].

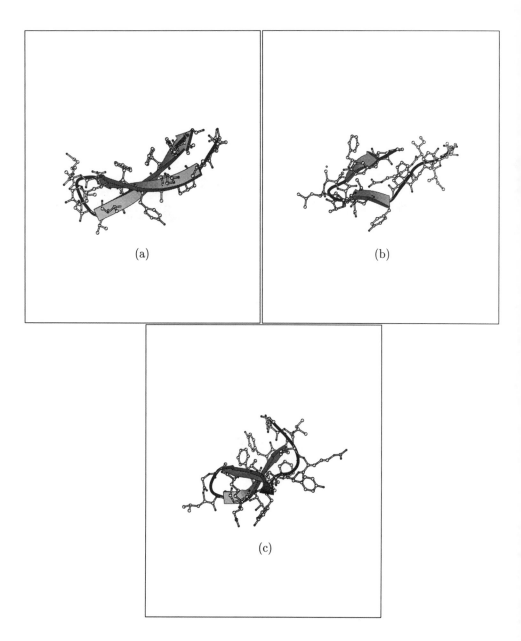

Figure 4: The X-ray structure (structure **X**) [43] of BPTI(16-36) (a) and the lowest-energy conformations of BPTI(16-36) obtained from Monte Carlo simulated annealing runs of 100,000 MC sweeps in gas phase with $\epsilon = 2$ (structure **V**) (b) and in aqueous solution represented by the term proportional to the solvent-accessible surface area (structure **S**) (c). The figures were created with Molscript [38].

Table 3: Intrachain hydrogen bonds of BPTI(16-36) found in structures **X**, **V**, and **S**.

Structure	Amino-acid Residues	
	Proton Donor (NH)	Proton Acceptor (CO)
X	Ile-18	Tyr-35
	Tyr-35	Ile-18
	Arg-20	Phe-33
	Phe-33	Arg-20
	Phe-22	Gln-31
	Asn-24	Leu-29
	Gly-28	Asn-24
	Ala-27	Asn-24
V	Phe-22	Tyr-35
S	Phe-33	Phe-22
	Ala-27	Cys-30
	Cys-30	Ala-27

The reverse turn is thus classified as a typical type II′ β-turn.

As is apparent from Figure 4 and Table 3, the lowest-energy structure of BPTI(16-36) in solvent (structure **S**) obtained by the present simulations is rather different from the one deduced from the X-ray diffraction experiments of the entire BPTI (structure **X**). However, this does not mean that our simulations are in failure. There is no reason to believe that the isolated peptide fragment should have the same structure as the corresponding segment in the whole protein molecule (if so, it should constitute a rather independent structural domain from the rest of the protein molecule, which is not the case here). In fact, recent NMR measurements of this peptide fragment have detected noticeable long-range NOE cross peaks. These include unambiguous correlations between C^{β}-H of Ala-27 and N^{α}H of Cys-30 and between N^{α}H of Tyr-23 and C^{β}-H (and C^{γ}-H) of Gln-31, supporting the existence of a reverse turn and the proximity of Phe-22 and Phe-33, respectively, in agreement with structure **S**. The hydrogen pairs that gave the NOE cross peaks and their corresponding distances in structures **V**, **S**, and **X** are listed in Table 4. Among the three structures, the remarkable agreement of structure **S** with the NMR experiment is obvious. The details of this NMR experiment will be presented elsewhere [44].

Finally, we study the role of solvent in the β-sheet formation. As shown in Figure 4, the lowest-energy conformation obtained from the simulations in gas phase (structure **V**) does not exhibit a conspicuous β-sheet structure compared to that obtained in solvent (structure **S**). These structures together with the structure deduced from the X-ray crystallographic experiments (structure **X**) are again compared in Figure 5, but this time only the main chain and the charged side chains (Arg-17, Arg-20, and Lys-26) are shown. The figures were created with RasMol [45]. One important difference between structures **S** (and **X**) and structure **V** is that the positively charged side chains are exposed to solvent in the former (namely, they point away from the main chain), while they are attracted to carbonyl oxygens (with negative partial charges) of the main chain in the latter. In particular, it is

Table 4: Long-range NOE correlations between pairs of residues of BPTI(16-36) detected by the NMR experiment [44] and the corresponding ^{1}H$-^1$H distances calculated from the atomic coordinates of structures **V**, **S**, and **X**.

NOE correlations detected		Distance (A–B; Å)^a		
Residue [proton A]	Residue [proton B]	**V**	**S**	**X**
Ile-19 [C$^\gamma$H$_3$]	Tyr-21 [N$^\alpha$H]	5.0	3.3	5.9
Ile-19 [C$^\delta$H$_3$]	Tyr-21 [N$^\alpha$H]	8.4	4.5	8.2
Ile-19 [C$^\gamma$H$_3$]	Tyr-21 [N$^\delta$H]	6.9	4.6	5.9
Ile-19 [C$^\delta$H$_3$]	Tyr-21 [N$^\delta$H]	9.3	4.8	8.2
Arg-20 [C$^\alpha$H]	Phe-22 [N$^\alpha$H]	6.2	5.0	5.7
Tyr-21 [C$^\delta$H]	Tyr-23 [N$^\alpha$H]	7.9	4.8	7.3
Phe-22 [N$^\alpha$H]	Asn-24 [C$^\alpha$H]	6.7	6.5	7.2
Tyr-23 [N$^\alpha$H]	Gln-31 [C$^\gamma$H$_2$]	8.3	6.2	7.2
Tyr-23 [C$^\alpha$H]	Ala-25 [C$^\beta$H$_3$]	8.5	5.3	6.4
Asn-24 [N$^\alpha$H]	Ala-25 [C$^\beta$H$_3$]	5.8	4.5	5.4
Ala-27 [C$^\alpha$H]	Cys-30 [N$^\alpha$H]	7.9	5.2	8.1
Ala-27 [C$^\beta$H$_3$]	Cys-30 [N$^\alpha$H]	8.5	4.6	7.2
Thr-32 [C$^\beta$H]	Val-34 [N$^\alpha$H]	6.3	4.0	6.1
Val-34 [C$^\alpha$H]	Gly-36 [N$^\alpha$H]	5.6	3.6	5.5

^{a} Note that the ^{1}H$-^1$H distance which allows NOE cross peaks to be detectable is less than about 6 Å at the present conditions of measurement and that only the distances in structure **S** are almost within this limit.

clearly observed that in structure **V** the positively charged guanidino group of Arg-20 is attracted to as many as five carbonyl oxygens of the main chain, which hinders the formation of the characteristic intrachain hydrogen bonds that connect a pair of β-strands in a β-sheet structure. Therefore, an important role of solvent on the β-sheet folding of BPTI(16-36) can be associated with its ability to extract the charged side chains from the interior of the molecule into solvent so that the formation of the intrachain hydrogen bonds between β-strands is enhanced, thus stabilizing the β-sheet structure.

5   Conclusions

In this article, we have presented the results of simulated annealing and multicanonical Monte Carlo simulations applied to study the α-helix formation in C-peptide of ribonuclease A and the β-sheet formation in the peptide fragment BPTI(16-36). The results were in good agreement with various implications of CD, NMR, and X-ray experiments. We demonstrated that the side-chain charges and solvent effects play important roles in the stabilities of both α-helix and β-sheet structures. It should be emphasized that the simulations were performed from completely random initial conformations and that no structural information from experiments was used as input.

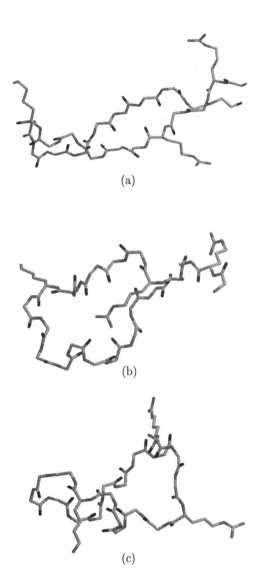

(a)

(b)

(c)

Figure 5: Structures **X**, **V**, and **S** of BPTI(16-36) redrawn from Figure 4 (seen from slightly different angles). Only the main chain and charged side chains (Arg-17, Arg-20, and Lys-26) are shown. The figures were created with RasMol [45].

Acknowledgements:
The author is grateful to his collaborators for discussions and suggestions. Especially he would like to thank U.H.E. Hansmann (for the work of C-peptide with multicanonical algorithm) and M. Masuya, M. Nabeshima, and T. Nakazawa (for the work of BPTI(16-36) with Monte Carlo simulated annealing). The simulations were performed on the computers at the Institute for Molecular Science. This work was supported, in part, by a Grant-in-Aid for Scientific Research from the Japanese Ministry of Education, Science, Sports and Culture, by a grant from the Research for the Future Program of Japan Society for the Promotion of Science (JSPS-RFTF98P01101).

References

[1] a) C. Levinthal, *J. Chim. Phys.* **65** (1968) 44.
b) D.B. Wetlaufer, *Proc. Natl. Acad. Sci. USA* **70** (1973) 691.

[2] C.J. Epstain, R.F. Goldberger, and C.B. Anfinsen, *Cold Spring Harbor Symp. Quant. Biol.* **28** (1963) 439.

[3] P.Y. Chou and G.D. Fasman, *Biochemistry* **13** (1974) 222.

[4] S. Kirkpatrick, C.D. Gelatt, Jr., and M.P. Vecchi, *Science* **220** (1983) 671.

[5] B.A. Berg and T. Neuhaus, *Phys. Lett.* **B267** (1991) 249.

[6] a) F.A. Momany, F.A., R.F. McGuire, A.W. Burgess, and H.A. Scheraga, *J. Phys. Chem.* **79** (1975) 2361.
b) G. Némethy, M.S. Pottle, and H.A. Scheraga, *J. Phys. Chem.* **87** (1983) 1883.
c) M.J. Sippl, G. Némethy, and H.A. Scheraga, *J. Phys. Chem.* **88**, (1984) 6231.

[7] a) H. Kawai, Y. Okamoto, M. Fukugita, T. Nakazawa, and T. Kikuchi, *Chem. Lett.* **1991** (1991) 213.
b) Y. Okamoto, M. Fukugita, T. Nakazawa, and H. Kawai, *Protein Eng.* **4** (1991) 639.

[8] N. Metropolis, A.W. Rosenbluth, M.N. Rosenbluth, A.H. Teller, and E. Teller, *J. Chem. Phys.* **21** (1953) 1087.

[9] a) B.E. Hingerty, R.H. Ritchie, T. Ferrell, and J.E. Turner, *Biopolymers* **24** (1985) 427.
b) J. Ramstein and R. Lavery, *Proc. Natl. Acad. Sci. USA* **85** (1988) 7231.

[10] Y. Okamoto, *Biopolymers* **34** (1994) 529.

[11] V. Daggett, P.A. Kollman, and I.D. Kuntz, *Biopolymers* **31** (1991) 285.

[12] T. Ooi, M. Oobatake, G. Némethy, and H.A. Scheraga, *Proc. Natl. Acad. Sci. USA* **84** (1987) 3086.

[13] L. Wesson and D. Eisenberg, *Protein Sci.* **1** (1992) 227.

[14] M. Masuya and Y. Okamoto, in preparation.

[15] Y. Okamoto, *Recent Research Devel. Pure Applied Chem.* **2** (1998) 1.

[16] U.H.E. Hansmann and Y. Okamoto, in *Annual Reviews in Computational Physics VI*, edited by D. Stauffer (Singapore, World Scientific, 1999) pp. 129–157.

[17] S.R. Wilson, W. Cui, J.W. Moskowitz, and K.E. Schmidt, *Tetrahedron Lett.* **29** (1988) 4373.

[18] H. Kawai, T. Kikuchi, and Y. Okamoto, *Protein Eng.* **3** (1989) 85.

[19] C. Wilson and S. Doniach, *PROTEINS: Struct. Funct. Genet.* **6** (1989) 193.

[20] A.T. Brünger, *J. Mol. Biol.* **203** (1988) 803.

[21] M. Nilges, G.M. Clore, and A.M. Gronenborn, *FEBS Lett.* **229** (1988) 317.

[22] S.R. Wilson and W. Cui, In: *The Protein Folding Problem and Tertiary Structure Prediction*, K.M. Merz, Jr. and S.M. Le Grand, eds. (Birkhäuser, 1994) pp. 43–70.

[23] U.H.E. Hansmann and Y. Okamoto, *J. Comp. Chem.* **18** (1997) 920.

[24] B.A. Berg, *Int. J. Mod. Phys.* **C3** (1992) 1083.

[25] U.H.E. Hansmann and Y. Okamoto, *J. Phys. Soc. Jpn.* **63** (1994) 3945; *Physica A* **212** (1994) 415.

[26] A.M. Ferrenberg and R.H. Swendsen, *Phys. Rev. Lett.* **61** (1988) 2635; *ibid.* **63** (1989) 1658(E).

[27] U.H.E. Hansmann and Y. Okamoto, *J. Comp. Chem.* **14** (1993) 1333.

[28] M.H. Hao and H.A. Scheraga, *J. Phys. Chem.* **98** (1994) 4940.

[29] U.H.E. Hansmann, Y. Okamoto, and F. Eisenmenger, *Chem. Phys. Lett.* **259** (1996) 321.

[30] N. Nakajima, H. Nakamura, and A. Kidera, *J. Phys. Chem.* **101** (1997) 817.

[31] C. Bartels and M. Karplus, *J. Phys. Chem. B* **102** (1998) 865.

[32] a) H.W. Wychoff, D. Tsernoglou, A.W. Hanson, J.R. Knox, B. Lee, and F.M. Richards, *J. Biol. Chem.* **245** (1970) 305.
b) R.F. Tilton, Jr., J.C. Dewan, and G.A. Petsko, *Biochemistry* **31** (1992) 2469.

[33] K.R. Shoemaker, P.S. Kim, D.N. Brems, S. Marqusee, E.J. York, I.M. Chaiken, J.M. Stewart, and R.L. Baldwin, *Proc. Natl. Acad. Sci. USA* **82** (1985) 2349.

[34] J.J. Osterhout, R.L. Baldwin, E.J. York, J.M. Stewart, H.J. Dyson, and P.E. Wright, *Biochemistry* **28** (1989) 7059.

[35] U.H.E. Hansmann and Y. Okamoto, *J. Phys. Chem. B* **102** (1998) 653.

[36] U.H.E. Hansmann and Y. Okamoto, *J. Phys. Chem. B* **103** (1999) 1595.

[37] a) Y. Okamoto, U.H.E. Hansmann, and T. Nakazawa, *Chem. Lett.* **1995** (1995) 391.
 b) Y. Okamoto and U.H.E. Hansmann, *J. Phys. Chem.* **99** (1995) 11276.

[38] P.J. Kraulis, *J. Appl. Cryst.* **24** (1991) 946.

[39] a) D. Bacon and W.F. Anderson, *J. Mol. Graphics* **6** (1988) 219.
 b) E.A. Merritt and M.E.P. Murphy, *Acta Cryst.* **D50** (1994) 869.

[40] T. Nakazawa, H. Kawai, Y. Okamoto, and M. Fukugita, *Protein Eng.* **5** (1992) 495.

[41] T. Nakazawa and Y. Okamoto, "Electrostatic effects on the α-helix and β-strand folding of BPTI(16-36) as predicted by Monte Carlo simulated annealing," *J. Peptide Res.* (1999), in press.

[42] Y. Okamoto, M. Masuya, M. Nabeshima, and T. Nakazawa, *Chem. Phys. Lett.* **299** (1999) 17.

[43] J. Deisenhofer and W. Steigemann, *Acta Crystallogr.* **B31** (1985) 238.

[44] T. Nakazawa, Y. Okamoto, Y. Kobayashi, Y. Kyogoku, and S. Aimoto, in preparation.

[45] R.A. Sayle and E.J. Milner-White, *TIBS* **20** (1995) 374.

Optimization in Computational Chemistry and Molecular Biology, pp. 91-105
C. A. Floudas and P. M. Pardalos, Editors
©2000 Kluwer Academic Publishers

Thermodynamics of Protein Folding - The Generalized-Ensemble Approach

Ulrich H.E. Hansmann
Department of Physics, Michigan Technological University
Houghton, MI 49931-1295 USA
hansmann@mtu.edu

Abstract

For many years the emphasis in protein-folding simulations has been laid as to how to predict the three dimensional structure of proteins. Only recently has there be a shift in interest towards the the thermodynamics of folding. We show that generalized-ensemble techniques are well suited to study these questions for realistic protein models.

Keywords: Generalized Ensembles, Monte Carlo Simulations, Thermodynamics of Protein Folding.

1 Introduction

A long standing goal of computational biochemistry is to understand folding of proteins solely from the amino-acid sequence information by means of computer simulations. For many years the emphasis in protein studies was on the structure prediction of proteins. Assuming that the native structure is thermodynamically stable, it is reasonable to identify the global-minimum conformation in the *free* energy at $T \approx 300$ K with the lowest *potential* energy conformation and to search for this conformation with powerful optimization techniques. Both deterministic methods (for instance, the α**BB** algorithm [1]) and stochastic algorithms (like simulated annealing [2]) are employed. However, with the recognition of energy landscape theory and funnel concept there has been an increased interest in the thermodynamics of folding. This "new view" asserts that a full understanding of the folding process requires a global knowledge of the free energy landscape of the protein system [3, 4, 5]. To probe these new ideas by computer simulations requires to go beyond global optimization techniques: one has to measure thermodynamic quantities, i.e. to sample a set of configurations from a canonical ensemble and take an average of the chosen quantity over this ensemble.

Unfortunately, such sampling has been proven to be notoriously difficult for realistic protein models. Simulations based on canonical Monte Carlo or molecular dynamics techniques will at low temperatures get trapped in one of the multitude of local minima separated by high energy barriers. Hence, only small parts of configuration space are sampled and physical quantities cannot be calculated accurately. One successful approach to overcome this problem and to enhance sampling in protein folding simulations is to perform simulations in so-called *generalized ensembles* which are defined in such a way that the probability to cross an energy barrier does not decrease exponentially with barrier heights. The first application of this approach to the protein folding problem can be found in Ref. [6], where a Monte Carlo technique was used. A formulation for the molecular dynamics method was also developed later [7, 8]. An overview on recent applications can be found in Ref. [9]. In the following we will present first a short review of the generalized-ensemble approach and demonstrate afterwards for two examples that the new approach allows indeed to study the thermodynamics of folding.

2   Generalized-ensemble techniques

A generalized-ensemble simulation is characterized by the condition that a Monte Carlo or molecular dynamics simulation shall lead to a uniform distribution of a pre-chosen physical quantity. Probably the earliest realization of this idea is *umbrella sampling* [10]. This idea was lately revived and a variety of new algorithms were developed whose usefulness for simulations of biological molecules has been increasingly recognized. Three prominent examples of these newer generalized-ensemble techniques are the multicanonical algorithm [11], $1/k$-sampling [12] and simulated tempering [13].

The underlying idea of all generalized-ensemble algorithms can be seen clearly for the example of the multicanonical algorithm [11]. Here, the weights $w(E)$ are chosen such the distribution of energies

$$P(E) \propto n(E)w(E) = \text{const}, \tag{1}$$

where $n(E)$ is the spectral density. A free random walk in the energy space is performed which allows the simulation to escape from any local minimum, and even regions with small $n(E)$ can be explored in detail. Similar, $1/k$-sampling [12] yields a uniform distribution in (microcanonical) entropy and simulated tempering [13] to an uniform distribution in temperature. We remark that there is no restriction of the approach to ensembles which lead to flat distributions in *one* variable. Extensions to higher number of variables are straight forward [14, 15]. In any case, from such a generalized-ensemble simulation one can calculate the thermodynamic average of any physical quantity $\mathcal{A}$ for a wide temperature range by the re-weighting technique [16]:

$$< \mathcal{A} >_T \; = \; \frac{\int dx \; \mathcal{A}(x) \; w^{-1}(x) \; e^{-E(x)/k_B T}}{\int dx \; w^{-1}(x) \; e^{-E(x)/k_B T}} \; . \tag{2}$$

Here x stands for configurations, $w(x)$ is the generalized-ensemble weight of configuration x and k_B the Boltzmann constant.

However, unlike in the canonical ensemble, the weights are not *a priori* known for simulations in these ensembles. For instance, in the multicanonical algorithm $w_{mu}(E) \propto n^{-1}(E)$, and knowledge of the exact weights is equivalent to obtaining the density of states $n(E)$, i.e., solving the system. Hence, one needs their estimators for a numerical simulation. The determination of the weight $w_{mu}(E)$ is usually based on an iterative procedure described in detail in Ref. [17]. In Figure 1 we display for the small peptide Met-enkephalin (see the following chapter) the microcanonical entropy

$$S(E) = \ln n(E) = -\ln w_{mu}(E) \qquad (3)$$

as obtained by the aboved described iterative procedure. Alternative methods rely on preliminary simulated annealing runs [18] or exploit a mean field approximation of the protein model [19].

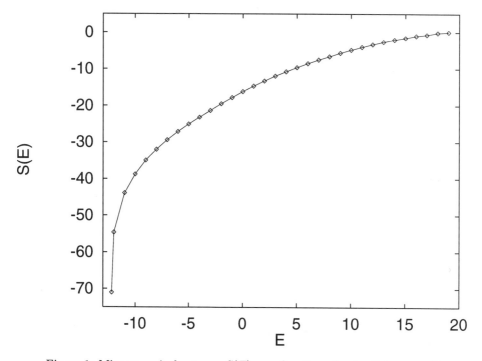

Figure 1: Microcanonical entropy $S(E)$ as a function of potential energy E.

It is obvious that the performance of the algorithm depends on how good the estimates for the weight factors are. For instance, the optimal performance for the multicanonical algorithm will be so that the autocorrelation time increases with the square of the energy range to be covered in the simulation (since a free random walk in energy is performed) while in a canonical simulation the autocorrelation time increases exponentially with system size. Since one has to use estimates for the weights in a multicanonical simulation, the autocorrelation time will scale like $\approx E^x$ with $x > 2$ instead of $\approx E^2$. However, the deviation is in general rather small. This is discussed in detail in Ref. [11] and was recently demonstrated for homopolymers in Ref. [20]. A detailed comparison of the various generalized-ensemble methods and their performance can be found in Ref. [21].

Some attempts were made to construct generalized ensembles where the determination of the estimators is simple and straightforward or where the weights are even *a priori* known. One possibility is *parallel tempering* [22] which was first introduced to the protein folding problem in Ref. [23]. More experience was gathered with the ensemble which was proposed in Ref. [24] and which is based on the following weight:

$$w(E) = \left(1 + \frac{\beta(E - E_0)}{n_F}\right)^{-n_F} . \qquad (4)$$

Here E_0 is an estimator for the ground-state energy and n_F is the number of degrees of freedom of the system. Obviously, the new weight reduces in the low-energy region to the canonical Boltzmann weight $\exp(-\beta E)$ for $\frac{\beta(E-E_0)}{n_F} \ll 1$. On the other hand, high-energy regions are no longer exponentially suppressed but only according to a power law, which enhances excursions to high-energy regions and therefore increases the probability to escape local minima. Note that this weight can be understood as a special case of the weights used in Tsallis generalized mechanics formalism [25] (the Tsallis parameter q is chosen as $q = 1 + 1/n_F$). In contrast to other generalized-ensembles the weights of the new ensemble are explicitly given by Eq. 4. One only needs to find an estimator for the ground-state energy E_0 which was found to be much easier than the determination of weights for other generalized ensembles.

3 Energy Landscape of Small peptides

To demonstrate that the generalized-ensemble approach is indeed well-suited for investigations of thermodynamics of folding, we first present results from a recent study on the energy landscape of a small peptide [26].

Our work was motivated by the "new view" of the protein folding which became increasingly popular over the last few years. Its framework is provided by energy landscape theory and the funnel concept, which assert that a full understanding of the folding process requires a global overview of the landscape. The folding landscape of a protein is assumed to resemble a partially rough funnel riddled with traps where the protein can transiently reside. There is no unique pathway but a multiplicity of convergent folding routes towards the native state [3, 4, 27].

However, these new ideas were derived from simulations of minimal protein models which capture only few, but probably dominant, interactions in real proteins. Because of the inherent limitations of minimal protein models it is desirable to check the above picture by comparison with more realistic energy functions. For reasons described in the introduction this became only possible with the recent development of new techniques such as the generalized-ensemble approach. Hence, the work in Ref. [26] is a first attempt to use the power of the new technique to check the above stated ideas for realistic protein models.

Our system of choice was the linear peptide Met-enkephalin which has the amino-acid sequence Tyr-Gly-Gly-Phe-Met. The generalized-ensemble algorithm used in this study is the one described in Ref. [24] and relies on the weight in Eq. 4. To actually observe the proposed folding funnel of the peptide, one has to study a projection of the energy landscape onto a set of suitable and appropriate order parameters. Using the ECEPP/2 force field [28] and excluding solvation effects, it was shown in a recent article [29] that Met-enkephalin undergoes a transition between extended and compact structures at a temperature $T_\theta = 295 \pm 20$ K. Above that temperature, the frequency of compact structures rapidly decreases while it increases below T_θ. Hence, our first order parameter is the volume V allowing us to distinguish between compact and extended conformations. In Ref. [29] it was also shown that by further lowering the temperature the peptide encounters a second transition. Below $T_f = 230 \pm 30$ K, the occupation of the ground-state conformations increases rapidly while it decreases for values of T above T_f. Hence, we chose as a second order parameters the overlap with the ground state, O_A, which allows us to distinguish between the various compact low-energy conformations.

Our simulation was started from a completely random initial conformation (Hot Start) and one Monte Carlo sweep updates every torsion angle of the peptide once. We fixed the peptide bond angles ω to their common value $180°$, which left us with 19 torsion angles (ϕ, ψ, and χ) as independent degrees of freedom (i.e., $n_F = 19$). In our simulations we did not explicitly include the interaction of the peptide with the solvent and set the dielectric constant ϵ equal to 2. However, we do expect some implicit solvent effect, since the various parameters for the energy function were determined by minimization of the potential energies of the crystal lattices of single amino acids, i.e., not in a vacuum. All thermodynamic quantities were then calculated from a single production run of 1,000,000 MC sweeps which followed 10,000 sweeps for thermalization. At the end of every fourth sweep we stored the energies of the conformation and our two "order parameters" (the corresponding volume and the overlap O_A of the conformation with the (known) ground state). Since large parts of the configuration space are sampled by our method, it is justified to calculate from this time series the thermodynamic quantities over a wide range of temperatures by Eq. (2).

Having defined the two order parameters, we tried to depict the folding funnel of Met-enkephalin by plotting the free energy $G(V, O_A)$ as a function of volume V and overlap O_A with the known ground state. Since the energy landscape for a folding protein depends strongly on temperature we have concentrated our analyses on four temperatures. The first one, $T = 1000$ K was chosen to probe the high-temperature regime where the peptide is fully unfolded and mostly in an extended form. In some early work [29], $T = 300$ K was identified as the collapse temperature T_θ and $T = 230$ K as the folding temperature T_f. The last temperature, $T = 150$ K, was chosen to study the low temperature behavior of the peptide where the glassy behavior is observed.

In Figure 2 we show the free energy landscape as a function of volume and overlap with the known ground state at the high-temperature region ($T = 1000$ K). Here, (as in the other free energy plots) we normalized the free energy in such a way that its observed minimum is set to zero. In the contour plots, the contour lines mark multiples of $k_B T$. We see that the free energy has its minimum at large volumes (≈ 1470 Å^3) and values of the overlap $O_A \approx 0.3$. Small volumes and larger values of the overlap are suppressed by many orders of $k_B T$. Hence, extended random coil structures are favored at this temperature. The picture changes dramatically once we reach the collapse temperature T_θ, shown in Figure 2. At this temperature a large part of the V-O_A space can be sampled in a simulation. The contour plot shows that regions with both small and large volumes and almost all values of O_A lie within the 2 $k_B T$ contour. This indicates that at this temperature the cross over between extended and compact structures happens with a small thermodynamic barrier between them. By lowering the temperature to $T_f = 230K$ (determined in ref. [29]), we now observe strong evidence for a funnel-like landscape (Figure 2). At this temperature the drive towards the native configuration is dominant and no long-lived traps exist. There is clearly a gradient towards the ground-state structure ($O_A \approx 1$), but other structures with similar volume (characterized by values of $O_A \approx 0.5$) are only separated by free energy barriers of order 1 $k_B T$. Below this temperature we expect that the ground state is clearly favored thermodynamically and separated from other low energy states by free energy barriers of many orders of $k_B T$. This can be seen in Figure 2 where at $T = 150K$ where other low energy states have free energies of 3 $k_B T$ higher than the ground state and are separated by an additional barrier of 2 $k_B T$. This kind of behavior has been predicted from simulations of minimalist models and now has been confirmed by our study [26] for simulations of real peptides.

4   The Helix-Coil Transition in Poly-Alanine

As mentioned above folding of proteins involves transitions between different thermodynamic states. The nature of these transitions is still not clearly understood. Is it possible to regard them as phase transitions or are they merely a crossover between the two states? Since such questions can be more easily investigated for homopolymers of amino-acids than for proteins (which are heteropolymers) we decided to look into more detail into the sharp transition between disordered coil conformers and an ordered helical phase observed by us for poly-alanine in earlier work [30]. To determine the nature of this helix-coil transition we studied in Ref. [20] the finite size scaling of this transition and extrapolated the results to the limit of an infinitely long polymer. The multicanonical algorithm was used to calculate various thermodynamic quantities as a function of temperature for poly-alanine of four different chain lengths. We concentrate on such quantities (like average number of helical residues or specific heat) where we expect to see the strongest signal for the helix-coil transition. The finite-size scaling of these quantities was studied and estimates for critical exponents were calculated. Again, use of a generalized-ensemble technique was crucial to avoid ergodicity problems in the low temperature phase (see the discussion in Ref. [30]).

Since one can avoid the complications of electrostatic and hydrogen-bond interactions of side chains with the solvent for alanine (a nonpolar amino acid), explicit solvent molecules were neglected for simplicity and the dielectric constant ϵ was set equal to 2. Again, the peptide-bond dihedral angles ω were fixed at the value 180° for simplicity, which leaves ϕ_i, ψ_i, and χ_i ($i = 1, \cdots, N$) as independent degrees Since alanine has only one χ angle in the side chain, the numbers of independent degrees of freedom are $3N$ where N is the number of residues. The multicanonical weight factors were determined by the iterative procedure described in Refs. [30, 17]. We needed between 40,000 sweeps (for $N = 10$) and 500,000 sweeps (for $N = 30$) for their calculation. All thermodynamic quantities were then calculated from one production run of N_{sw} Monte Carlo sweeps following additional 10 000 sweeps for equilibrization. We chose $N_{sw} = 200,000$ for $N = 10$, $N_{sw} = 250,000$ for $N = 15$, $N_{sw} = 500,000$ for $N = 20$, and $N_{sw} = 1,000,000$ for $N = 30$. In all cases, each

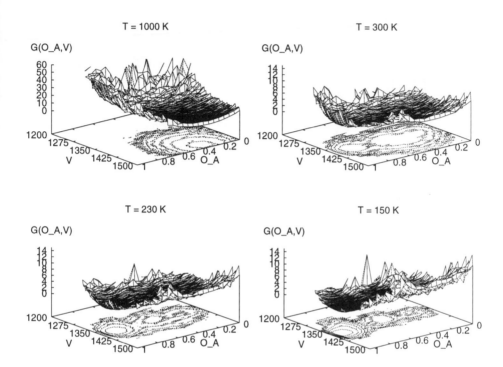

Figure 2: Free energy $G(V, O_A)$ (in kcal/mol) as a function of both peptide volume V (in Å^3) and overlap O_A for a) $T = 1000$ K, b) $T = 300$ K, c) $T = 230$ K, and d) $T = 150$ K. Both the free energy surface and the contour plot are shown. The contour lines are multiples of $k_B T$. $G(V, O_A)$ was normalized such that $\min(G(V, O_A)) = 0$.

simulation started from a completely random initial conformation.

The steep helix-coil transition for poly-alanine can be seen in Figure 3 where the order parameter

$$q = \frac{\tilde{n}_H}{N - 2} \qquad (5)$$

is plotted as a function of temperature. Here $\tilde{n}_H$ is the number of helical residues in a conformation, however, without counting the first and last residues. We chose this definition in order to have $q = 1$ for a completely helical conformation. Since the residues at the end of the polymer chain can move freely, they will not be part of a helical segment and therefore should not be counted. In Figure 4, where we display the average number of helical segments as a function of temperature and chain length, we find evidence that this transition indicates indeed a phase transition. It is clear from this plot that for each chain length the low-temperature region is dominated by a single helix. This indicates the the existence of long-range order, since we find no indications that helical segments become unstable once they reach a critical length. On the other hand, around the critical temperature T_c the number of helical segments is maximal and its average number increases with the size of the chain. This is consistent with a second-order phase transition where one would also expect fluctuations on all length scales at T_c.

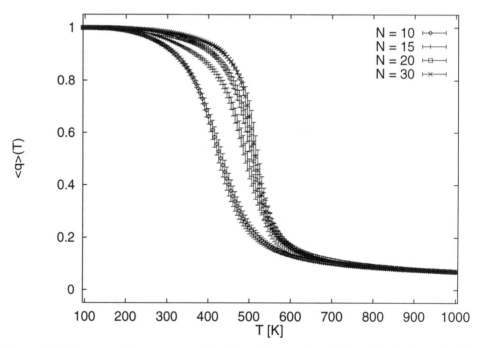

Figure 3: Order parameter $< q >_T$ as a function of temperature T for poly-alanine molecules of chain length $N = 10, 15, 20,$ and 30.

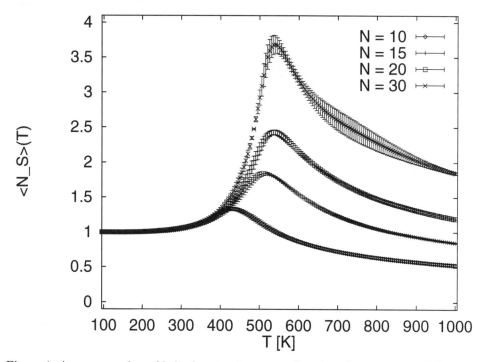

Figure 4: Average number of helical segments n_S as a function of temperature T for poly-alanine molecules of chain length $N = 10, 15, 20,$ and 30.

To further characterize this phase transition we tried to determine the critical exponents. Conventional arguments for finite-size scaling for a second-order transition are based on the assumption that the singular part of the free energy depends only on the system size N and the correlation length ξ. The critical exponents can be extracted from the finite-size scaling of the heights and width of the peaks in specific heat

$$C_N(T) \equiv \frac{1}{N\,k_B} \frac{d\left(<E_{tot}>_T\right)}{dT} = \beta^2 \frac{<E_{tot}^2>_T - <E_{tot}>_T^2}{N} . \tag{6}$$

and susceptibility

$$\chi_N(T) = \frac{1}{N-2}(<q^2>_T - <q>_T^2) . \tag{7}$$

which are plotted as a function of temperature in Figure 5, and Figure 6, respectively. For details, see Ref. [20]. Here, we only remark that we do not find in Figure 5 any indications for another peak in the specific heat at lower temperature $T < T_c$, which, if existed, could be interpreted as a transition between two helix states. Such a solid-solid transition was observed in a recent study on wormlike polymer chains.[31] Similarly, we do not see a shoulder in the specific heat for $T > T_c$. Hence, we conjecture that no other transitions but helix-coil one exist for poly-alanine.

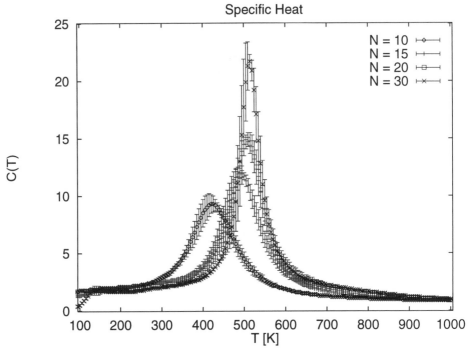

Figure 5: Specific heat $C(T)$ as a function of temperature T for poly-alanine molecules of chain length $N = 10, 15, 20$, and 30.

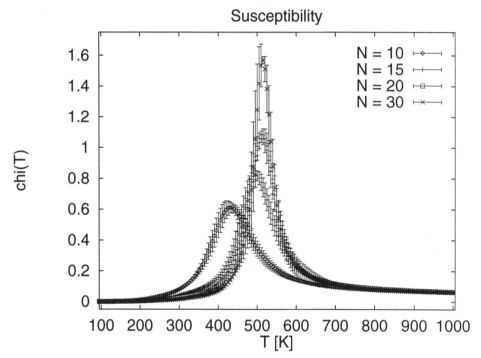

Figure 6: Susceptibility χ as a function of temperature T for poly-alanine molecules of chain length $N = 10, 15, 20,$ and 30.

With the critical temperature $T_c(N)$ as the position where the peak in the specific heat has its maximum, and $T_1(N)$ and $T_2(N)$ (with $T_1(N) < T_c(N) < T_2(N)$) chosen such that $C(T_1) = 1/2C(T_c) = C(T_2)$, we have

$$\Gamma_C(N) = T_2(N) - T_1(N) \propto N^{-\frac{1}{\nu}}, \tag{8}$$

and

$$C_N(T_c) \propto N^{\frac{\alpha}{\nu}} . \tag{9}$$

Similarly we find from the heights of the peak in the susceptibility

$$\chi_N(T_c) \propto N^{\frac{\gamma}{\nu}} , \tag{10}$$

and from the temperatures where $\chi(T) = 1/2\chi(T_c)$ we get a second, independent estimate for the critical exponent ν by

$$\Gamma_\chi(N) = T_2(N) - T_1(N) \propto N^{-\frac{1}{\nu}} . \tag{11}$$

Using the above equations the following estimates for the critical exponents were obtained in Ref. [20]: Eq. (8) yields an estimate for $1/\nu = 0.54(5)$ which is comparable with $1/\nu = 0.51(6)$, the value we obtained from the fitting of Eq. (11). Combining both values, we have as our final estimate for the correlation length exponent for the helix-coil transition in poly-alanine:

$$\nu = 1.9(2) . \tag{12}$$

With a value of $\alpha/\nu = 0.79(9)$, obtained by fitting Eq. (9), we find the following specific heat exponent:

$$\alpha = 1.5(2) . \tag{13}$$

Similarly, from Eq. (10) we obtain a value $\gamma/\nu = 0.88(7)$, from which we get our estimate for the susceptibility exponent:

$$\gamma = 1.7(1) . \tag{14}$$

The non-trivial values we obtained for these critical exponents give further evidence for the second-order phase transition. However, note that the above critical exponents do not obey the hyperscaling relation:

$$\nu = 2 - \alpha . \tag{15}$$

We are currently investigating the reasons for the violation of hyperscaling [32].

5   Conclusions

To summarize, we reported results from two different generalized-ensemble simulations of peptides where the interactions among all atoms were taken into account. Our results support pictures for the kinetics of protein folding which were developed from the study of simplified protein models. We could determine the nature of the helix-coil transition for poly-alanine and presented estimates for the critical exponents which characterizes this transition. These two examples demonstrate that the generalized-ensemble algorithms are well-suited for investigations of thermodynamics of proteins. Hence, it can be expected that the new approach will lead to an increased understanding of the protein folding problem not only in minimal models but also in realistic protein systems.

Acknowledgements:
The presented work was done in collaboration with J. Onuchic (UCSD, San Diego) and Y. Okamoto (IMS, Okazaki, Japan). Financial supports from a Research Excellence Fund (E27448) of the State of Michigan is gratefully acknowledged.

References

[1] Androulakis I.P., Maranas C.D., Floudas C.A. (1997), " Prediction of oligopeptide conformations via deterministic global optimization," *Journal of Global Optimization*, *Vol. 11*, 1-34.

[2] Kirkpatrick S., Gelatt C.D. Jr., Vecchi M.P. (1983), "Optimization by simulated annealing," *Science, Vol. 220*, 671-680.

[3] Onuchic J.N., Luthey-Schulten Z., Wolynes P.G. (1997) "Theory of protein folding: the energy landscape perspective," *Annual Reviews in Physical Chemistry, Vol. 48*, 545-600.

[4] Dill K.A., Chan H.S. (1997), " From Levinthal to pathways to funnels," *Nature Structural Biology, Vol. 4*, 10-19.

[5] Dobson C.M., Sali A., Karplus M. (1998), "Protein folding: a perspective from theory and experiment," *Angewante Chemie*, in press.

[6] Hansmann U.H.E., Okamoto Y. (1993), "Prediction of peptide conformation by multicanonical algorithm: a new approach to the multiple-minima problem," *Journal of Computational Chemistry, Vol. 14*, 1333-1338.

[7] Hansmann U.H.E., Okamoto Y., Eisenmenger F. (1996) "Molecular dynamics, Langevin and hybrid Monte Carlo simulations in a multicanonical ensemble," *Chemcal Physics Letters, Vol. 259*, 321-330.

[8] Nakajima N., Nakamura H., Kidera A. (1997), " Multicanonical ensemble generated by molecular dynamics simulation for enhanced conformational sampling of peptides," *Journal of Physical Chemistry, Vol. 101*, 817 -824.

[9] Hansmann U.H.E., Okamoto Y. (1999) " The generalized-ensemble approach for protein folding simulations." In *Annual Reviews in Computational Physics VI*. Edited by Stauffer D. Singapore: World Scientific; 1999, 129-157.

[10] Torrie G..M, Valleau J.P. (1977), " Nonphysical sampling distributions in Monte Carlo free-energy estimation: umbrella sampling," *Journal of Computational Physics, Vol. 23*, 187-199.

[11] Berg B.A., Neuhaus T. (1991), " Multicanonical algorithms for first order phase transitions," *Physics Letters, Vol. B267*, 249-253.

[12] Hesselbo B., Stinchcombe R.B. (1995), "Monte Carlo simulation and global optimization without parameters," *Physical Review Letters, Vol. 74*, 2151-2155.

[13] Lyubartsev A.P., Martinovski A.A., Shevkunov S.V., Vorontsov-Velyaminov P.N. (1992), "New approach to Monte Carlo calculations of the free energy: method of expanded ensembles," *Journal of Chemical Physics, Vol. 96*, 1776-1783; Marinari E., Parisi G. (1992), "Simulated tempering: a new Monte Carlo scheme," *Europhysics Letters, Vol. 19*, 451-458.

[14] Kumar S., Payne P., Vásquez M. (1996), "Method for free-energy calculations using iterative techniques," *Journal of Computational Chemistry, Vol. 17*, 1269-1275.

[15] Higo J., Nakajima N., Shirai H., Kidera A., Nakamura H. (1997), "Two-component multicanonical Monte Carlo method for effective conformational sampling," *Journal of Computational Chemistry, Vol. 18*, 2086-2092.

[16] Ferrenberg A.M., Swendsen R.H. (1988), "New Monte Carlo technique for studying phase transitions" *Physical Review Letters, Vol. 61*, 2635-2638.

[17] Hansmann, U.H.E. and Okamoto, Y. (1994), " Comparative Study of Multicanonical and Simulated Annealing Algorithms in the Protein Folding Problem," *Physica A, Vol. 212*, 415-437.

[18] Hansmann, U.H.E (1997), "An Effective Way for Determination of Multicanonical Weights", *Physical Review E, Vol. 56*, 6200-6203.

[19] Hansmann, U.H.E. and de Forcrand, Ph. (1997), "Simple Ansatz to Describe Thermodynamic Quantities of Peptides and Proteins at Low Temperatures" *International Journal for Modern Physics C, Vol. 8*, 1085 -1094.

[20] Hansmann, U. H. E., Okamoto, Y. (1999), "Finite-size scaling of helix-coil transitions in poly-alanine studied by multicanonical simulations," *Journal of Chemical Physics, Vol. 110*, 1267-1276.

[21] Hansmann, U. H. E., Okamoto, Y. (1997), "Numerical Comparisons of Three Recently Proposed Algorithms in the Protein Folding Problem," *Journal of Computational Chemistry, Vol. 18*, 920–933.

[22] Hukushima K., Nemoto K. (1996), "Exchange Monte Carlo method and application to spin glass simulations," *Journal of the Physical Society (Japan), Vol. 65*, 1604-1608.

[23] Hansmann U.H.E. (1997), "Parallel tempering algorithm for conformational studies of biological molecules," *Chemical Physics Letters, Vol. 281*, 140-150.

[24] Hansmann, U.H.E. and Okamoto, Y. (1997), " Generalized-Ensemble Monte Carlo Method for Systems with Rough Energy Landscape" *Physical Review E, Vol. 56*, 2228 -2233.

[25] Tsallis C. (1988), "Possible generalization of Boltzmann-Gibbs statistics," *Journal of Statistical Physics, Vol. 52*, 479-487.

[26] Hansmann U.H.E., Okamoto Y., Onuchic J.N. (1999), "The folding funnel landscape for the peptide Met-enkephalin," *Proteins, Vol. 34*, 472-483.

[27] Bryngelson, J.D., Wolynes, P.G. "Spin glasses and the statistical mechanics of protein folding," *Proceedings of the National Academy of Sciences (USA), Vol. 84*, 524-7528.

[28] Sippl, M. J., Némethy, G., Scheraga, H. A. (1994), "Intermolecular potentials from crystal data. 6. Determination of empirical potentials for O-H$\cdots$O=C hydrogen bonds from packing configurations," *Journal of Physical Chemistry, Vol. 88*, 6231–6233; and references therein.

[29] Hansmann U.H.E., Masuya, M. and Okamoto, Y. (1997), "Characteristic Temperatures of Folding of a Small Peptide", *Proceedings of the National Academy of Sciences (USA), Vol. 94*, 10652 - 10656.

[30] Okamoto Y. and Hansmann,U.H.E. (1995), "Thermodynamics of Helix - Coil Transitions Studied by Multicanonical Algorithms", *Journal of Physical Chemistry, Vol. 99*, 11276 - 11287.

[31] Kemp J.P. and Chen Z.Y. (1998), "Formation of helical states in wormlike polymer chains", *Physical Review Letters, Vol. 81*, 3880-3883.

[32] Alves, N. and Hansmann, U.H.E. (1999), manuscript in preparation.

Optimization in Computational Chemistry and Molecular Biology, pp. 107-129
C. A. Floudas and P. M. Pardalos, Editors
©2000 Kluwer Academic Publishers

An approach to detect the dominant folds of proteinlike heteropolymers from the statistics of a homopolymeric chain

Erik D. Nelson
Center for Advanced Computational Science and Engineering
San Diego Supercomputer Center 0505,
University of California at San Diego,
La Jolla, CA 92093
enelson@sdsc.edu

Peter G. Wolynes
School of Chemical Sciences,
University of Illinois,
Urbana, Illinois 61801
wolynes@scs.uiuc.edu

Jose' N. Onuchic
Department of Physics
University of California at San Diego,
La Jolla, CA 92093
jonuchic@ucsd.edu

Abstract

Statistical optimization of proteins is interpreted from the standpoint of optimal memory storage in neural networks. This approach results in the concept of an intrinsic
learning rule which occurs in place of the usual learning rule in neural networks and
incorporates geometric, topological and statistical constraints which make one folded
shape kinetically more accessible than another. As a first step to extract this learning
rule from the behavior of model chains, we approximate the free energy of proteinlike
heteropolymers by an expansion about the free energy of an "equivalent homopolymer",
the coefficients of which determine a potential for heteropolymer sequences. Expansion
coefficients are computed for a simple bead chain homopolymer model and the results
are compared to a hydrophobic and polar (HP) model of proteins for which the optimal
folds are already known.

Keywords: protein evolution, topology, gauge invariance.

1 Folding and Hard Problems

The size of the sequence search space for even biologically short proteins ($N \sim 100$) is astronomically large (20^N), however, there is growing evidence to suggest that a complete set of 20 residue types is not essential to accomplish protein folding on physiological timescales ($\sim msec$) under typical biochemical conditions [2, 1, 3, 4, 7, 5, 6]. Sequences that do fold posess a specific type of energy landscape [8, 9, 11, 10, 12] qualitatively similar to the landscape of a single stored memory in a neural network [15, 16, 17, 18, 19, 20, 21] textured by an ensemble of local energy minima [13, 14]. In the ideal case, this funnel shaped landscape kinetically attracts a protein to its folded structure from any initial configuration of the chain [8].

The energy landscape of a non−folding (frustrated) sequence is, on the other hand, similar to that of a neural network with strongly overlapping memories [15, 16, 17, 18, 19, 20]. For frustrated sequences different starting configurations of the chain tend to become trapped within structurally dissimilar basins of attraction because the configurational energy landscape of such a sequence is textured by many deep local minima.

This rough analogy between spin glass and neural network optimization problems [20, 21, 22, 23] is the core approach in many theoretical attempts to model protein folding and to predict protein structure In such approaches [24, 25, 26, 27, 28, 29] one associates the configurational variables of a protein with neuron connections (synapses) and the sequence variables (residue types) with states of the neurons. Crudely speaking, folding can then be pictured as annealing the connections for a fixed state of the neurons [20, 21], while sequence evolution can be viewed as annealing the neurons [24] for a fixed set of connections (determined by a set of optimal structures), although in reality both sequence and structure evolve simultaneously [24, 25, 26].

In proteins, a randomly chosen sequence generally does not fold, in other words, protein sequences are "special instances" [30] of the folding problem which posess funnel like configurational landscapes [8]. Accordingly, to each foldable structure corresponds a family of sequences − each family being associated with a sequence funnel, or attractor in sequence space [24]. Consequently, the emergence of a configurational funnel upon sequence design or through evolutionary selection [31] signals a transition [8, 32, 33] between two completely different types of behavior − (i) frustrated [32, 33] and (ii) minimally frustrated [34, 35, 36].

The occurrence of a particular pattern of attractors in sequence space [24, 25] for a given model of proteins may depend strongly on the internal constraints of the chain (chain connectivity, excluded volume, and non−crossing). These constraints result in a type of potential for the sequences − somewhat analogous to the Hebb learning rule in the Hopfield model − which acts to connect sequences to kinetically accessible folded structures. In this paper we consider the possibility that the landscape of a sequence averaged (and therefore homopolymeric) chain, in which only these (uniform) constraints operate, may already contain some number of deep valleys [37], and that these valleys correspond to the most designable shapes of the heteropolymer model.

To investigate this possibility, we propose using a perturbation expansion for the heteropolymer free energy about the free energy of an equivalent homopolymer. We apply this description to a simple bead chain heteropolymer model for which we have already determined the most designable shapes for sequences with hydrophobic and polar residue types and a fixed proteinlike hydrophobicity. By measuring the parameters of the heteropolymer expansion, we observe a rough signature of the dominant folding motifs determined for the hydrophobic—polar (HP) model. We later speculate as to how this free energy expansion may be used as a type of potential to determine sequence families that select one of the statistically dominant collapsed shapes of an equivalent homopolymer.

The paper is organized as follows. First we look more closely at the relationship between associative memory neural networks and protein evolution under a foldability constraint. This helps to illustrate that frustration results from a conflict between the attractive interactions which, because of the constraints on dynamical variables (i.e. chain connectivity, excluded volume, and non—crossing restrictions), can never be completely satisfied. The fact that these interactions are maximally satisfied [34] (i.e. frustration is minimized) for some set of highly accessible [8, 38, 39, 40] collapsed homopolymer shapes [24] leads to the concept of an intrinsic learning rule, or potential, "contained" in the homopolymer free energy. In remaining sections we discuss the free energy expansion and apply our methods to the simple bead—chain model. The the sequence free energy is discussed in a short Appendix.

2   Proteins as Evolutionarily Stored Memories

As is well known, there is a limit to the number of memories that can be stored in a neural network [18]. This limit depends not just on the size of the system (the number of neurons) but also on the types of patterns stored [18, 17]. If the patterns are dissimilar (uncorrelated) the storage capacity of the network is maximized — or alternatively, the noisy overlaps between stored patterns (which texture the energy landscape) are minimized [15, 16, 18, 17].

In an optimally trained network [15, 16], the energy landscape of the neurons contains a set of deep valleys (attractors), each corresponding to one of the stored memories. Below a certain "pattern temperature" [18] an incomplete or damaged input pattern of the neurons is kinetically attracted to the most similar pattern stored in memory. However, when the memory capacity of the network is exceeded, the overlaps reach a critical value, which can not be reduced by pushing the patterns apart energetically [18]. Consequently, the attractors blend together and the network loses its functionality.

As we noted above, this situation closely resembles the storage of foldable protein structure memories by sequence, and suggests why a limiting number of foldable families should occur for a given chain length [24]. In earlier work we established a precise connection between pattern attractors in a neural network and sequence attractors on the sequence landscape of proteinlike heteropolymers [24]. Using a simple off—lattice model (Figs. 1−3) we showed that sequences which fold to the most accessible structures of a homopolymeric chain exhibit a feature called pseudo—orthogonality [15] enforced by the synapse constraints (learning rule) in the Hopfield network. This rule pushes apart the attractors so that input neuron patterns which are energetically favored in one attractor are disfavored in all the others.

To be more precise one makes the analogy between a sequence and a frozen (fixed, or specific) state of the neurons. Then the valleys which appear on the sequence landscape [24] are analogous to pattern attractors in a neural network. In addition, because the sequence

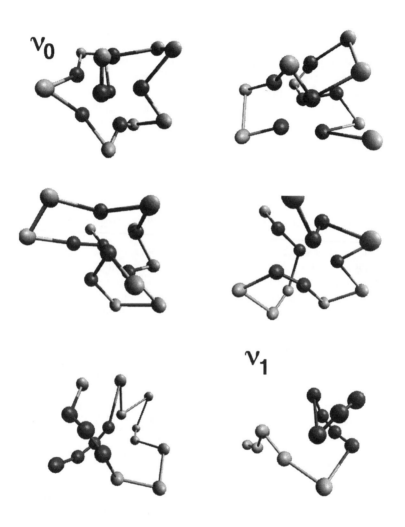

Figure 1: Snapshots of 6 representative sequences illustrating the two optimal hydrophobic core geometries in the HP−model chains (hydrophobic residues are dark beads). The top 4 structures have the highest symmetry, pentagonal bipyrimid core structure. The lower two structures have a core geometry with two planes of symmetry. Each structure represents 1 of 15 local gauge invariance classes − i.e. each structure typifies a large number of sequences with the same ground state configuration of hydrophobic monomers shown in the figure.

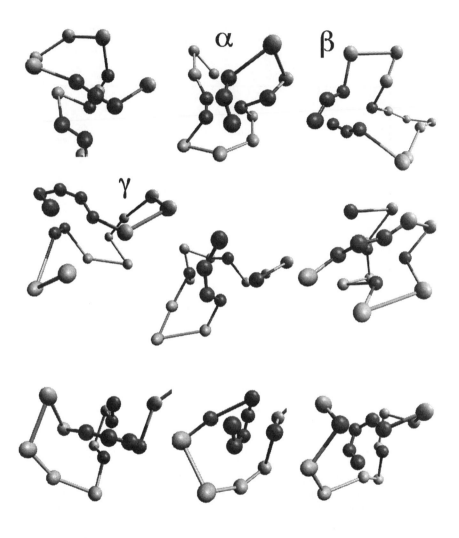

Figure 2: Snapshots of the remaining 9 allowed core structures. Again, each structure represents an invariance class of sequences folding to the same core monomer configuration, however, these sequences are much more frustrated, and less symmetrical than those in Fig. 1.

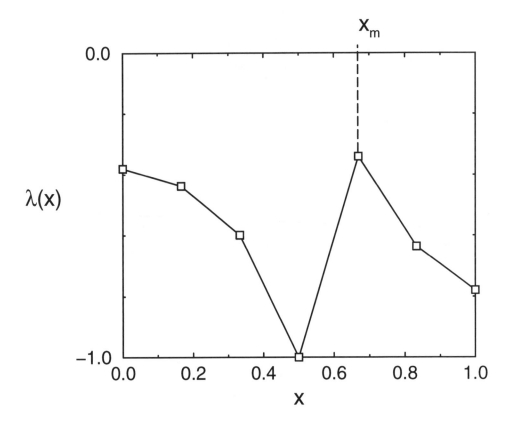

Figure 3: Illustration of a cut through the foldability landscape showing two dominant valleys (attractors) in sequence space correponding to structure types ν_0 (left valley) and ν_1 (right valley) in Fig. 1. Specifically, the figure shows *minus the maximum folding ability,* $-\lambda(x)$ for sequences with domain parameter x. In contrast to the remaining figures, the domain parameter measures the *period* of domain alternation between H and P residues along the sequences.

landscape is determined by a set of dominant folding motifs [20], one can compare the residue pair contacts made for this whole collection of motifs to the neuron connections (synapses) determined by an optimally stored set of memories in the Hopfield model (in other words, the dominant folding motifs play the role of memories stored in the synapses of Hopfield's model [20, 21]). Crudely speaking, when a sequence evolves it is like fixing the memories [1] and allowing the neurons to evolve toward the base of an attractor [2]. However, since there is no artificially imposed learning rule in proteins, we expect a similar rule to emerge from factors that define the kinetic accessibility of collapsed chain topologies.

This was strongly indicated by the results of our small ($N = 16$) 2−letter heteropolymer model [5, 24]. Briefly, for hydrophobic and polar interactions, two preferred core geometries emerge (ν_0, ν_1 in Fig. 1) having symmetric arrangements of hydrophobic (core) monomers (dark beads). The seqences connected with core types ν_0, ν_1 are the fastest folding, and thermally most stable. The highest symmetry structure ν_0 is the most robust to sequence mutations [5, 24]. These two structures define two minimally frustrated sequence families, each of which is embedded in a sea of frustrated sequences folding to the core structure types shown in Fig. 2. As a result, two deep valleys (attractors) emerge in the sequence landscape (Fig. 3) separated by a barrier connected with frustrated (non folding) sequences having ground state core structures like α, β, γ in Fig. 2. As noted above, the sequence space valleys are the pattern attractors of the heteropolymer model. In the following we show that a collapsed homopolymer with the same internal architecture can collapse easiest into two types of core structure similar to the dominant heteropolymer modes ν_0, ν_1. The emergence of attractors in the heteropolymer model is therefore a consequence of an intrinsic potential defined by the homopolymer statistics.

The folding and memory problems now appear very similar, except for this one crucial difference which characterizes the storage features of each system − in the neural network, the energetic dissimilarization of memories must be enforced by an artificial learning rule [18], while in proteins the learing rule is inherent in the mechanism for folding [38]. Consequently, the dominant folds of proteinlike heteropolymers may, in a coarse grained sense, reflect some universal (shared) features characteristic of polymers [44, 45].

3 Hierarchy of Geometry and Topology

To investigate these issues more concretely, we can construct various explicit heteropolymer models for computer simulation [5, 10, 7, 46, 47, 48, 6] such as the one above. In simple models the "residues" posess spherical symmetry, and the present model grew out of considering the rules (magic numbers) for assembly [49, 50, 51] of such monomers into clusters

---

[1] As we have mentioned, both sequence and structure evolve during protein evolution. In this example we have in mind a steady state (neutral evolution [43]) situation for structure memories.

[2] Although we shall not approach the problem exactly this way, one can consider an ensemble of independent, identical heteropolymers (replicas) in which both structure $\hat{C}$ and sequence $\mathbf{p}$ temporally evolve according to the same energy function $H(\hat{C}, \mathbf{p})$ but at separate temperatures T and T' (respectively) [25, 21, 27]. At low temperatures the replicas can all condense, resulting in the allowed folded motifs and sequence families of the polymer model (this signifies the memory retrieval phase in neural networks [17, 18]). After replica condensation, both sequence and structure continue to fluctuate, but in such a way that the emergent motifs are preserved. However, in the usual spin glass approach [25, 27, 21] the replicas are averaged together in the $n \longrightarrow 0$ limit of the replica trick [19] producing the characteristic free energy of a proteinlike polymer.

— the chain bonded interactions acting as a perturbation.

The model considered here is a continuum realization of the hydrophobic–polar (HP) model [7]. The hydrophobic residues are strongly attractive while all other pairs of residues (HP and PP) interact as hard beads. The residues are joined together by immaterial string, and each string bond has a fixed maximum and minimum extension length. As a result, the core interactions are just as they would be in a homopolymeric chain. Figures 1 and 2 show all the allowed core geometries of this model for sequences with length $N = 16$ and 7 hydrophobic beads (proteinlike hydrophibicity). As one can clearly see, the chain perturbations cause various natural subunits of chain structure to emerge (i.e. larger units than a single monomer) within the core structures. Although, the smaller subunits are usually incorporated into a pentagonal bipyrimid structure (similar to that for disconnected monomers, ν_0), occasionally the native core is an assembly of "oblong" subunits, (such as in ν_1 Fig. 1), which illustrates the perturbative effect of the chain bonds on the core structure for identical disconnected monomers. As noted above, the core geometries ν_0 and ν_1 are the most optimal — sequences yielding alternative core geometries (Fig. 2) are strongly frustrated, folding approximately $10 - 100$ times slower than those in Fig. 1. The two optimal sequence families, folding to ν_0 and ν_1, together comprise the main fraction of the sequence distribution.

In the above model, the most symmetric core structure permits the highest number of threaded topologies [52, 53] [3]. For example, a chain with single monomer subunits folding to structure class ν_0 sterically allows about 100 separate chain threadings (determined by the core monomer contact matix) with the same core energy. Kinetically only about 20 percent of these threadings are occupied at the folding temperature [5], however, even in less symmetric structures HP interactions are not sufficient to restrict the chain to a unique threading. For example, the core geometry ν_1, even with its oblong subunits, still has two distinguishable threadings.

In simple bead chain models, topological degeneracy is related to *local gauge invariance* [32, 56] of the energy function — specifically, to the fact that the Hamiltonian is invariant under specific simultanous local changes in sequence and structure (which leave the pair separations between monomers unchanged [5]). To illustrate this, suppose we index the 15 "local invariance classes" (typified by the representative structures in Fig. 1 and 2) as $n = 1, \ldots, 15$. To each class corresponds a set of sequences $(p^{(n)}, q^{(n)}, \ldots)$, and each sequence folds to one of the structures $(x^{(n)}, y^{(n)} \ldots)$ all typified by one of the representative structures in Fig. 1–2 (in other words, each of the structures $(x^{(n)}, y^{(n)} \ldots)$ has the *same* configuration of core monomers [5, 24]). Therefore, any mutation from a sequence $p^{(n)}$ to another sequence $q^{(n)}$ in the same class must fold to the same geometry, in other words, the geometry is "conserved". In our model, the largest invariance class ν_0 also has the greatest topological degeneracy, while the smallest invariance class ν_1 is nearly unique. Consequently, the sizes of valleys on the homopolymer landscape indicate the degree of mutational invariance (robustness to mutations [5]).

From these results, the following picture emerges. The configurational energy landscape of a uniformly attractive chain already contains some number of deep valleys which

---

[3]Throughout our discussion, we understand a chain topology to mean a distinguishable bending, or threading of the chain (with fixed sequence) through a connected graph of core vertices, such that the "colors" of the vertices match the colors of the residues.

correspond to the dominant geometries, probably of some repeated natural unit of chain structure [57, 58, 59], for folding a heteropolymer. Within each valley are sub−valleys [60] assigned to the set of topologies that correspond to a threading of the chain of subunits [59] through the subunit geometry. If we increase the heterogeneity of the residue dictionary in the right way (without changing the homogeneous component of the interactions), it should become possible to "select" a dominant geometry, and further, to reduce topological degeneracy to the point of favoring just one threading of the chain through the selected geometry.

4   Projection method

In this section we describe an approach to design sequences that select one of the dominant sub−ensembles on the configurational energy landscape of a collapsed homopolymer. To accomplish this, we approximate the free energy of a proteinlike heteropolymer as a collapsed homopolymer with heterogeneous sequence perturbations. The idea being that in proteins, folding involves the cooperative attraction of many residues so the heterogeneous perturbations do not need to be very strong to select one of the maximally accessible structures.

To separate off the heterogeneous interaction component of an arbitrary heteropolymer, we decompose its energy in terms of a complete basis of copolymer sequences. This results in a decomposition of the energy in terms of projection matrices (sequence product matrices) that connect the basis sequences to the families of chain structure they stabilize. The perturbation expansion, and *all quantitative estimates in this section are computed using only homopolymer statistics.*

We first adopt a coarse grained approximation to the Hamiltonian by assuming short ranged interactions [4]. The topologies of any such chain model are partitioned according to contact matrices

$$C_{ij} = \theta(b - x_{ij}) \qquad (1)$$

where x_{ij} is the distance between monomers i and j non−adjacent in sequence, b is the range of the potential, and $\theta(x)$ is the step function

$$\theta(x) = 1 \qquad x \geq 1 \ , \qquad (2)$$

$\theta(x) = 0$ otherwise. To this level of accuracy, the total energy $H(C)$ of a structure C_{ij} is

$$H(C) = \frac{1}{2} \sum_{ij} E_{ij} C_{ij} \qquad (3)$$

where E_{ij} defines the energy of a contact between two residues (monomers). The contact matrix is of course time dependent, $C_{ij}(t) \equiv C_{ij}$ (where t is the time) and incorporates all the chain constraints (excluded volume, non−crossing etc.). It is helpful to recall that these constraints put hard limits on the number and type of residue contacts (for example, there are no self or nearest neighbor contacts, $C_{jj} = C_{jj\pm 1} = 0$.).

---

[4]For comparison with proteins see, for example, references [61, 62]

The most general form of the contact energy is a symmetric matrix of real numbers. Such a matrix can always be expressed as the superposition of a complete set of matrices composed of products between orthgonal (basis) sequences $\mathbf{p}^\nu$, i.e.

$$E_{ij}(\{E^{\mu\nu}\}) = -\sum_{\mu\nu} E^{\mu\nu} p_i^\mu p_j^\nu . \tag{4}$$

where $E^{\mu\nu}$ is the amplitude (in units of energy) of the matrix component $p_i^\mu p_j^\nu$, and the (as yet unspecified) basis sequences $\mathbf{p}^\nu$, $\nu = 0, \ldots, N-1$ are members of a group of N orthogonal vectors, i.e. such that $\sum_{j=1}^N p_j^\mu p_j^\nu = N\Delta(\mu - \nu)$ where $\Delta(\mu - \nu)$ is the Kronecker delta function. The free energy $F(E_{ij}) = F(E^{\mu\nu})$ determined by (4) is now interpreted in terms of the amplitudes $E^{\mu\nu}$ which result from expanding E_{ij} using a particular set of basis vectors.

In equation (4) we are completely free to choose any basis we want. However, to be effective in the free energy perturbation approximation, we want to choose a basis that is as unbiased as possible with respect to missing information [63], which in this case corresponds to the most accessible collapsed structures. Assuming these structures to be unknown at the begginning, there is no a priori reason to have a particular polarity of the basis residues p_j^μ (for example, more positive than negative). For simplicity we choose a discrete (copolymer, or Ising) model $p_j^\mu = \pm 1$. Moreover, since there is no a priori reason to choose a particular composition or patterning of residues in the basis sequences, we choose sequences with an equal composition of $p_j^\mu = 1$ and $p_j^\mu = -1$ residues and an unbiased patterning of the domains overall. A basis whose heterogeneous sequences satisfy all these constraints is defined by the mnemonic device in Fig. 4 [64].

For proteinlike (minimally frustrated) heteropolymers the diagonal elements $\mu = \nu$ of equation (4) play an important role, and in the following derivation we will take the approximation of considering only these terms

$$E_{ij}(\{E^\nu\}) = -\sum_{\nu=1}^N E^\nu p_i^\nu p_j^\nu \tag{5}$$

for which the pair interactions are cooperative [32, 28] like those of a copolymer sequence. The sequences are alternating (rougly periodic [65, 66]) arrangements of $+1$ and -1 domains. Consequently, the alignment matrices $p_i^\nu p_j^\nu$ resemble chessboard patterns of $+1$ and -1 blocks seen at N different magnifications. The strongest magnification corresponding to the all ones (homopolymer) matrix ($\nu \equiv 0$), the weakest, a chessboard of unit sized $+1$ and -1 blocks ($\nu = N - 1$).

At the most basic level, this reduced model (with $E^\nu > 0$) describes a mixture of the energetics for N sequences with proteinlike (cross−chain ferromagnetic) symmetry interactions (like residues attract, opposites repel). In the opposite case, $E^\nu < 0$, the cross chain interactions would be caused to have an anti−ferromagnetic (opposites attract) symmetry.

To approximate the free energy of any proteinlike heteropolymer we therefore start from the folowing assumptions − (i) individual pair interactions between residues are relatively weak and (ii) optimal proteins are minimally frustrated. Accordingly, we can expand $F(\{E^\nu\})$ to second order in E^ν about the collapsed homopolymer free energy using the

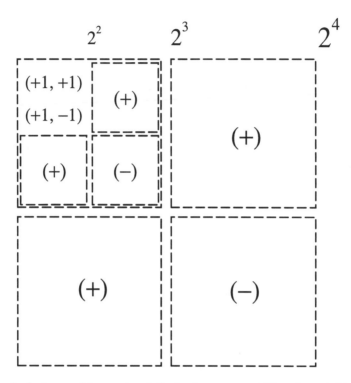

Figure 4: Mnemonic device used to construct the basis sequences. The elemental basis for $N = 2$ is written in the upper left corner. To generate the basis for $N = 4$, this basis is first *replicated* and placed exactly as shown into the small dashed boxes directly below and to the right of it — the negative of this basis is replicated and placed in the small dashed box below and along the diagonal. The sign of the replication is indicated by the symbols $(+)$ and $(-)$. Next, these length 2 vectors are *polymerized* along a vertical line extending below the number 2^2 which indicates the vector length after the polymerization process. This process results in 4 mutually perpendicular vectors $(+1, +1, +1, +1)$, $(+1, +1, -1, -1)$, $(+1, -1, +1, -1)$, and $(+1, -1, -1, +1)$ of length $N = 4$. The procedure can be iterated once more to obtain the basis for $N = 8$, using the 4, length 4 vectors now contained in the large dashed box in the upper left corner of the figure. The next ($N = 8$) polymerization line extends vertically below the number 2^3. Bases of length 2^n are generated by $n - 1$ iterations.

model (5). We obtain the following coefficients in a Taylor series

$$Q^\nu = -\frac{\partial F}{\partial E^\nu} \tag{6}$$

and

$$Q^{\mu\nu} = -\frac{\partial^2 F}{\partial E^\mu \partial E^\nu} \tag{7}$$

where $F \equiv F(\{E^\nu\}, T)$ is the configurational free energy of any proteinlike *heteropolymer* with energy described according to (5), i.e.

$$F(E, T) = -T \log \left[\sum_C \exp -H(E, C)/T \right] \tag{8}$$

and $\{E^\nu\} \equiv E$. The parameters (6) and (7) are evaluated at "zero field" (i.e. at $E^{\nu \neq 0} = 0$), specifically, for a homopolymer chain near the collapse temperature. In this case, one obtains

$$Q^\nu = \langle A^{\nu\nu} \rangle_0 \tag{9}$$

and

$$Q^{\mu\nu} = \frac{1}{T} [\langle A^{\mu\mu} A^{\nu\nu} \rangle_0 - \langle A^{\mu\mu} \rangle_0 \langle A^{\nu\nu} \rangle_0] \tag{10}$$

where

$$A^{\mu\nu}(t) = \frac{1}{2} \sum_{ij} C_{ij}(t) p_i^\mu p_j^\nu \tag{11}$$

T is the temperature, t is the time, and the subscript 0 signifies the homopolymer thermal average. In a Monte Carlo simulation, $C_{ij}(t)$ is a time dependent quantity, and the true thermal average is an infinite time average at thermal equilibrium [67].

Besides the parameters Q^ν and $Q^{\mu\nu}$ it is necessary to measure off–diagonal projections in equation (11) to check the validity of the approximation (5) to the Hamiltonian. All quantities of interest in this paper can be obtained by recording histograms of the two quantities $A^{\mu\nu}$ and $A^{\mu\mu} A^{\nu\nu}$ during a Monte Carlo simulation.

To describe these two quantities it is helpful to notice that equations (9) and (10) are analogous to the zero field polarization and susceptibility tensor of a magnet [67, 68] [5]. At the most basic level, the parameter $A^{\mu\nu}(t)$ measures the projection of an instantaneous chain configuration $C_{ij}(t)$ against the pattern matrix $p_i^\mu p_j^\nu$ – its average $< A^{\mu\nu} >_0$ measures the "zero field" polarization of the chain along the "direction" $p_i^\mu p_j^\nu$. The average projections $< A^{\mu\mu} A^{\nu\nu} >_0$ measure the simultaneous projection of the chain against two patterns.

The two expansion parameters Q^ν and $Q^{\mu\nu}$ are coefficients in the approximation for the heteropolymer free energy under the assumption of minimally frustrated energetics (see Appendix). In the following section we show how these parameters indicate the connection between basis sequences and compact structures in the homopolymer ensemble.

The parameters Q^ν and $Q^{\mu\nu}$ are somewhat reminiscent of the usual Edwards–Anderson and Parisi order parameters in spin glass theory [41, 19]. We can define a simultaneous projection function $q^{\mu\nu} = |\langle A^{\mu\mu} A^{\nu\nu} \rangle|$ in terms of the $A^{\nu\nu}$ which substitutes more closely for the Parisi overlaps. Like the Parisi overlaps, we find (following the approach of Bhatt and Young [70]) that $q^{\mu\nu}$ also exibits a strong *ultrametric* signature just as obtained for finite sized spin glasses [70, 19, 71].

---

[5]This is reminiscent of the random phase approximation in quantum statistical mechanics [69].

5 Results

To gather histograms for $A^{\mu\nu}$ and $A^{\mu\mu}A^{\nu\nu}$, we simulate a homopolymer in the neighborhood of its theta and collapse temperatures according to standard Monte Carlo dynamics [72]. All the results are presented in Fig.s 5–8. To interpret these figures it is important to remember that the basis sequences are indexed according to increasing frequency of sequence domain alternation.

In Fig. 5 we plot the diagonal elements of the projection function $< A^{\nu\nu} >_0$ for a short chain $(N = 16)$ at the theta temperature $T_\vartheta = .6$ and a temperature $T_{\vartheta-\delta} = .4$ just above the collapse temperature $T_c = .35$, where $T_c(N)$ is defined according to the specific heat peaks obtained in [73]. The figure exibits two regimes (high and low frequency sequence domain alternation) for which the chain can fold itself in order to produce a strong projection. The tails of the histograms for high and low frequency also extend to the largest positive projections, which signifies the two types of mechanisms observed in the HP model. To illustrate this, we have retrieved the coordinates of the structures that produce maximal projections, **max** $A^{\nu=1\,\nu=1}$ and **max** $A^{\nu=N-1\,\nu=N-1}$, for high $(\nu = N - 1)$ and low $(\nu = 1)$ frequency sequence domain alternation (the simulation was conducted at $T = .5$). As illustrated in Fig. 6, these two structures utilize folding mechanisms very similar to those in Fig. 1. Finally, as shown in Fig. 7, the off–diagonal projections $< A^{\mu\nu} >_0$ associated with partially frustrated interactions are statistically unimportant.

For long chains, each basis sequence is increasingly compatable with a greater variety of monomer geometries. Consequently, the connection between a basis sequence and a particular geometry of sub–structures is becoming less defined. In this case it is necessary to examine the corellations and anti–corellations between pairs, triples, etc. of projections against alignment matrices to focus the energetics on a particular geometry, and it may even be necessary to include off–diagonal projections in the free energy perturbation expansion given in the Appendix.

To understand whether a free energy perturbation expansion can be reduced to something less complex, such as (9)–(12), it is important to calculate all elements of the projection function $< A^{\mu\nu} >$ for longer chains. [6]. The results for $N = 32, 64$ are shown in Fig. 8. Again we observe a predominance of the diagonal projections over the off–diagonal projections, and the same characteristic shape of the curve appears for $Q^\nu = < A^{\nu\nu} >$ [7]. Although, a greater number of statistically significant off–diagonal terms emerge (i.e. with $|< A^{\mu\nu} >| > \sigma$), the overall number of such terms still scales with N.

6 Discussion

To complete this approach we would like to provide a general scheme for assigning mixtures of the amplitudes E^ν to the most accessible topologies of homopolymer. The sketch of a possible approach for accomplishing this is presented in a short Appendix directly below. In this Appendix we interpret the configurational free energy $F(\{E^\nu\})$ defined by the perturbation expansion as a potential (energy) function in the sequence free energy. The

---

[6]In the figures we actually compute a symmetrized version of the projections, $a^{\mu\nu} = \frac{1}{2} < A^{\mu\nu} + A^{\nu\mu} >$, since it is easier to interpret symmetric matrices.

[7]It is important to point out that similar results are obtained for the standard Gaussian chain model

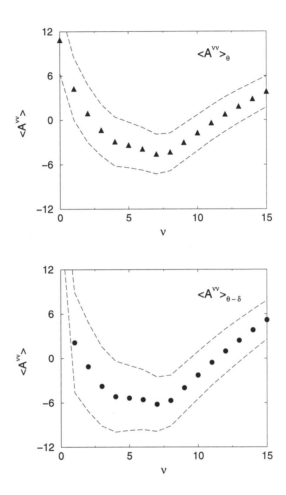

Figure 5: Diagonal elements of the projection function $Q^\nu = < A^{\nu\nu} >$ for an $N = 16$ homopolymeric chain computed at the theta temperature $T_\vartheta \simeq .6$ (triangles) and a lower temperature $T_{\vartheta - \delta} = .4$ near $T_c \simeq .35$ (circles). The projection function is plotted against the number of domain boundaries $\nu - 1$. The dashed lines indicate the standard deviation of histograms for $A^{\nu\nu}$. Inclusion of the homopolymeric term ($\nu - 1 = 0$) is an artifact of our perterbation method. Both plots exhibit peaks for high and low frequency domain alternation. The energetically favorable contacts for such sequences, defined by $K_{ij}^\nu = \frac{1}{2}p_i^\nu p_j^\nu + \frac{1}{2}$ are characteristic of the structures in Fig. 1, as discussed in the next figure.

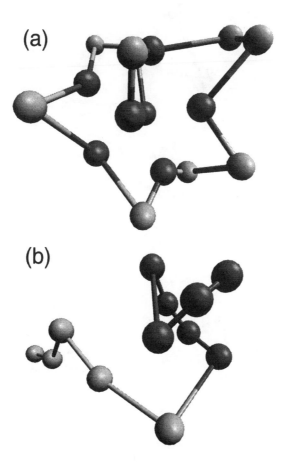

Figure 6: Enlargement of the two folded structures from Fig. 1 which have the highest and lowest frequency of sequence domain boundaries between H and P monomers

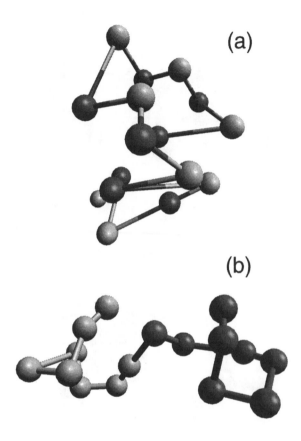

Figure 7: Snapshot of two structures, recorded during a homopolymer simulation at $T = .5$, which produce the maximum projections $A^{\nu\nu}$ for high ($\nu - 1 = 15$, top) and low ($\nu - 1 = 1$, bottom) frequency domain alternation. The beads have been colored to reflect the sign of residues in the corresponding basis pattern $p_i^{\nu} p_j^{\nu}$ (dark beads; $p_j^{\nu} = +1$, light beads; $p_j^{\nu} = -1$). The top structure has a core shape similar the pentagonal bipyrimid, while the lower structure contains two domains similar to the single H core in Fig. 6 (recall that only H monomers are attractive in Fig. 6).

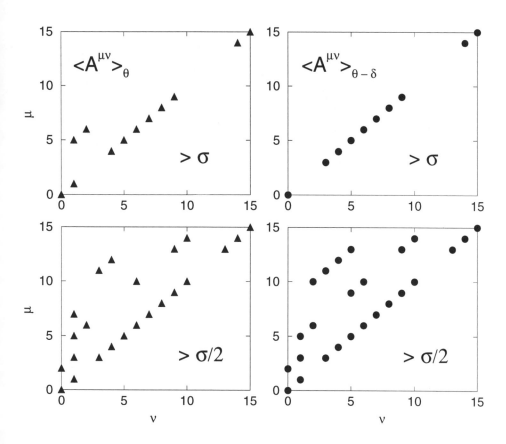

Figure 8: Threshold plots of the projection function $< A^{\mu\nu} >$ at T_ϑ and $T_{\vartheta-\delta}$. The data points indicate when the mean value $< A^{\mu\nu} >$ exceeds the mean standard deviation $\sigma = \overline{< A^{\mu\nu} A^{\mu\nu} > - < A^{\mu\nu} >< A^{\mu\nu} >}$ of the off–diagonal terms. The lower figures are threshold plots for exceeding half the fluctuation, $\sigma/2$. The plots illustrate that the dominant signal comes primarily from the diagonal terms $< A^{\mu\nu} >$ corresponding to no energetic frustration. Similar results are obtained for $N = 32, 64$ as shown in the next figure, and for an unweighted ensemble of dissimilar protein domains.

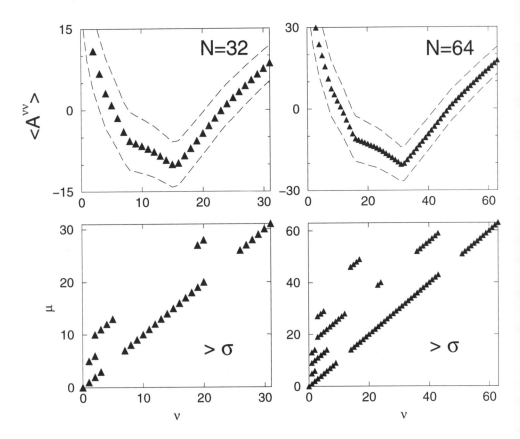

Figure 9: Graphs analogous to Fig.s 3 and 5 for $N = 32, 64$ homopolymers at T_ϑ with interaction parameters defined as in Milchev, Paul, and Binder. Again the diagonal terms contribute the dominant projections, and the curves $< A^{\nu\nu} >$ have a shape similar to Fig. 3. However, here, more of the off−diagonal terms exceed the fluctuation threshold σ. For these longer chains, mixtures of the patterns become necessary to specify a particular folded geometry, so the peak at high (or low) frequency domain alternation can not be associated with a single folding mechanism or folded geometry.

above results, and particularly the ultrametric signature of the overlaps $q^{\mu\nu}$ indicate that this sequence free energy may contain minima corresponding to mixtures of the E^ν that select the accessible motifs. Preliminary results of this approach are also mentioned in the Appendix.

In closing, we should remark that local gauge invariance is connected with the conserved properties of particle–like objects in other systems (for example, defects in ordered media [32, 42], and the properties of elementary particles [75]). In the future it may be worthwhile to look for specific connections between these extremely different regimes of matter.

7 Appendix

Throught this paper, we view the the sequence–structure assignment problem in terms of an approximation to the free energy of a homopolymeric chain F_0. In equations (9) and (10) we define expansion coefficients in a second order perturbation series of the free energy. Specifically, $F(E) - F_0 =$

$$-\sum_\nu Q^\nu E^\nu - \frac{1}{2} \sum_{\mu\nu} Q^{\mu\nu} E^\mu E^\nu \tag{12}$$

where again $E \equiv \{E^\nu\}$ is shorthand for a particular configuration of the amplitudes in equation (4). In the following, we write this perturbation as $\Delta F(E) = F(E) - F_0$.

The free energy $\Delta F(E)$ can be interpreted as a sequence potential that determines which different sets $E = \{E^\nu\}$ select which of the many dominant collapsed configurations of the homopolymer. Thus, we interpret $\Delta F(E)$ as a potential for the E degrees of freedom in the sequence free energy, similar to the neural network model of Dotsenko, Franz, and Mezard [17]. This results a partition function for the amplitudes,

$$Z' = \sum_E P(E) \exp -\Delta F(E)/T' \tag{13}$$

resulting in a free energy

$$F' = -T' \log Z' \tag{14}$$

and T' is the sequence pseudo–temperature [63, 57]. In equation (14) the whole problem has been renormalized, because the structure free energy $\Delta F(E)$ in (12) is just the energy function for a neural network. Consequently, the intrinsic learning rule of a sequence aver-aged (homopolymeric) chain is now absorbed into the free energy of a neural network with continuous neurons.

The distribution function $P(E)$ in (13) takes into account any compositional constraints on the sequence amplitude E^ν. For example, if the interactions E_{ij} are constrained to have $\overline{|E_{ij}|} = 1$ as, for example, in a Gō model, a short calculation shows that $\overline{|E^\nu|} = 1/\sqrt{N}$. In this case, we could constrain the amplitudes according to a spherical model $\sum_\nu (E^\nu)^2 = const.$ [76]), however, in practice we find it more appropriate to use

$$\log P(E) = -(\sum_\nu E^\nu - 1)^2/\Delta \tag{15}$$

where Δ determines the flexibility (half width) of the constraint. It is known that under certain conditions (uncorrelated, weakly ferromagnetic $Q^{\mu\nu}$) the spherical model leads to a phase with multiple free energy minima [8] [76]. Thus, we have reason to expect multiple minima in the sequence free energy F' corresponding to different mixtures of amplitudes which fold the chain into its dominant configurations. It is important to see whether such a drastic approximation as this can lead to the sequence free energy minima noted above. We intend to explore different approximations [77] to the free energy functions F and F' in a future article. Results of the present model are in agreement with the homopolymer collapse model of de Gennes [38].

References

[1] H. Xiong, B. L. Buckwalter, H. Sheih, and M. H. Hecht, *Proc. Natl. Acad. Sci.* **92**, 6349 (1995).

[2] B. I. Dahiyat and S. L. Mayo, *Science* **278**, 82 (1997).

[3] D. D. Axe, N. W. Foster, and A. R. Fersht, *Proc. Natl. Acad. Sci.* **93**, 5590 (1997).

[4] C. R. Robinson and R. T. Sauer, *Proc. Natl. Acad. Sci.* **95**, 5929 (1998).

[5] E. D. Nelson, L. F. Ten Eyck, and J. N. Onuchic, *Phys. Rev. Lett.* **79**, 3534 (1997).

[6] N. D. Socci and J. N. Onuchic, *J. Chem. Phys.* **103**, 4732 (1995).

[7] K. F. Lau, and K. A. Dill, *Macromolecules* **22**, 3986 (1989).

[8] P. E. Leopold, M. Montal, and J. N. Onuchic, *Proc. Natl. Acad. Sci. USA* **89**, 8721 (1992).

[9] N. D. Socci, J. N. Onuchic, and P. G. Wolynes, *Proteins* **32**, 136 (1998).

[10] H. Nemyer, A. Garcia, and J. Onuchic, Onuchic, *Proc. Natl. Acad. Sci.* **95**, 5921 (1998).

[11] J. N. Onuchic, Z. Luthey – Schulten, and P. G. Wolynes *Annu. Rev. Phys. Chem.* **48**, 539 (1997).

[12] K. A. Dill and H. S. Chan, *Nature Struct. Biol.* **4**, 10 (1997).

[13] J. N. Onuchic, N. D. Socci, Z. Luthey – Schulten and P. G. Wolynes *Folding and Design* **1**, 441 (1996).

[14] S. S. Plotkin, J. Wang, and P. G. Wolynes, *J. Chem. Phys.* **106**, 2932 (1997).

[15] J. J. Hopfield, *Proc. Natl. Acad. Sci. USA* **79**, 2554 (1982).

[16] J. J. Hopfield, D. I. Feinstein, and R. G. Palmer, *Nature* **304**, 158 (1983).

[17] V. S. Dotsenko, S. Franz, and M. Mezard, *J. Phys.* **A27**, 2351 (1994).

---

[8]The spin glass phase appears in the $N \longrightarrow \infty$ limit of the spherical model.

[18] Dotsenko, V. S., *An introduction to the theory of spin glasses and neural networks*, World Scientific, Singapore (1994).

[19] Mezard, M., Parisi, G., and Virasoro, M., *Spin glass theory and beyond*, World Scientific, Singapore (1987).

[20] M. S. Friedrichs, and P. G. Wolynes, *Science* **246**, 371 (1989).

[21] M. Sasai, and P. G. Wolynes, *Phys. Rev. Lett.* **65**, 2740 (1990).

[22] P. G. Wolynes, in *Spin glass ideas and the protein folding problem*, D. L. Stein ed., World Scientific, Singapore, 225−259 (1992).

[23] N. Qian and T. J. Sejnowski, *J. Mol. Bio.* **202**, 865 (1988).

[24] E. D. Nelson, and J. N. Onuchic, *Proc. Natl. Acad. Sci. USA* **95**, 10682 (1998).

[25] E. I. Shakhnovich, and A. M. Gutin, *Biophys. Chem.* **34**, 187 (1989).

[26] V. S. Pande, A. Y. Grosberg and T. Tanaka, *Folding and Design* **7**, 109 (1997).

[27] Pande, V. S., Grosberg, A. Yu., and Tanaka, T., *Phys. Rev. E* **51**, 3381 (1995).

[28] T. Garel and H. Orland, *Europhys. Lett.* **6**, 597 (1988).

[29] A. V. Finkelstein, A. Y. Bedretdinov, and A. M. Gutin, *Proteins 23* 142 (1995).

[30] Yaotian Fu, and P. W. Anderson, *J. Phys. A* **19**, 1605 (1986).

[31] Gillespie, J. H., *The causes of molecular evolution*, Oxford University Press, New York (1991).

[32] G. Toulouse, *Commun. Phys.* **2**, 115 (1977).

[33] P. W. Anderson, *J. Less Common Met.* **62**, 291 (1978)

[34] J. D. Bryngelson, and P. G. Wolynes, *Proc. Natl. Acad. Sci. USA* **84**, 7524 (1987).

[35] J. D. Bryngelson, J. N. Onuchic, N. D. Socci, and P. G. Wolynes, *Proteins* **21**, 167 (1995).

[36] N. Gō, *Annu. Rev. Biophys. Bioeng.* **12**, 183 (1983).

[37] C. Anfinsen *Science* **181**, 223 (1973).

[38] de Gennes, P.−G., *J. Physique Lett.* **46**, L−639 (1985).

[39] V. I. Abkevich, A. M. Gutin, E. I. Shakhnovich, *Proteins* **4**, 335 (1998).

[40] D. Thirumalai, D. K. Klimov, and S. Woodson, *Theor. Chem. Acct.* **1**, 23 (1997).

[41] K. Binder, *Z. Physik* **B26**, 339 (1977).

[42] K. Binder and A. P. Young, *Rev. Mod. Phys.* **58**, 801 (1986).

[43] M. Kimura, *Neutral Theory of Evolution*, Cambridge University Press, New York, (1983).

[44] de Gennes, P.–G., *Rev. Nuovo Cimemto* **7**, 363 (1977).

[45] Y. Zhou, C. Hall and M. Karplus, *Phys. Rev. Lett.* **77**, 2822 (1996).

[46] E. I. Shakhnovich, *Phys. Rev. Lett.* **72**, 3907 (1994).

[47] D. Thirumalai, and Z. Guo, *Biopolymers* **35**, 137 (1989).

[48] H. Li, R. Helling, and C. Tang, *Science* **273**, 666 (1996).

[49] P. A. Braier, R. S. Berry, and D. J. Wales, *J. Chem. Phys.* **93**, 8745 (1990).

[50] D. J. Wales, M. Miller, and T. Walsh, *Nature* **394**, 758 (1998).

[51] D. J. Wales, *Science* **271**, 925 (1996).

[52] I. M. Lifshitz, *Sov. Phys. JETP* **28**, 1280 (1969).

[53] A. Y. Grosberg, *Phys. Usp.* **40**, 125 (1997).

[54] Grosberg, A. Y., and Kokhlov, A. R. *Statistical physics of macromolecules*, AIP Press, New York (1994).

[55] K. W. Plaxco, K. T. Simons, D. Baker, em J. Mol. Bio. **277**, 985 (1998).

[56] G. Toulouse and M. Kléman, *J. Physique Lett.* **37**, L–149 (1976).

[57] P. G. Wolynes, *Proc. Natl. Acad. Sci* **93**, 14249 (1996).

[58] A. Murzin, and A. Finkelstein, *J. Mol. Bio.* **204**, 749 (1988).

[59] M. Levitt, and C. Chothia, (1976) *Nature* **261**, 552 (1976).

[60] R. Baviera, and M. A. Virasoro, in *Landscape paradigms in physics and biology: Concepts and dynamics*, H. Frauenfelder et. al. eds., North–Holland, Amsterdam (1996).

[61] Y. Zhou and M. Karplus, *Proc. Natl. Acad. Sci.* **94**, 14429 (1997).

[62] E. M. Boczko and C. L. Brooks III, *Science* **269**, 393 (1995).

[63] E. T. Jaynes, *Physical Review* **106**, 620 (1957).

[64] The construction rule is similar to that used in Walsh functions, J. L. Walsh, *Amer. J. Math.* **45**, 5 (1923), Harmuth, H. F., *Information theory applied to space time physics*, World Scientific, Singapore (1992).

[65] A. Irbäck, C. Peterson, and F. P. Pottast, *Proc. Natl. Acad. Sci* **93**, 9533 (1996).

[66] S. Rackovsky, *Proc. Natl. Acad. Sci.* **95**, 8580 (1998).

[67] K. Binder and D. Heerman, *Monte Carlo simulation in statistical physics*, Springer Verlag, New York (1988).

[68] Toda, M., Kubo, R., and Saitô, N., *Statistical physics*, Springer Verlag, New York (1983).

[69] Kadanoff, L. P. and Baym, G., *Quantum statistical mechanics*, Benjamin Inc., New York (1962).

[70] R. N. and A. P. Young, *J. Magn. Matt.* **54–57**, 191 (1986).

[71] For ultrametricity in neural networks see, for example N. Parga and M. A. Virasoro, *J. Physique* **47**, 1857 (1986). For protein sequences see for example, M. Farach, S. Kannan, T. Warnow, *Proceeding of the twenty fifth annual ACM symposium on the theory of computing*, ACM digital library, (1993). which cites many of the important references on protein evolutionary trees.

[72] N. Metropolis, A. W. Rosenbluth, M. N. Rosenbluth, A. H. Teller, and E. Teller, *J. Chem. Phys.* **21**, 1088 (1953).

[73] A. Milchev, W. Paul, and K. Binder, *J. Chem. Phys.* **99**, 4786 (1993).

[74] E. D. Nelson, unpublished calculations.

[75] C. N. Yang and R. L. Mills *Physical Review* **96**, 191 (1954).

[76] J. M. Kosterlitz, D. J. Thouless, and R. C. Jones, *Phys. Rev. Lett.* **36**, 1217 (1976).

[77] D. J. Thouless, P. W. Anderson, and R. G. Palmer, *Phil. Mag.* **35**, 593 (1977).

Optimization in Computational Chemistry and Molecular Biology, pp. 131-140
C. A. Floudas and P. M. Pardalos, Editors

Gene Sequences are Locally Optimized for Global mRNA Folding

William Seffens
Department of Biological Sciences
Center for Theoretical Study of Physical Systems
Clark Atlanta University
Atlanta, GA 30134 USA
wseffens@cau.edu

David Digby
Department of Biological Sciences
Clark Atlanta University
Atlanta, GA 30134 USA
ddigby@mindspring.com

Abstract

An examination of 51 mRNA sequences in GENBANK has revealed that calculated mRNA folding free energies are more negative than expected. Free energy minimization calculations of native mRNA sequences are more negative than randomized mRNA sequences with the same base composition and length. Randomization only of the coding region of genes also yields folding free energies of less negative magnitude than the original native mRNA sequence. Examination of the predicted basepairing within the coding sequence finds an unequal distribution between the three possible frames. The wobble-to-1 frame, which is "in-frame", is preferred significantly compared to randomized sets of mRNA sequences. This suggests that evolution may bias or adjust the local selection of codons to favor the global formation of more mRNA structures. This would result in greater negative folding free energies as seen in the 51 mRNAs examined.

Keywords: mRNA, folding free energy, coding sequence, global optimization, codon-choice optimization.

1 Introduction

Control of gene expression is known to occur at any of the events from promotion of transcription to stabilization of the mature polypeptide product. The role of RNA structure in gene expression is not well understood. Conserved RNA secondary structures in families of viruses are known to be important to the viral life cycle [3]. It is thought that the 3' UTRs of certain developmental genes in Drosophila are similarly important for localization [10]. RNA secondary structure is thought to be important in the coding regions of certain RNA viruses. Motifs such as pseudoknots have been shown to be responsible for frame-shifting. It is not clear if RNA structure has any importance for non-viral protein coding mRNAs. Several studies have demonstrated that mRNA stability may be an important factor in gene expression for certain non-viral genes [1][4][9]. Structural RNA features are suspected to be involved in the regulation of mRNA degradation in those cases. Several authors have suggested that the choice of codons in eukaryotic genes may be constrained by effects other than the frequency of codons in the whole gene [6][8]. Major influences on codon usage have been shown to result from amino acid residue preferences and di-residue associations in proteins. Such biases could be related to replication and repair processes, and/or to DNA structural requirements. For a message coding 100 amino acids, there are approximately 3^{100} or 10^{48} different combinations of bases using synonymous codons coding for the same polypeptide. Evolution then has great freedom to tinker with mRNAs subject to the constraint of the amino acid sequence. This work tested the hypothesis that codon choice is biased to generate mRNAs with greater negative folding free energies, and results in base-pair frame patterns within the coding sequence.

2　Methods

mRNA sequences were selected from GENBANK using programs in the Wisconsin Group GCG software package [2]. mRNA sequence files were randomly selected from GENBANK with short Locus descriptors (limited to 8 or 9 characters) and which possessed sufficient information in the Features annotation to reconstruct the sequence of the mRNA. Fifty-one mRNA sequences were selected possessing the following properties identified in the Features annotation: 1) mRNA +1 start site identified, 2) MET start codon identified, 3) termination codon identified, 4) poly-A site or signal identified, and 5) the mRNA sequence must be less than 1200 bases long (due to computational constraints). A variety of sequence files were examined from diverse species including prokaryotes, plants, invertebrates and higher animals (Table 1 in [12]). These mRNA sequences were *in silico* folded using Zuker's MFOLD program from GCG using a VAXStation 4000 or SUN Ultra computer [14].

Each *in silico* folded native mRNA was compared with folded mRNA sequences randomized by one of five different procedures. In the first procedure, each native mRNA sequence was randomized at least ten times using the SHUFFLE program in GCG. SHUFFLE randomizes the order of bases in a sequence keeping the composition constant. These randomized sequences (termed "whole-random" sequences) were folded and the free energies averaged. In the second randomization procedure, the native sequences were randomized only within the coding region, yielding "CDS-random" sequences. These sequences contained unmodified 5' and 3' UTRs. In the third randomization procedure, codons were shuffled within the coding sequence only, yielding "codon-shuffled" sequences. These contain unmodified UTR sequences of the respective native mRNA, and code for a polypeptide with identical amino acid composition yet different amino acid sequence. A program (RNAshuffle) was written in FORTRAN using the GCG software library that randomized only the codon choice to produce "codon-random" sequences for the fourth randomization procedure. Codon-random sequences have the same nucleotide base composition and translated polypeptide product as the respective native mRNA. The last randomization procedure was a modification of the previous codon-random algorithm without constraining the base composition. All codons for each amino acid were allowed to be equally likely. The resulting sequences also coded for the same polypeptide as the native mRNA, but the base composition was generally more G+C rich. These sequences were labeled as "codon-flat".

A C program was written to analyze the GCG PlotFold output files of the most stable secondary structure found in MFOLD. The program FrameCount counted the number of predicted CDS base pairings for each of the three possible frames.

Statistical significance was tested for the biases observed in calculated folding free energy between native and randomized sequences, and for the frame counts. The statistical significance of the differences in free energy was measured in standard deviation units, termed the segment score from [7]. Large sets of randomized mRNA folding free energies were found to be normally distributed. Standard hypothesis tests were employed using statistical analysis software in Excel 97 (Microsoft). All thermodynamic energies are free energies expressed as kcal/mol. A greater negative free energy indicates that a more stable folding configuration is possible.

3 Results

Fifty-one mRNA sequences were selected from a variety of plant, animal and bacteria sequences in GENBANK [12]. These sequences were *in silico* folded using Zuker's MFOLD program. The single most stable configuration was examined for the basepairing occurring within the coding sequence of the native mRNAs. Counts of the predicted base-pairing occurring within the CDS were totaled according to the three possible frames. A wobble-to-one (W-1) frame is "in-frame", such that a codon would be exactly matched up with one reverse-complement codon. The other two possible frames, wobble-to-wobble (W-W) and wobble-to-two (W-2) are frame shifted to the left or the right by one nucleotide compared to the W-1 frame. In these later two frames, a codon in a RNA stem structure would be opposite to two codons, rather than the perfectly aligned one codon of the W-1 frame. One would predict that the distribution of base-pair framing relationships should be equally likely for the three frames. Surprisingly in Table 1 for the native mRNA set, there are 7.4% more bases paired in the W-1 frame compared to the W-W frame, and 21% more compared to the W-2 frame (2737 vs 2549 and 2262). The two frames, W-W and W-1 are about 12% different from each other in terms of native frame counts. This suggests the local arrangement or selection of codons is optimized for in-frame basepairing within the CDS of mRNAs. The significance of this framing bias can be assessed by examining control sets of randomized sequences. The controls here are whole, CDS, and codon randomized sets of mRNAs.

Counts of the base-pairing occurring within ten randomized sequences were averaged. The position where the CDS would have been located is used for counting in the randomized sets shown in Table 1. From the ten different randomized sets examined for each random-

Table 1.

mRNA Set	W-W			W-1			W-2		
	Frame	SD	FS	Frame	SD	FS	Frame	SD	FS
Native	2549			2737			2262		
Whole-Random	2370	65.4	2.7	2409	56.4	5.8	2400	96.6	-1.4
CDS-Random	2442	75.1	7.0	2457	113.6	9.6	2487	153.9	-5.9
Codon-Shuffle	2497	90.9	2.3	2672	111.3	1.9	2210	103.6	2.2
Codon-Random	2653	101.8	-4.1	2579	131.2	5.2	2140	95.3	5.6
Codon-Flat	2515	119.0	1.0	2400	133.1	9.8	2357	127.7	-2.9

Counts of the predicted base pairs for the most stable secondary structure for each of the three frames are shown. The native set is the sum of the base pair counts for all of the 51 mRNAs in Table 1 of Seffens&Digby (1999). The randomized sets are reported as the mean of 10 sets, each the same size as the native set. The standard deviation (SD) is also shown for each randomized set. The Frame Score (FS) is calculated as [(native count)-(randomized count)]/SD , and represents the significance of the difference in frame counts of the native set from a randomized set.

ization scheme, each the same size as the native set, a standard deviation is reported for the framing mean values. The difference between the native counts and the mean of the randomized set is divided by the standard deviation, to report as a "frame score" (FS). This method of assessing the significance of mRNA framing counts is similar to the segment score of Le and Maizel for differences in folding free energies of mRNAs[7]. Using this terminology, the frame score of the W-1 frame is 5.8 based on whole-randomized mRNA sequences. This indicates that the count of native W-1 frame bases is 5.8 standard deviations greater than the count in the whole-randomized set. The native W-W frame is also larger than the whole-randomized set with a frame score of 2.7, while the native W-2 frame is not as significantly different (frame score is -1.4). The significance level of the enrichment of the W-1 frame suggests that frame W-1 possesses unique properties. The significance to a graph theory representation or underlying structure of the genetic code will be considered later below.

In addition to summing the frame-count information from the MFOLDed native sequences, the free energy of folding for each mRNA structure was also collected. Only the free energy of folding for the most stable configuration was examined. These native free energies can be compared to the mean of the free energies of ten randomized sequences from each native sequence. From each group of 10 randomized sequences a standard deviation (SD) is also computed. The significance of the difference of the mean from the native free energy is then reported as the segment-score according to Le and Maizel [7]. For the different randomization schemes, native mRNA sequences are generally more stable than the corresponding randomized sequences as seen from segment scores. From Table 2 the average segment score for the whole-randomized set is -1.23 with a 95% confidence interval of 0.45, indicating there is a significant bias in folding free energy.

To determine if this observed bias toward greater negative folding free energies of mRNAs resides in the coding region or the flanking untranslated regions of the mRNAs, native se-

Table 2.

mRNA Set	Segment Score	CI(95%)	% Difference	CI(95%)
Whole-Random	-1.23	0.45	4.9	2.1
CDS-Random	-0.87	0.48	2.9	1.9
Codon-Shuffle	-0.63	0.47	1.5	1.6
Codon-Random	0.97	NA	2.2	1.8
Codon-flat	-1.81	NA	6.6	3.4

Segment score as calculated by Le and Maizel from the average folding free energies of 51 mRNAs. CI is the confidence interval (P=0.05) of the segment score or percent difference columns. Entries with NA are sets with extremely small standard deviations, and resultant large segment scores. The percent difference is a better measure in those cases.

quences were randomized only within the coding region, yielding "CDS-random" sequences. These sequences contain identical 5' and 3' untranslated regions of the respective native mRNA. Again the native mRNA sequences when folded are usually more negative than the corresponding CDS-random sequences (Table 2). The average segment score is -0.87 with a 95% confidence interval of 0.48. This suggests there is a significant difference in folding free energies between native and partially randomized mRNA sequences. Of the 51 mRNAs examined, 37 or 73% are more negative than the CDS-randomized sequences [12]. The 51 mRNAs possessed a total 5'UTR length of 3014 nucleotides (nt), a total coding length of 27306 nt, and a total 3' UTR length of 8533 nt. Therefore the coding regions comprise 70% of the total mRNA nucleotides, yet randomization of the coding region does not substantially alter the number of mRNAs observed with a bias toward more negative folding free energies. The significance level of this bias is also only slightly reduced by CDS-randomization compared to whole-sequence randomization. The frame scores under CDS-randomization are only slightly reduced compared to the whole randomization scores (Table 1). The trend remain unchanged, a strong enrichment of the W-1 frame, a smaller bias for the W-W frame, and an avoidance of the W-2 frame.

The above randomization procedure was further modified to shuffle the codons while preserving the native amino acid composition and UTR sequences. These "codon-shuffled" mRNAs again are generally less stable than the respective native mRNA sequence (Table 2). Thirty-two of the 51 mRNAs (63%) have negative segment scores, with 13 (or 25%) being greater than -1 [12]. The mean of the segment scores is not as great as the other two randomization procedures, yet the 95% confidence interval (0.47) still does not include zero (Table 2). Under this randomization procedure, the frame scores are considerably reduced (Table 1). This indicates that the codon-shuffle set of mRNAs has retained characteristics that give rise to the frame bias observed in the native set, yet absent from the whole- or CDS-random set. The only codon-shuffle characteristic that satisfies these constraints is codon composition. This implies that the sequence of codons has less influence on frame scores than codon choice.

To determine if the observed bias in frame scores resides in the choice of codons within the coding sequence, a fourth randomization procedure was performed. Native sequences were randomized only by codon choice within the coding region, yielding "codon-random" sequences with unmodified base composition. These sequences contained identical 5' and 3' untranslated regions of the respective native mRNA and coded for the same polypeptide.

Again the native mRNA sequences tend to fold more negative in free energy than the corresponding codon-random sequences (Table 2). Since only a relatively small number of bases will change under this randomization procedure, the resulting sequences in the randomized set will be similar. As a consequence the folding free energies are very close in each set of randomized sequences, resulting in a very low standard deviation. This results in a very large segment score for several mRNAs, so instead the percent difference from the native free energy becomes a more appropriate measure of randomization effects. The mean of the percent difference of native from codon-random free energies is -2.2 percent, with a 95% confidence interval from -4.0 percent to -0.4 percent. The confidence interval does not include zero, demonstrating a significant difference in folding free energies between native and codon-randomized mRNA sequences. Of the 51 mRNAs examined, 35 or 69% are more negative than their codon-randomized mRNAs [12]. The frame scores under codon-randomization again show a significant difference from the native sequences. This indicates that the codon-randomization procedure has destroyed some characteristic in the native sequences that is giving rise to the larger-than-expected W-1 base pair counts in the native set. Since only the degenerate bases can change under this randomization, codon-choice appears to be a major cause of the enhancement of the W-1 frame counts in the native set.

To further investigate the effect of the choice of codons within the coding sequence, a fifth randomization procedure was performed. Native sequences were randomized by an equal selection of codons within the coding region, yielding "codon-flat" sequences. These sequences contained identical 5' and 3' untranslated regions of the respective native mRNA and coded for the same polypeptide as with the codon-random sets. Since the choice of codons was unbiased, the resulting base composition was usually different from the native sequences. Again the native mRNA sequences tend to fold more negative in free energy than the corresponding codon-flat sequences (Table 2). The mean of the percent difference of native from codon-flat free energies is -6.6 percent, with a 95% confidence interval from -10.0 percent to -3.2 percent. The confidence interval does not include zero, demonstrating a significant difference in folding free energies between native and codon-flat mRNA sequences. Of the 51 mRNAs examined, 37 or 73% are more negative than their codon-flat mRNAs [12].

The frame scores from codon-flat randomization are most similar to the whole- and CDS-random set due to the large W-1 frame scores. The frame scores are quite different when comparing codon-random with codon-flat, yet the number of bases actually changed between the two sets should be small. Approximately one-third of CDS bases are degenerate, and each has a probability from $\frac{1}{2}$ to $\frac{1}{4}$ of being changed. Therefore this data is consistent with the hypothesis that local codon choice is optimized to result in a global preference for W-1 base pairing. The relationship to the biased negative free energies in Table 3 is probably related to the total number of predicted base pairing. The sum of the three frame counts for the native sequences is larger than the sum of any of the randomized sequences, but is closest to the sum for the codon-shuffle case.

To analyze frame W-1 a graph-drawing exercise can be performed by listing the codons of each amino acid in the genetic code, then drawing a line to link codon: reverse-complement codons together. The codons then are grouped into a vertex for each amino acid. The lines then represent in-frame (or W-1) base pairing that could occur within the CDS of mRNAs. Surprisingly this exercise results in three independent graph components or families, each

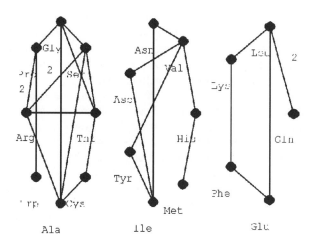

Figure 1: Graph theory representation of the standard genetic code. Multi-lines are shown with numbers indicating their valency.

comprising a subset of the twenty amino acids (Figure 1).

The three amino acid families can be identified by the member with the greatest degeneracy or graph vertex degree [13]. The largest amino acid family, Serine, contains a C or G at the second codon position, while the Valine and Leucine families contain A or T at this second codon position. This decomposition of the twenty amino acids into three families is a structural feature of the genetic code. Since the second codon position is constant (except for Serine) among all synonymous codons, the twenty amino acids must group into at least two families based on the pairing of reverse complement codons. One family contains C and G at the second codon position, and the other family contains A and T. It is interesting that the A/T family has decomposed further into two smaller graphs as shown in Figure (1).The alternate genetic codes follow a similar pattern of possessing either three or two graph components or families [13]. Others have also reported the graph property of the genetic code from different viewpoints. Zull and Smith [15] observed this graph while considering the properties of "anti-proteins" to explain receptor/ligand biochemistry. This graph theory representation may be related to the frame counts since the W-W and W-2 frames are symmetrically related (shift left or right), while the W-1 frame is not shifted. Hence the shifted-frames form a related group different from the W-1 group.

4   Discussion

The biases in calculated mRNA folding free energies observed in Table 2 are small yet significant. For a 400 nt mRNA that is 50% basepaired and 50% G+C, a 5% increase in folding free energy could be caused by changes in only 7-20 basepairs. Since the CDS comprises about 70% of the mRNA examined, more of the bias is due to amino acid sequence and codon choice than due to the UTR sequences alone. If the global amino acid sequence is constrained by considerations of protein function, then the bias is most likely due to subtle local arrangements in codon choice. The effect observed here may be due to a selective advantage for mRNAs to be more basepaired, perhaps to resist degradation or modification. If indeed RNA was the original genetic material as suggested by the research of Joyce [5], then the genetic code may be arranged in such a manner as to encourage intermolecular bonding for single-stranded RNA. This certainly would have been of advantage for any protoorganism with genetic information stored in RNA.

The fact that these biases are observed from an empirical molecular calculation also suggests that local secondary structures are the causative agents. Most algorithms for predicting RNA secondary structure from base sequence are based on a nearest neighbor model of interaction [16]. Experimental evidence indicates short-ranged stacking and hydrogen bonding are important determinants of RNA stability while hydrophobic bonding is of lesser importance [11]. Numerous algorithms have been developed to predict RNA secondary structure by minimizing the configurational free energy. The quality of these predictions depends upon 1) the accuracy of the thermodynamic data which describe the free energies of various secondary structural features, 2) the folding rules that an algorithm uses to find the lowest free energy structure, and 3) the degree to which environmental conditions stabilize alternate structures of equivalent or higher energy.

Free energy minimizing algorithms such as Zuker's MFOLD program output a family of structures that have the same or nearly the same free energy. This study has compared the optimal free energy of folding with a reference set of optimal free energies obtained by sequence randomization. Thus what are being compared are the locations of the minima in the configurational energy profiles for folding. Different regions of mRNA sequence were randomized, including codon choice, resulting in destabilization of the folding free energy. This suggests that local secondary structure interactions are causing the observed bias in folding free energies and frame counts.

5   Conclusion

A survey of 51 mRNA sequences reveals a bias in the coding and untranslated regions that allows for greater negative folding free energies than predicted by sequence length or nucleotide base content. A free energy reference state is taken to be a large enough set of randomized mRNA sequences. Randomization can be implemented over the whole sequence or over sections such as the CDS, UTR, or codon choice. Randomization of most regions of the mRNA sequences display lower folding stability as measured by calculated free energy values. Randomization of codon choice while still preserving original base composition also results in less stable mRNAs. This suggests that a bias in the selection of codons favors mRNA structures which contribute to folding stability. More predicted base-pairs are in the W-1 frame than the other two possible frames. Codon composition appears to be the major determinant for this behavior

6 Acknowledgment

This work was supported (or partially supported) by NIH grant GM08247, by a Research Centers in Minority Institutions award, G12RR03062, from the Division of Research Resources, National Institutes of Health, and NSF CREST Center for Theoretical Studies of Physical Systems (CTSPS) Cooperative Agreement #HRD-9632844.

References

[1] de Smit, M.H and van Duin, J. (1990),"Control of translation by mRNA secondary structure in Escherichia coli. A quantitative analysis of literature data," J Mol Biol 1994 Nov 25;244(2):144-50 .

[2] Devereux,J., Haeberli,P., and Smithies,O. (1984),"A comprehensive set of sequence analysis programs for the VAX," *Nuc. Acids Res.*,12, 387-395.

[3] Hofacker IL, Fekete M, Flamm C, Huynen MA, Rauscher S, Stolorz PE, Stadler PF (1998),"Automatic detection of conserved RNA structure elements in complete RNA virus genomes," Nucleic Acids Res 1998 Aug 15;26(16):3825-36.

[4] Jacobson,A.B.,Arora, R., Zuker, M., Priano,C., Lin, C.H., &Mills, D.R. (1998),"Structural plasticity in RNA and its role in the regulation of protein translation in coliphage Q beta," *J. Mol. Biol.* 275, 589-600.

[5] Joyce, G. (1997),"Evolutionary chemistry: getting there from here," Science 13;276(5319):1658-9.

[6] Karlin, S. and Mrazek, J. (1996), "What drives codon choices in human genes?," *J. Mol. Biol.* 262:459-472.

[7] Le, S.-Y. and Maizel, J.V., Jr. (1989), "A method for assessing the statistical significance of RNA folding,"*J. Theor. Biol.*, 138,495-510.

[8] Lloyd,A.T. and Sharp,P.M. (1992), "Evolution of codon usage patterns: the extent and nature of divergence between Candida albicans and Saccharomyces cerevisiae. ," *Nuc. Acids Res.*,20,5289-5295.

[9] Love Jr.,H.D., Allen-Nash,A., Zhao,Q., and Bannon,G.A. (1988), "mRNA stability plays a major role in regulating the temperature-specific expression of a Tetrahymena thermophila surface protein ," *Mol. Cell. Biol.*,8,427-432.

[10] Macdonald PM, Kerr K (1997), "Redundant RNA recognition events in bicoid mRNA localization," RNA 1997 Dec;3(12):1413-20.

[11] Mathews DH, Sabina J, Zuker M, Turner DH (1999), "Expanded sequence dependence of thermodynamic parameters improves prediction of RNA secondary structure," J Mol Biol 1999 May 21;288(5):911-40

[12] Seffens, W. and Digby, D., (1999), "mRNAs Have Greater Calculated Folding Free Energies Than Shuffled or Codon Choice Randomized Sequences," *Nucleic Acids Research*, 27,1578-1584.

[13] Seffens, W. (1999), "Graph generation and properties of the standard genetic code," *Mathematica World*, submitted.

[14] Zuker,M. and Stiegler,P. (1981),"Optimal computer folding of large RNA sequences using thermodynamics and auxiliary information.," *Nuc. Acids Res.*,9,133-148.

[15] Zull, J.E. and Smith, S.K. (1990), "Is genetic code redundancy related to retention of structural information in both DNA strands?," Trends Biochemical Sci., 15,257-261.

[16] Walter, A.E., Turner, D.H., Kim, J.,Lyttle, M.H.,Müller, P.,Mathews, D.H., and Zuker, M. (1994),"Coaxial stacking of helixes enhances binding of oligoribonucleotides and improves predictions of RNA folding," *Proc. Natl. Acad. Sci. USA* 91, 9218-9222.

Optimization in Computational Chemistry and Molecular Biology, pp. 141-156
C. A. Floudas and P. M. Pardalos, Editors

Structure Calculations of Symmetric Dimers using Molecular Dynamics/Simulated Annealing and NMR Restraints: The Case of the RIIα Subunit of Protein Kinase A

Dimitrios Morikis, Marceen Glavic Newlon and Patricia A. Jennings
Department of Chemistry and Biochemistry
University of California, San Diego
La Jolla CA 92093-0359 USA
dmorikis@ucsd.edu; mnewlon@ucsd.edu; pajennin@ucsd.edu

Abstract

A comparative study of two different molecular dynamics/simulated annealing (MD/SA) protocols for the structure determination of protein dimers using NMR-derived distance (NOE) and dihedral angle restraints has been performed. The solution structure determination of the dimeric regulatory subunit of the type IIα (RIIα) protein kinase A (PKA) has been used as a test case. An asymmetrically isotopically labeled sample and an X-filtered NOE experiment were important for the crucial assignment of inter-monomer NOEs. The first protocol is an *ab initio* MD/SA calculation using starting structures with random ϕ- and ψ-dihedral angles (Nilges, M., 1993, *Proteins: Structure, Function, and Genetics, Vol. 17*, 297-309). The second protocol is an MD/SA calculation using starting structures "reasonably well-defined" monomers (O'Donoghue, S.I., King, G.F. and Nilges, M., 1996, *Journal of Biomolecular NMR, Vol. 8*, 193-206), in which monomer structures were first calculated from a subset of the available NOEs. Two calculations with variable number of intra-monomer long range NOEs were made to determine what a "reasonably well-defined" monomer is. RIIα consists of two helix-loop-helix patterns that form an X-type four helix bundle structural motif. A strong hydrophobic core is observed which contributes to the structural stability of the protein. In addition, RIIα possesses a solvent accessible hydrophobic surface which is involved in protein-protein interactions that mediate signal transduction pathways.

Keywords: Protein kinase A, PKA, RIIα, NMR, NOE, molecular dynamics, simulated annealing, MD/SA, distance geometry.

1   Introduction

Protein kinases catalyze the addition of phosphate groups from ATP to specific sites on target proteins. This simple change directs subsequent protein-protein interactions and modulates diverse cellular processes including cell differentiation and death (apoptosis). The correct intracellular targeting of protein kinases and phosphatases confers specificity to the enzymes, in part, by placing them in close proximity to their preferred substrates. Targeting of these enzymes occurs via association with specific proteins which are found in different locations in the cell. Perhaps the best characterized of the mammalian proteins which serve a similar function is the family of A-Kinase Anchoring Proteins (AKAPs) identified by Scott and coworkers.[1, 2, 3] AKAPs maintain the cyclic AMP (cAMP) dependent protein kinase (PKA) in specific subcellular compartments, thereby ensuring accessibility of the kinase to a limited number of substrates in a particular location.

The enzyme is an inactive tetramer composed of a regulatory dimer (R_2) and two catalytic (2C) subunits (CR_2C) which make up the inactive PKA holoenzyme complex. Binding of the second messenger, cAMP, dissociates the CR_2C tetramer to expose a catalytically active monomeric C-subunit that phosphorylates protein substrates on serine or threonine residues according to the scheme:

$$CR_2C \ + \ 4cAMP \ \rightarrow \ R_2 \cdot cAMP_4 \ + \ 2C$$
$$C \ + \ ATP \ + \ target \ \rightarrow \ C \ + \ ADP \ + \ phosphorylated \ \ target$$

Localization of PKA occurs through the N-terminal dimerization domain of the regulatory subunit. N-terminal constructs spanning the first 44 residues of the enzyme demonstrate that this small region is necessary and sufficient for both stable dimer formation and for high affinity AKAP interaction.[4, 5]

We have determined the solution structure of the N-terminal dimerization domain (residues 1-44) of the type IIα regulatory subunit (RIIα) of PKA [5, 6] using two molecular dynamics-based simulated annealing (MD/SA) methodologies and experimental NMR restraints.[7, 8] The input restraints were NOE-derived distance restraints, J-coupling constant derived ϕ-dihedral angle restraints and J-coupling constant/NOE-derived χ_1-dihedral angle restraints. In addition, NMR-deduced hydrogen bond restraints were incorporated as distance restraints. ^{1}H, ^{15}N and ^{13}C chemical shift assignments, backbone sequential assignments and spin system identification were made using standard multi-dimensional and multi-nuclear NMR spectra.[9]

There is an inherent simplification in assigning the NMR spectra of symmetric dimers; the equivalent chemical environment of equivalent protons within the two monomers produces resonances with identical chemical shifts (symmetry degeneracy).[7, 8, 10] In this sense, only one monomer needs to be assigned. However, the symmetry degeneracy produces a significant difficulty in extracting the necessary inter-monomer NOE assignments that define the relative topology (and symmetry) of the two monomers during the structure calculation.[7, 8, 10] Conventional 2D and 3D NOESY spectra do not discriminate between intra-monomer, inter-monomer and mixed NOEs. In order to reduce the symmetry degeneracy of symmetric dimers, an asymmetrically labeled protein sample should be available to perform an X-filtered NOE experiment (Fig. 1). In our case the asymmetrically labeled sample is prepared by mixing equivalent and equimolar quantities of a ^{15}N/^{13}C-labeled with an unlabeled sample, by unfolding them in the presence of a denaturant and then by slowly refolding them using a dilution method followed by sample concentration (Fig. 1a). The final sample contains a statistical mix of labeled-labeled, labeled-unlabeled, unlabeled-unlabeled protein dimers (Fig. 1a) of which only the labeled-unlabeled portion will be detectable by the X-filtered NOE experiment (Fig. 1a). Thus, the X-filtered NOESY spectrum allows identification of inter-monomer NOEs (Fig. 1b). Conventional NOESY spectra provide all possible combinations of intra-, inter-monomer and mixed NOEs (Fig. 1b). Comparison of conventional NOESY spectra and X-filtered NOESY spectra, usually allows the classification of a sufficient number of NOEs as intra-monomer, inter-monomer, mixed and ambiguous. The ambiguity arises from the inherent low sensitivity and the possibility of artifacts in the X-filtered NOE experiments which may hinder the identification of certain inter-monomer NOEs.

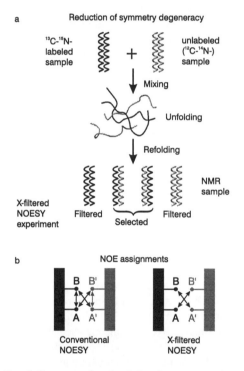

Figure 1: (a) Schematic of the steps involved in the preparation of the asymmetrically isotopically labeled sample used to collect the X-filtered NOE experiment. This sample preparation involves mixing, unfolding and refolding of a labeled and an unlabeled sample. Only the labeled-unlabeled portion of the final sample is detected in the X-filtered NOE experiment. (b) Example of NOE assignments in a symmetric protein dimer. Possible NOEs among protons A, B of one monomer and their equivalent protons A', B' of the other monomer are shown in the case of conventional and X-filtered NOE spectra.

In order to succeed in the solution structure determination of RIIα, we prepared an asymmetrically ^{15}N/^{13}C-labeled sample using standard protein expression and folding techniques.[5] This sample was used to perform an X-filtered NOE experiment [11, 12] [3D ^{13}C-edited (ω_2)-^{12}C-filtered (ω_1)/^{13}C-filtered (ω_3) NOESY], which allowed the assignment of purely inter-monomer NOEs. In addition, conventional NOE experiments [13, 12] were performed using unlabeled, ^{15}N-labelled or ^{15}N/^{13}C-doubly labeled samples (2D NOESY, 3D ^{15}N-edited NOESY-HSQC, 3D ^{13}C-edited NOESY-HMQC, respectively), which were used to assign intra-monomer, inter-monomer and mixed NOEs.[5, 6]

2 Methodology

2.1 General strategy

For this study we used an RIIα expression construct encompassing residues 1-44. Two additional residues were also added to the N-terminus in the construction of the expression system.[9, 5, 6] These two residues do not affect dimerization or AKAP interactions. Thus the residue numbering that is used herein is 1-46, with 3-46 corresponding to residues 1-44 of the native protein. The sample preparation of RIIα, the experimental conditions, the types of experiments used and the derivation of restraints from NMR data are described elsewhere.[9, 5, 6] Since there are multiple approaches to the calculation of dimer structures with NMR and it is not clear which is the best strategy we undertook a comparative study. We focus on different methodologies used for the structure calculation of RIIα.[5, 6] All calculations were performed using the program X-PLOR 3.851 [14, 15] with the topology and parameter files TOPALLHDG.PRO and PARALLHDG.PRO, respectively.[14, 15] Three different dimer structure calculations are presented:

(1) An iterative *ab initio* MD/SA calculation using initial structures with random ϕ- and ψ-angle dihedral angles.[7] Structure calculation iterations were performed to resolve NOE assignment ambiguities, to assign new NOEs based on structural arguments and to aid in the classification of intra-, inter-monomer, mixed and ambiguous NOEs.[5]

(2) An iterative MD/SA calculation with starting structures "reasonably well-defined" monomers.[8] The "reasonably well-defined" monomer structures were calculated using a hybrid distance geometry-simulated annealing protocol [16, 14] and a subset of the NOEs used for the dimer calculation. Structure calculation iterations were performed using different numbers of intra-monomer long range NOEs, which are responsible for the packing of the monomer elements of secondary structure during the calculation. The number of intra-monomer long range NOEs is important in determining what is a "reasonably well-defined" monomer structure that can be used for a subsequent successful dimer calculation. A minimal number of intra-monomer long range NOEs was used for the monomer calculations. These NOEs were unambiguously assigned mainly from the NMR spectra and to a lesser extent from structural arguments.[6] In the past this approach was used only for dimers where homologous structures were available.

(3) An MD/SA calculation with starting structures "reasonably well-defined" monomers similar to (2).[8] However, the "reasonably well-defined" monomer structures were calculated with a larger number of intra-monomer long range NOEs. The additional NOEs were assigned from structural arguments, based on the results of calculations (1) and (2).[6]

2.2 NOE restraint classification

NOEs were classified as strong ($1.8 \leq r_{ij} \leq 2.7$Å), medium ($1.8 \leq r_{ij} \leq 3.3$Å), weak ($1.8 \leq r_{ij} \leq 5.0$Å) and very weak ($1.8 \leq r_{ij} \leq 6.0$Å), where r_{ij} is the inter-proton distance and 1.8 Å is the van der Waals contact distance for hydrogen atoms. A correction of 0.5 Å was added to the upper bound of the distances involving methyl groups to account for the averaging of the three methyl protons. Also a correction of 0.2 Å was added to the upper bound of distances involving amide groups to account for the higher apparent intensity of these types of NOEs. During dimer structure calculations, mixed NOEs and NOEs that could not be unambiguously assigned as intra- or inter-monomer were treated as ambiguous NOEs with the dynamic NOE assignment method,[7, 14] similar to the one used for non-stereospecifically assigned methyl or methylene protons. In this method the sum of the intra- and inter-monomer distances is used to enter an "effective distance" into the calculation, according to $(\sum r^{-6})^{-1/6}$. NOEs were classified as intra-residue ($|i - j| = 0$) NOEs, sequential intra-monomer ($1 \leq |i - j| \leq 4$)NOEs, long range intra-monomer ($|i-j| > 4$) NOEs, inter-monomer NOEs and ambiguous NOEs. During <u>monomer</u> structure calculations [(2) and (3)] only intra-monomer (intra-residue, sequential and long range) NOEs were used. During <u>dimer</u> structure calculations [(1), (2) and (3)] all five classes of intra-monomer (intra-residue, sequential and long range), inter-monomer and ambiguous NOEs were used. In all three <u>dimer</u> structure calculations, the number and type of intra-monomer (intra-residue, sequential and long range), inter-monomer and ambiguous NOEs was kept the same. In the <u>monomer</u> structure calculations [(2) and (3)], the number and type of intra-residue and sequential NOEs was kept the same but the number of long range NOEs was varied (*vide infra*). In all three calculations (monomer or dimer) 30 dihedral angle restraints and 19 intra-monomer hydrogen bond restraints were used. The dihedral angle restraints consisted of 25 ϕ- and 5 χ_1-dihedral angles. Hydrogen bond restraints were incorporated in the calculations as distance restraints, two restraints per hydrogen bond. For the O-N distance we used 3.3 Å (lower bound of 2.5 Å and upper bound of 3.5 Å) and for the O-HN distance we used 2.3 Å (lower bound of 1.5 Å and upper bound of 2.5 Å).

2.3 Calculation (1)

The structure of RIIα was determined [5] using an *ab initio* MD/SA protocol (REFINE for dimers [17]) proposed by Dr. Nilges.[7] A template random structure with correct local geometries and no non-bonded contacts was used to generate initial structures with random ϕ- and ψ-dihedral angles. These structures were used as input in the calculation to generate an ensemble of structures of RIIα. A total of 505 NOE restraints were utilized of which 185 were intra-residue NOEs, 231 were sequential intra-monomer NOEs, 25 were long range intra-monomer NOEs, 38 were inter-monomer NOEs and 26 were ambiguous NOEs. A total of 100 dimer structures were calculated and 30 of them with no NOE violation > 0.3 Å, no dihedral angle violation > 5°, no bond length violation > 0.05 Å and no bond angle or improper angle violation > 5° were accepted for the final ensemble of structures. All 30 accepted structures converged to the same structural motif.

2.4   Calculation (2)

The structure of RIIα was determined in two stages. First, an ensemble of monomer structures were calculated using a subset of the observed NOEs which were unambiguously characterized as intra-monomer NOEs. Then, the dimer structures were calculated using as initial input structures the previously determined monomer structures. The monomer structures were calculated using a standard X-PLOR hybrid distance geometry-simulated annealing protocol (DG_SUB_EMBED, DGSA, REFINE [15]).[16, 14] The dimer structures were determined using an MD/SA protocol (MDSA-SO-WDMR-1_0 [17]) also proposed by Dr. Nilges group.[8]

A total of 441 NOE restraints were used. The NOE restraints consisted of 185 intra-residue NOEs, 231 sequential NOEs and 25 long range NOEs. A total of 100 monomer structures were calculated of which 24 showed no NOE violation > 0.5 Å, no dihedral angle violation $> 5°$, no bond length violation > 0.05 Å and no bond angle or improper angle violation $> 5°$. These 24 monomer structures were used as input for the subsequent dimer calculation.

The dimer structure calculation used a total of 505 NOE restraints of which 185 were intra-residue NOEs, 231 were sequential intra-monomer NOEs, 25 were long range intra-monomer NOEs, 38 were inter-monomer NOEs and 26 were ambiguous NOEs. A total of 24 dimer structures were generated and 10 of them with no NOE violation > 0.3 Å, no dihedral angle violation $> 5°$, no bond length violation > 0.05 Å and no bond angle or improper angle violation $> 5°$ were accepted for the final ensemble of structures. All 10 accepted structures converged to the same structural motif.

2.5   Calculation (3)

The structure of RIIα was also determined using the same protocol as in calculation (2),[8, 10] except that a larger number of long range intra-monomer NOE restraints were used for the monomer structure calculations.[6] During monomer structure calculations a total of 457 experimental NOE restraints were used of which 185 were intra-residue NOE restraints, 231 were sequential intra-monomer NOE restraints and 41 were long range intra-monomer NOE restraints [as opposed to 25 in calculation (2)]. The additional 16 NOE restraints were classified as long range intra-monomer NOE restraints based on structural arguments, following calculations (1) and (2). However these 16 restraints were treated as ambiguous NOE restraints during the dimer calculation because a mixed character could not be excluded, based again on structural arguments. A total of 100 monomer structures were generated and 49 of them with no NOE violation > 0.5 Å, no dihedral angle violation $> 5°$, no bond length violation > 0.05 Å and no bond angle or improper angle violation $> 5°$ were used as input for the dimer calculation.

During dimer structure calculations a total of 505 experimental NOE restraints were used, of which 185 were intra-residue NOEs, 231 were sequential intra-monomer NOEs, 25 were long range NOEs, 38 were inter-monomer NOEs and 26 were ambiguous NOEs. A total of 49 dimer structures were generated and 30 of them with no NOE violation > 0.3 Å, no dihedral angle violation $> 5°$, no bond length violation > 0.05 Å and no bond angle or improper angle violation $> 5°$ were accepted for the final ensemble of structures. All 30 accepted structures converged to the same structural motif.

2.6 X-PLOR target function

The X-PLOR target function for Calculation (1) consisted of quadratic harmonic energy terms for covalent geometry (bond lengths, bond angles, planes, chirality), a quartic van der Waals repulsion term for non-bonded contacts, soft-square energy terms for all classes of experimental distance (NOE) restraints (intra-monomer, inter-monomer and ambiguous) and a quadratic square well energy term for experimental dihedral angle restraints. Also, to enforce symmetry, a quadratic harmonic energy term for non-crystallographic symmetry (NCS) restraints and a symmetry soft-square energy term for global symmetry NOE restraints were used. The NCS restraints produce two nearly superimposamble monomers by minimizing the atomic RMSD between the two monomers, without taking into account their relative orientation. The global symmetry NOE restraints arrange the monomers in a symmetric way without imposing a symmetry axis, by restraining a set of equivalent inter-monomer distances to be equal.[7, 14]

The X-PLOR target function for monomer calculations (2) and (3) consisted of quadratic harmonic energy terms for covalent geometry, a quartic van der Waals repulsion term for non-bonded contacts and quadratic square well energy terms for experimental NOE and dihedral angle restraints.[14]

The X-PLOR target function used for dimer calculations (2) and (3) consisted of the same energy terms as in monomer calculations for covalent geometry, van der Waals repulsion, experimental intra-monomer NOE restraints and experimental dihedral angle restraints. Additional soft-square energy terms were used for experimental inter-monomer and ambiguous NOE restraints, a quadratic harmonic energy term was used for non-crystallographic symmetry (NCS) restraints and a symmetry soft-square energy term was used for global symmetry NOE restraints. Also, a quadratic harmonic energy term was used for packing to prevent the monomers from drifting apart. The packing term restrains all atoms to the reference coordinates.[8, 10, 14]

No explicit hydrogen bonding, 6-12 Lennard-Jones or electrostatic energy terms were used.

The input force constants during minimization were 1000 kcal mol^{-1} $\mathring{A}^{-2}$ for bond lengths, 500 kcal mol^{-1} rad^{-2} for bond angles and improper angles, 4 kcal mol^{-1} $\mathring{A}^{-4}$ for van der Waals repulsion in Calculation (1), 1 kcal mol^{-1} $\mathring{A}^{-4}$ for van der Waals repulsion in Calculations (2) and (3), 50 kcal mol^{-1} $\mathring{A}^{-2}$ for experimental NOE restraints, 200 kcal mol^{-1} rad^{-2} for experimental dihedral angle restraints, 10 kcal mol^{-1} $\mathring{A}^{-2}$ for NCS restraints in Calculation (1), 2 kcal mol^{-1} $\mathring{A}^{-2}$ for NCS restraints in Calculations (2) and (3), 1 kcal mol^{-1} $\mathring{A}^{-2}$ for symmetry NOE restraints in Calculation (1), 2 kcal mol^{-1} $\mathring{A}^{-2}$ for symmetry NOE restraints in Calculations (2) and (3), and 0.3X10^{-6} kcal mol^{-1} $\mathring{A}^{-2}$ for packing restraints. The input force constants were varied during simulated annealing according to the standard X-PLOR protocols.[7, 8, 14, 17, 15]

2.7 Computational requirements

A total of 1526 atoms (763 per monomer) where used in the dimer calculations corresponding to 92 residues (46 per monomer). An Indigo 2 SGI workstation was used for all calculations with a 200 MHz R4400 processor. The CPU time required for the calculation of 100 structures was 64 hours for Calculation (1), 67 hours for dimer Calculations (2) and (3) and 9.4 hours for monomer Calculations (2) and (3).

3 Results and Discussion

3.1 Analysis of calculated structures

Figure 2a shows a backbone representation of the final ensemble of the 30 dimer structures of RIIα determined with Calculation (1). Figure 2b shows the ensemble of the 10 monomer structures that were used as initial structures for the accepted dimer structures determined in Calculation (2). Figure 2c shows the final ensemble of the 10 accepted dimer structures of RIIα. Figure 2d shows the ensemble of the 30 monomer structures that were used as initial structures for the accepted dimer structure determined in Calculation (3). Figure 2e shows the final ensemble of the 30 accepted dimer structures of RIIα. The ensembles of the dimer structures are in excellent agreement with each other as it is indicated by visual inspection (Figs. 2a,c,e). RIIα is a four helix bundle which consists of two helix-loop-helix monomers. Furthermore, the structure of RIIα is classified as an X-type four helix bundle structural motif, with alternating nearly anti-parallel and orthogonal packing of α-helices.[5, 6] Each monomer of RIIα possesses a disordered amino terminal region, an α-helix (I, I'), a small loop, a second α-helix (II, II') and a small disordered carboxy terminal region. All three accepted dimer ensembles of RIIα have the same relative topology of the four helices. More specifically, helices I, II are on top of helices I', II'. This has been called a "top-top" configuration.[6] A smaller ensemble consisting of rejected structures with high total energy terms and with large numbers of NOE and dihedral angle violations was also observed during the structure calculations. This alternative configuration (called "top-bottom") consisted of a packing arrangement in which helices I, II and I', II' were interwoven.[6]

Table 1 presents a comparison of the agreement criteria of the structures determined using Calculations (1), (2) and (3) and the convergence criteria within each structural ensemble. These criteria are secondary, tertiary and quaternary structure formation and the precision of each structural ensemble. In all three dimer ensembles and the two monomer ensembles helices I, I' start at residue 11 and end at residue 24, and helices II, II' start at residue 30 and end at residues 43-44 (Table 1). The tertiary and quaternary packing of the four helices in the dimer structures is indicated by their inter-helical angles which are in excellent agreement within the standard deviations (Table 1) and by the relative topology of the four helices (they all conform with the "top-top" configuration). The precision of each individual ensemble is indicated by the root mean square deviation (RMSD) from the mean structure (Table 1).

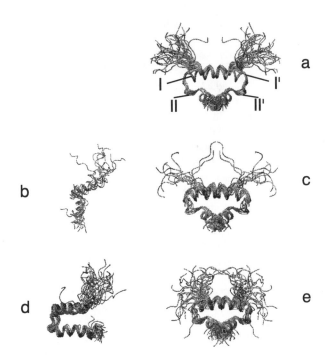

Figure 2: (a) The ensemble of 30 dimer structures of RIIα generated with Calculation (1). (b) The ensemble of 10 (out of 24) accepted monomer structures used as input for the accepted ensemble of 10 dimer structures of RIIα generated with Calculation (2). (c) The ensemble of 10 accepted dimer structures of RIIα generated with Calculation (2). (d) The ensemble of 30 (out of 49) accepted monomer structures used as input for the accepted ensemble of 30 dimer structures of RIIα generated with Calculation (3). (c) The ensemble of 30 accepted dimer structures of RIIα generated with Calculation (3). In each individual ensemble structures are superimposed by fitting the coordinates of the backbone heavy atoms (N, Cα, C) in the ordered region [residues 11-24 for monomer Calculations (2) and (3), 11-44 for dimer Calculation (1) and 11-43 for dimer Calculations (2) and (3); Table 1].

Table 1. Structural comparison of Calculations (1), (2) and (3).

Calculation	(1)	(2)	(2)	(3)	(3)
	Dimer	Monomer	Dimer	Monomer	Dimer
Number of structures	30	10	10	30	30
Secondary structure					
Helix I	11-24	11-24	11-24	11-24	11-24
Helix II	30-44	30-43	30-43	30-43	30-43
Inter-helical angles[a] (°)					
I, II	125±2	53±22	126±6	140±9	130±6
I, I'	160±2	-	164±9	-	165±7
II, II'	150±3	-	145±4	-	145±4
RMSD[b] (Å)					
Backbone heavy atoms	0.44	2.65	0.95	0.96	0.85
All heavy atoms	0.93	3.54	1.42	1.68	1.41

[a] Inter-helical angles were calculated with the program MolMol.[18] Parallel and anti-parallel arrangement of the NH-CO vectors of two helices correspond to 0° and 180°, respectively. [b] Residues 11-44, 11-43, 11-43 were used to fit the backbone heavy atoms and to calculate the RMSDs of the ensembles resulted from calculations (1), (2) and (3), respectively, in agreement with the secondary structure of each calculation. RMSDs are calculated from the mean structure of each individual ensemble.

Figure 2b-e also addresses the issue of what is a "reasonably well-defined" monomer structure, a term that has been loosely used in the past. Figure 2b,d and Table 1 demonstrate that the secondary structure of the monomer is identical to that of the dimer. This is expected because the secondary structure is primarily defined by the sequential ($1 \leq |i - j| \leq 4$) NOEs.[13] We have used the same number and type of sequential NOEs for both monomer and dimer calculations. However, the monomer structures of Calculations (2) and (3) show significant differences in the angle between helices I and II and in the standard deviation (Fig. 2b,d; Table 1). Monomer Calculation (3) has produced better defined inter-helical angle with smaller standard deviation and with value closer to the dimer inter-helical angle than Calculation (2) (Table 1). This is attributed to the fact that a larger number of long range ($|i - j| > 4$) intra-monomer NOE restraints have been used in Calculation (3) compared to Calculation (2). These NOEs are responsible for the tertiary packing of elements of secondary structure.[13] Despite their significant packing differences, monomer structures from both Calculations (2) and (3) yielded secondary and tertiary structure for the dimers in excellent agreement with each other and with Calculation (1) (Fig. 2; Table 1). Based on these observations, the term "reasonably well-defined" monomer structures refers mainly to the correct definition of the two elements of secondary structure (helices I, II) rather than the correct formation of super-secondary or tertiary structure which is defined by the inter-helical angle.

Calculation (3) has resulted in significantly smaller RMSD for the backbone heavy atoms (N, Cα, C) and all heavy atoms than Calculation (2) at the monomer structure level (Fig. 2b,d; Table 1). The large RMSD variation of the monomer structures of Calculation (2) is again attributed to the small number of intra-monomer long range NOEs used. At the dimer level, Calculation (3) shows smaller RMSD than Calculation (2) which reflects the tighter packing of the input monomer structures (Fig. 2c,e; Table 1). Finally, the *ab initio* Calculation (1) shows significantly smaller RMSD than Calculations (2) and (3) (Fig. 2a,c,e; Table 1). Use of a van der Waals repulsion force constant of 4 kcal mol^{-1} Å^{-4} for Calculation (3) has resulted in a smaller RMSD (0.71 Å) and a reduced number of structures (23) in the accepted ensemble, compared to Calculation (3) with a van der Waals repulsion force constant of 1 kcal mol^{-1} Å^{-4} (everything else being kept the same). This is still significantly higher than the RMSD of the accepted structures in Calculation (1). The higher RMSD of Calculations (2) and (3) may be attributed to biasing the dimer structure calculation towards the initial monomer structures.

Figure 3 shows a comparison of the lowest energy structures of the final ensembles of 30, 10 and 30 structures determined using calculations (1), (2) and (3), respectively. The three structures are found to be in excellent agreement with each other. The RMSD for the backbone heavy atoms of residues in the ordered region 11-43 is 0.45 Å.

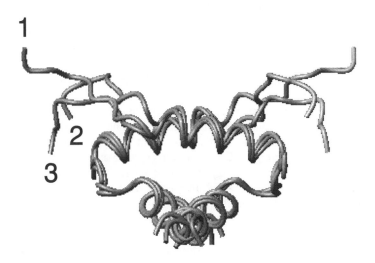

Figure 3: The lowest energy structures of RIIα determined in Calculations (1), (2) and (3). The structures have been superimposed by fitting the coordinates of the backbone heavy atoms (N, Cα, C) in the ordered region 11-43.

MD/SA dimer structure calculations using NMR restraints and protocols similar to the ones described in Calculations (2) and (3) have been presented in the past in cases where previous knowledge of homologous monomer or dimer structures was available. In these cases assessment of the symmetry axis or dimer interface is simplified and inter-monomer NOE assignments can be facilitated with this *a priori* structural knowledge. Here, we demonstrate that even in the absence of a previously known homologous structure, a subset of the available NOEs can be used to calculate "reasonably well-defined" monomer structures which result in the determination of well-defined dimers.

In comparing Calculations (2) and (3) in view of future work, an initial conservative approach in selecting the input NOEs is preferred in the early rounds of structure calculations [like in Calculation(2)]. The resulted structures can be used to back-calculated all expected NOEs which can then be compared to the actual experimental NOE data to make additional assignments. This can be a manual or semi-automated process. Inclusion of these additional NOEs in later rounds of Calculations should produce better defined structures [like in Calculation(3)]. In comparing Calculations (1) and (3), it appears that Calculation (1) produces better defined structures as judged by the RMSD of the accepted ensemble. While this is true for our simple structural motif (symmetric packing of two helix-loop-helix patterns), it is not obvious that this will be the case for more complicated structural motifs or in cases of symmetric oligomers with higher order of symmetry.

3.2   Modeling of RIIα-AKAP interactions

Figure 4a shows the final ensemble of 30 structures [Calculation (3)] of RIIα. Only a backbone and hydrophobic side chain representation is shown, while all other side chains have been deleted for clarity. RIIα possesses a central hydrophobic core which is responsible for the packing of the two monomers in an X-type four helix bundle structural motif and for maintaining the intra-monomer, inter-helical packing in a helix-loop-helix structural motif.[5, 6] The interaction between RIIα and a model of the amphipathic helical peptide segment of the human thyroid anchoring protein (Ht31, residues 493-515) is also illustrated (Fig. 4a). Ht31 is a member of large family of AKAPs found in diverse cell types and locations.[3] The peptide is positioned above a solvent accessible hydrophobic surface of RIIα in an orientation that permits maximal protein-peptide hydrophobic interactions. There is no significant charged character along the interface between RIIα and Ht31 peptide; however, there is significant charge distribution in RIIα and Ht31, at their solvent exposed faces opposite to the protein-peptide interface.[5, 6] Figure 4b shows a ribbon representation of the RIIα-Ht31 complex in the same orientation as in Fig. 4a. Only the lower energy structure of RIIα is shown for clarity. Figure 4c shows the AKAP binding surface of the lowest energy structure of RIIα. A characteristic knob and hole pattern [19] is observed that allows optimal packing of hydrophobic residues between RIIα and AKAPs.

The RIIα-Ht31 interaction model based on structural arguments alone is further supported by chemical shift analysis of NMR experiments.[5] A sample of ^{15}N-labeled RIIα complexed with an unlabeled Ht31 peptide was prepared and ^{1}H-^{15}N heteronuclear single quantum coherence (HSQC) NMR spectra were collected.[5] The HSQC spectra of the complex when compared to the HSQC spectrum of free RIIα, showed chemical shift doubling only in residues of RIIα that participate in the protein-peptide interface.[5] This was attributed to asymmetries in the local chemical and magnetic environments of residues in the RIIα-Ht31 interface, induced by the asymmetric nature of the peptide. Small structural changes in the same region of RIIα, induced by the peptide binding cannot be excluded.[5] A full structural analysis of the complex in underway.

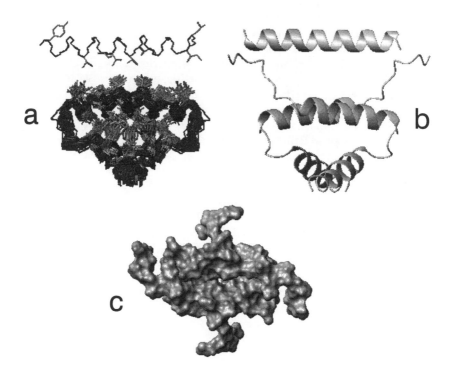

Figure 4: (a) A backbone and hydrophobic side chain representation of the ordered region (residue 11-43) of the ensemble of 30 structures of RIIα and the amphipathic AKAP peptide Ht31 derived from the human thyroid anchoring protein. The backbone is drawn in black and the hydrophobic side chains are drawn in gray. All other side chains are deleted for clarity. The relative orientation of RIIα and Ht31 is chosen to demonstrate optimal hydrophobic interactions in the bound state. (b) A ribbon model of Ht31 and the lowest energy structure of RIIα in the same orientation as in (a). (c) A representation of the AKAP binding surface of the lowest energy structure of RIIα. This view is a 90° rotation of the RIIα view in (a) or (b). The RIIα structures shown in this Figure are from Calculation (3). Individual panels of this Figure have been generated with the program MolMol.[18]

4 Conclusions

We have shown that structure determination of symmetric protein dimers is possible using the MD/SA protocol of Nilges [8] in which the initial structures are "reasonably well-defined" monomers even in the absence of previously known homologous protein structures or the dimer symmetry axis or dimer interface. A single asymmetrically isotopically labeled sample was used to perform a crucial X-filtered NOE experiment that was helpful in distinguishing inter-monomer from intra-monomer NOEs. The monomer structures were calculated utilizing a subset of the NMR restraints used for the dimer structure calculation. We have also demonstrated that in the case of RIIα, the term "reasonably well-defined" monomers means (1) correct secondary structure, (2) not necessarily correct relative topology of elements of secondary structure, (3) possibly large deviation of inter-helical angles from those of the final dimer structure and (4) sometimes significantly larger RMSDs from those of the final dimer structure.

We have also used RIIα to compare two MD/SA protocols proposed by Dr. Nilges, an *ab initio* calculation from templates with random ϕ-, ψ-dihedral angles [7] and the calculation from initial structures of "reasonably well-defined" monomers.[8] The outputs of the two calculations were in excellent agreement; however, the *ab initio* calculation generated an ensemble of structures with higher precision.

RIIα forms an X type four helix bundle structural motif.[5, 6] The central core of the protein is formed by strong hydrophobic packing of side chains. In addition, a solvent accessible hydrophobic patch is observed.[5, 6] We have used structural modeling to identify the binding site of RIIα with an AKAP peptide, using the calculated structures and a peptide model. Hydrophobic interactions between the solvent accessible hydrophobic surface of RIIα and the hydrophobic face of the amphipathic helix of the AKAP peptide are essential for binding. This observation is in agreement with earlier NMR and NMR-based modeling data.[5, 6]

5 Acknowledgments

This work was supported in part by National Institutes of Health Grants GM54038, DK54441 (P.A.J); The Cancer Research Coordinating Center (P.A.J.); The American Cancer Society (P.A.J); National Institutes of Health Training Grants DK07233 (D.M), GM07313 (M.G.N.) and American Heart Association Grant 97-425 (M.G.N.). We thank Drs. John Scott, Melinda Roy and Sean O'Donoghue for helpful discussions.

References

[1] Faux, M.C. and J. D. Scott, J.D. (1996) "More on target with protein phosphorylation: conferring specificity by location," *Trends in Biochemical Sciences Vol. 21, No 8*, 312-315.

[2] Coghlan, V.M., Perrino, B.A., Howard, M., Langeberg, L.K., Hicks, J.B., Gallatin, W.M., and Scott, J.D. (1995) "Association of Protein Kinase A and Protein Phosphatase 2B with a Common Anchoring Protein," *Science, Vol. 267, No 5194*, 108-111.

[3] Dell'Acqua, M.L. and Scott, J.D. (1997) "Protein kinase A anchoring," *Journal of Biological Chemistry, Vol. 272, No 20*, 12881-12884.

[4] Hausken, Z. E., Coghlan, V.M., Schafer Hastings, C.A., Reimann, E.M., and Scott, J.D. (1994) "Type II Regulatory Subunit (RII) of the cAMP-dependent Rotein Kinase Interaction with A-kinase Anchor Proteins Requires Isoleucines 3 and 5," *Journal of Biological Chemistry, Vol. 269, No 39*, 24245-24251.

[5] Newlon, G.M., Roy, M., Morikis, D., Hausken, Z.E., Coghlan, V., Scott, J.D. and Jennings, P.A. (1999a) "The molecular basis of protein kinase A anchoring revealed by solution NMR," *Nature Structural Biology, Vol. 6, No 3*, 222-227.

[6] Newlon, G.M., Morikis, D., Roy, M., Scott, J.D. and Jennings, P.A. (1999b) "Solution structure determination of the anchoring domain of protein kinase A IIα," submitted for publication.

[7] Nilges, M. (1993) "A calculation strategy for the structure determination of symmetric dimers by ^1H NMR," *Proteins: Structure, Function, Genetics, Vol. 17*, 297-309.

[8] O'Donoghue, S.I., King, G.F. and Nilges, M. (1996) "Calculation of symmetric multimer structures from NMR data using a priori knowledge of the monomer structure, co-monomer restraints, and interface mapping: The case of leucine zippers," *Journal of Biomolecular NMR, Vol. 8*, 193-206.

[9] Newlon, G.M., Roy, M., Hausken, Z.E., Scott, J.D. and Jennings, P.A. (1997) "The A-kinase anchoring domain of type IIα cAMP-dependent protein kinase is highly helical," *Journal of Biological Chemistry, Vol. 272, No 38*, 23637-23644.

[10] O'Donoghue, S.I. and Nilges, M. (1999) "Calculation of symmetric oligomer structures from NMR data," in *Biological Magnetic Resonance, Vol. 17, pp. 131-161: Structure Computation and Dynamics in Protein NMR*, Berliner, J.L. and Krishna, N.R. (eds.), Plenum, New York.

[11] Ikura, M., Clore, G. M., Gronenborn, A. M., Zhu, G., Klee, C. B. and Bax, A. (1992) "Solution structure of a calmodulin-target peptide complex by multidimensional NMR," *Science, Vol. 256, No 5057*, 632-638.

[12] Clore, G. M. and Gronenborn, A. M. (1998) "Determining the structures of large proteins and protein complexes by NMR," *Trends in Biotechnology, Vol. 16*, 22-34.

[13] Wuthrich, K. (1986), *NMR of Proteins and Nucleic Acids*, John Wiley and Sons, New York.

[14] Brunger, A. (1992), *X-PLOR*, Yale University Press, Yale, New Haven.

[15] *http://pauli.csb.yale.edu/xplor-info/*.

[16] Nilges, M., Clore, G.M. and Gronenborn, A.M (1988) "Determination of three-dimensional structures of proteins from interproton distance data by hybrid distance geometry-dynamical simulated annealing calculations," *FEBS Letters, Vol. 229*, 317-324.

[17] *http://www.NMR.EMBL-Heidelberg.DE/nilges/.*

[18] Koradi, R., Billeter, M. and Wuthrich, K. (1996) "MOLMOL: A program for display and analysis of macromolecular structures," *Journal of Molecular Graphics, Vol. 14*, 51-55.

[19] Crick, F.H.C. (1953) "The packing of α-helices: Simple coil-coils," *Acta Crystallographica, Vol. 6*, 689-697.

Optimization in Computational Chemistry and Molecular Biology, pp. 157-189
C. A. Floudas and P. M. Pardalos, Editors
©2000 Kluwer Academic Publishers

Structure Prediction of Binding Sites of MHC Class II Molecules based on the Crystal of HLA–DRB1 and Global Optimization

M.G. Ierapetritou
Department of Chemical and Biochemical Engineering,
Rutgers University, Piscataway, NJ 08854

I.P. Androulakis
Corporate Research Science Laboratories,
Exxon Research & Engineering Co., Annandale, NJ 08801

D.S. Monos
Department of Pediatrics, University of Pennsylvania,
The Children's Hospital of Philadelphia, Philadelphia, PA 19104

C.A. Floudas
Department of Chemical Engineering
Princeton University, Princeton, N.J. 08544-5263
floudas@titan.princeton.edu

Abstract

Class II histocompatibility molecules are cell surface molecules that form complexes with self and non-self peptides and present them to T-cells that activate the immune response. A number of class II histocompatibility molecules have been analyzed by crystallography and include the molecules HLA-DR1 [59], HLA-DR3 [22], and I-E^k [21].

A novel theoretical predictive approach is presented that can determine three dimensional structures of the binding sites of the HLA–II molecules based on the crystallographic data of previously characterized HLA class II molecules. The proposed approach uses the ECEPP/3 detailed potential energy model for describing the energetics of the atomic interactions in the space of substituted residues dihedral angles and employs a rigorous deterministic global optimization algorithm αBB [1, 6, 2, 3, 4] to obtain the global minimum energy conformation of the binding site. The binding sites of the HLA–DR3 and I-E^k molecules are predicted based on the crystallographic data of HLA-DR1 [59]. The predicted structures of the binding sites of these molecules exhibit small root mean square differences that range between 1.09-2.03Å (based on all atoms) in comparison to the reported crystallographic data [21, 22]. The energetic driving forces for binding of the predicted structures are also studied using the decomposition-based approach of Androulakis et al. [28] and found to provide very good agreement with the results of the crystallographically obtained binding sites.

Keywords: Structure prediction, Global optimization, MHC class II molecules, Binding sites, Peptide docking

1   Introduction

Class II histocompatibility molecules are polymorphic cell surface glycoproteins that form complexes with self and non-self peptides. These complexes are recognized by the T cell antigen receptor of the CD4 positive T cells. This interaction initiates the activation of the antigen specific immune response. Crystallographic analysis of the class II molecules has been reported for the human alleles HLA-DR1 [59], HLA-DR3 [22], and the mouse I-E^k [21]. Detailed knowledge of the structural characteristics of these molecules is very essential for the understanding of molecular mechanisms in normal and pathogenic processes that involve HLA-peptide interactions.

These pioneering crystallographic studies provide the basis for theoretical investigations concerning both structure prediction and binding affinity of different MHC molecules for various peptide antigens. Among the theoretical and computational contributions are the works of Lee and Richard [35], Connolly [15], Bacon and Moult [8], Jiang and Kim [29], Kuntz et al. [27] that follow shape-based methods and the works of Goodsell and Olson [23], Hart and Read [24] and Calfisch [11] following energy-based methods. Recently, Androulakis et al. [28] presented a decomposition based approach that allows the independent study of the different binding sites (i.e., pockets) of the HLA-DR1 molecule using deterministic global minimization of intra and inter energetic interactions modeled via the ECEPP/3 [51, 57] force field. Based on the crystallographic data of HLA-DR1 [59], that allows the description of the different pockets, their theoretical approach predicts a rank ordered list of the amino-acids with regard to their binding to pocket 1 of HLA-DR1. These theoretical results are in excellent agreement with the results of competitive binding assays [50].

On the other hand, the determination of high quality models of protein structure for which no experimentally determined coordinates exist has received considerable attention in the literature. A commonly used approach is based on the homology modeling in which a model for a target protein is generated using the known structure of a homologous protein. A backbone model is constructed typically for the structurally conserved regions, and loops and side chain are then added [9, 60]. For the prediction of side-chain conformation there exist many approaches based on homology modeling that differ from each other regarding (a) the rotamer library used, (b) the energy function and (c) the search strategy. When composing the sampling of conformational space through rotamer libraries, many different approaches have been used, including backbone independent rotamer libraries [12], or rotamer sets that incorporates the backbone-sidechain interactions [18]. Extended libraries derived form the cluster analysis of the experimentally determined database [56] and augmented sets using discrete values around observed χ angles values $\pm 10°$ [63] are also used. Regarding the energy function used, simplistic local interactions are typically limited to van der Waals or hard-sphere energies [13, 34]. Finally, the employed search strategies are mainly heuristic methods involving Monte Carlo techniques [25] Genetic Algorithms [61], Neural Networks [26], Mean-Field optimization [34], and combinatorial search [18].

A recent review of global optimization approaches for protein folding and peptide docking can be found in Floudas et al. [20]. The main disadvantages of these approaches are:

(a) a very limited conformational space is only considered since usually less than 10 rotamers are used for each residue,

(b) the simplicity of the energy functions used are not able to capture realistic description of the molecular system, and

(c) no systematic search methodology exists to guarantee the determination of the global optimal solution even by utilizing simplified energy functions.

The objective of this paper is to propose a systematic and rigorous approach for the determination of the three-dimensional structure of the polymorphic surface of an HLA type II molecule based on the crystallographic data of HLA-DR1 molecule [59], with the correct conformation of the binding sites residues and same binding affinities as the crystallographicaly obtained structures. The HLA-DR3 and I-E^k molecules have been selected as the benchmark studies in this work since their crystals have been determined recently [22, 21], and hence provide the basis for evaluating the predictive approach. The proposed theoretical approach is based on detailed atomistic-level modeling and deterministic global optimization of the energetic interactions. The results are verified by the superposition of the crystallographic data with the predicted binding sites. Further justification of the proposed approach is provided by reversing the problem and predicting the structure of binding sites of HLA-DR1 molecule based on the crystallographic data of HLA-DR3 and furthermore by being able to evaluate the affinity of the predicted pockets. Finally, the binding studies of the predicted pockets are in very good agreement with the binding studies of the crystallographic pockets for all the examined systems for HLA-DR1, HLA-DR3 and I-E^k.

2 Problem Definition

The recent crystallographic studies of class II HLA molecules [59, 22, 21], suggest an overall similarity in their structures. The conformation of HLA-DR3 in the HLA-DR3-CLIP complex is only slightly different from that of HLA-DR1 in HLA-DR1-HA [22], and a comparison of two I-E^k structures with HLA-DR1 identifies that only a few differences in β chain amino acids exist between I-E^k and both the HLA-DR1 and HLA-DR3 sequences. However, these few variable residues are sufficient to explain antigenic differences without recourse to allosteric transitions or novel conformations.

Consequently, specific information about the structure of the histocompatibility molecules is needed in order to be able to analyze their specificity. Because crystal structures of class

II molecules are not available except for the human crystals of HLA-DR1-HA and HLA-DR3-CLIP and the murine crystals I-E^k-HB, I-E^k-Hsp, we propose a novel approach based on decomposition and deterministic global optimization that enables the prediction of the three dimensional structure of the binding sites of class II molecules and can be efficiently used for the qualitative assessment of their binding affinities.

The question that is addressed in this paper is stated as follows:

> "Given the (x,y,z) coordinates of the atoms in pockets 1,4,6,7 and 9 of the HLA-DR1 [59], can we predict the three dimensional structures of the corresponding pockets of HLA-DR3 and I-E^k that exhibit the same binding characteristics ?"

In the next section the basic steps of the proposed approach are presented. The presentation focuses on the structure prediction of the HLA-DR3 and I-E^k binding site utilizing the crystallographic data of the HLA-DR1 molecule [59]. It should be noted that the proposed methodology is generalizable and can be used for the prediction of unknown HLA structures.

3 Proposed Approach

3.1 System Definition

The geometric shape of a protein under the assumption of rigid bond lengths and bond angles is uniquely determined by its dihedral angles. If more than one polypeptide is involved then the relative orientations, and locations of these different chains must be defined. This can most easily be accomplished by defining a translation vector and a rotation matrix. The translation is achieved through the cartesian coordinates of the initial nitrogen atom of each independent chain. The Euler angles specify the rotations necessary to orient a particular polypeptide and are defined as the angles between the coordinate axes defined by the initial hydrogen, nitrogen, and alpha carbon of each residue.

The system under study involves all the residues of the binding site. The substituted amino acids constitute the problem variables, whereas the residues that remain the same are treated as fixed based on the crystallographic data. Since there may be multiple amino acid substitutions, the problem variables are the amino coordinates their euler angles and their dihedral angles of all substituted residues.

It should be highlighted here that in contrast to the existing approaches the euler angles and the dihedral angles are considered to span the whole feasible range [-180°, +180°] and not restricted to specified discrete values.

3.2 Potential Energy Function

The most accurate representation of the potential energy of a molecule is the *ab initio* quantum mechanical approach. Using the Born-Oppenheimer approximation, one can determine the energy for fixed atomic nuclei from the smallest eigenvalue of the Hamiltonian of the electron system. However, due to their computational complexity, such calculations are limited to extremely small molecules.

As a result, tractable potential energy models have been derived which adequately capture the energy contributions resulting from various types of atom interactions. Molecular potential functions include ECEPP [47, 48, 49], AMBER [66, 67], CHARMM [10], DIS-COVER [16], GROMOS [62], MM3 [5], ENCAD [36], ECEPP/2 [52] and ECEPP/3 [51].

In this work, the ECEPP/3 (Empirical Conformational Energy Program for Peptides) potential model is utilized. In this force field, it is assumed that the covalent bond lengths and bond angles are fixed at their equilibrium values. It has been observed that variations in bond lengths and bond angles depend mostly on short range interactions; that is, those between the side chain and backbone of the same residue. Under this assumption, all residues of the same type have essentially the same geometry in various proteins [49]. Therefore, a chain of any sequence can be generated using the fixed geometry specific to each type of amino acid residue in the sequence.

Based on these approximations, the conformation is only a function of the dihedral angles. That is, ECEPP/3 accounts for energy interaction terms which can be expressed solely in terms of the dihedral angles. The total conformational energy is calculated as the sum of the electrostatic, nonbonded, hydrogen bonded, and tortional contributions. Loop closing contributions, if the polypeptide contains two or more sulfur–containing residues, are a fixed internal conformational energy of the pyrolidine ring for each prolyl and hydroxyprolyl residue contained in the peptide chain, are also represented. The main energy contributions (electrostatic, nonbonded, hydrogen bonded) are computed as the sum of terms for each atom pair (i,j) whose interatomic distance is a function of at least one dihedral angle.

Let M denote the total number of substitutions, then based on the description of our system, the set of variables include the $N_x^m, N_y^m, N_z^m, m = 1, ..., M$ vector of all N coordinates for which bounds can be obtained $\pm \delta$ from the coordinates of HLA-DR1 molecule, the euler angles, and the dihedral angles ϕ^m, ψ^m, ω^m and χ_k^m which vary between $[-\pi, +\pi]$. The constraints on the translation vector of each substituted residue are required to assure that the residue remains within the vicinity of the binding site. The contributing terms to the total potential energy of ECEPP/3 are:

$$
\begin{aligned}
E = & \sum_{(i,j) \in \mathcal{ES}} 332.0 \frac{q_i q_j}{D r_{ij}} \quad \text{(Electrostatic)} \\
& + \sum_{(i,j) \in \mathcal{NB}} F \frac{A}{r_{ij}^{12}} - \frac{C}{r_{ij}^6} \quad \text{(Nonbonded)} \\
& + \sum_{(hx) \in \mathcal{HX}} F \frac{A'}{r_{hx}^{12}} - \frac{B}{r_{hx}^{10}} \quad \text{(Hydrogen bonding)} \\
& + \sum_{k \in \mathcal{TOR}} (\frac{E_o}{2})(1 \pm \cos n_k \theta_k) \quad \text{(Torsional)} \\
& + \sum_{l \in \mathcal{LOOP}} B_L \sum_{i_l=1}^{i_l=3} (r_{i_l} - r_{i_o})^2 \quad \text{(Cystine Loop–Closing)} \\
& + \sum_{l \in \mathcal{LOOP}} A_L (r_{4_l} - r_{4_o})^2 \quad \text{(Cystine Torsional)}
\end{aligned}
\tag{1}
$$

Note that in the above energy function all the interactions of the atoms belonging to a single substituted residue (**intra-interactions**), as well as the interactions between the atoms of the different substituted residues (**inter-interactions**) are simultaneously considered in an explicit systematic way. In the next section the complete description of the total energy of the polypeptide system under consideration along with the interactions with the solvent is presented.

3.3 Solvation Model

The explicit incorporation of solvation effects involves calculating solvent-peptide and solvent-solvent interactions using potentials similar to those previously described. Although these methods are conceptually simple, explicit inclusion of solvent molecules greatly increases the computational time needed to simulate the polypeptide system. Therefore, most simulations of this type are limited to restricted conformational searches. In addition, the interactions between the protein molecule and the surrounding water molecules are not fixed for a given peptide configuration. In reality, a large number of solvent configurations must be considered, and the free energy of hydration can then be calculated by averaging over these configurations. Average values for peptide-solvent interaction energies can also be calculated using a mean-field approximation for the free energy of solvation. In this case the interactions are expressed as an average over all positions of solvent molecules for a given protein configuration. Limitations in these methods have led to the development of simpler implicit solvation models [64] .

There are a number of empirical hydration models which can be used to implicitly predict hydration effects. The main assumption of these models is that, for each functional group of the peptide, a hydration free energy can be calculated from an averaged free energy of interaction of the group with a layer of solvent known as the hydration shell. In addition, the total free energy of hydration is expressed as a sum of the free energies of hydration for each of the functional groups of the peptide, that is, an additive relationship is assumed.

In this paper, a solvent accessible surface area method [14], which is based on the assumption that the free energy of hydration is proportional to the solvent-accessible surface area of the peptide is used:

$$E_{HYD} = \sum_{i=1}^{N}(A_i)(\sigma_i) \tag{2}$$

In Equation (2), an additive relationship for N individual functional groups is assumed. (A_i) represents the solvent-accessible surface area for the functional group, and (σ_i) are empirically derived free energy density parameters.

The development of the peptide surface is made in the following way. First the peptide surface is represented by a union of spheres, with the radii of the spheres set by the van der Waals radii of the constituent atoms. A spherical test probe is then rolled over these spheres, thereby tracing out a surface. The molecular surface is set by direct contact between the probe sphere and the peptide spheres. In areas where the probe cannot make direct contact, the closest part of the probe is used. The solvent–accessible surface is defined by the surface traced by the center of the probe as the probe rolls over the peptide spheres.

Once the solvent–accessible surface areas have been calculated, these values must be multiplied by the appropriate (σ_i) parameters as shown in Equation (2).

There are a number of models available, including JRF, OONS, SRFOPT, which provide estimates for these parameters based on interactions between water and the functional groups of peptides. Detailed studies of solvation models in connection with deterministic global optimization approaches for oligopeptide folding have been conducted [32, 33]. The JRF parameter set is used in this paper [65]. Since this parameter set was developed from minimum energy conformations of peptides, the surface–accessible solvation energies are only included at local minimum conformations.

Consequently, the total energy function is defined as:

$$E_{TOT} = E_{UNSOL}^{MIN} + E_{SOL} \tag{3}$$

3.4   Mathematical Formulation

Based on the above description the mathematical formulation can be posed in the following way:

$$\min \quad E_{Total}\left(\phi^m, \psi^m, \omega^m, \chi_k^m, N_x^m, N_y^m, N_z^m, \varepsilon_1^m, \varepsilon_2^m, \varepsilon_3^m\right) \tag{4}$$

$$\text{s.t.} \quad -\pi \leq \phi^m, \psi^m, \omega^m, \chi_k^m, \varepsilon_1^m, \varepsilon_2^m, \varepsilon_3^m \leq \pi \tag{5}$$

$$(N_x^m)^L \leq N_x^m \leq (N_x^m)^U \tag{6}$$

$$(N_y^m)^L \leq N_y^m \leq (N_y^m)^U \tag{7}$$

$$(N_z^m)^L \leq N_z^m \leq (N_z^m)^U \tag{8}$$

$$(C_x'^m)^L \leq C_x'^m\left(\phi^m, \psi^m, \omega^m, \chi_k^m, N_x^m, N_y^m, N_z^m, \varepsilon_1^m, \varepsilon_2^m, \varepsilon_3^m\right) \leq (C_x'^m)^U \tag{9}$$

$$(C_y'^m)^L \leq C_y'^m\left(\phi^m, \psi^m, \omega^m, \chi_k^m, N_x^m, N_y^m, N_z^m, \varepsilon_1^m, \varepsilon_2^m, \varepsilon_3^m\right) \leq (C_y'^m)^U \tag{10}$$

$$(C_z'^m)^L \leq C_z'^m\left(\phi^m, \psi^m, \omega^m, \chi_k^m, N_x^m, N_y^m, N_z^m, \varepsilon_1^m, \varepsilon_2^m, \varepsilon_3^m\right) \leq (C_z'^m)^U \tag{11}$$

where m=1,...,M corresponds to total number of substitutions.

Note that the additional constraints (6-11) are considered to represent the bounds on the N and C' coordinates expressing the binding of the specific residue with the rest of pocket [28], since the substituted residue is part of a longer polypeptide and consequently is not allowed to rotate freely. Since the C' coordinates can be evaluated as functions of the independent variables the restrictions on C' position is implemented by the incorporation of a penalty function, P, in the objective function.

$$P = \beta\{ \ \langle C_x'^l - C_x'\rangle + \langle C_x' - C_x'^u\rangle + $$
$$\langle C_y'^l - C_y'\rangle + \langle C_y' - C_y'^u\rangle + $$
$$\langle C_z'^l - C_z'\rangle + \langle C_z' - C_z'^u\rangle \ \}$$

The $\langle \rangle$ function is defined as follows: $\langle \mathcal{A} \rangle$ equals $\mathcal{A}$ if $\mathcal{A}$ is greater than zero, otherwise $\langle \mathcal{A} \rangle$ equals zero. Thus, any coordinate value beyond the specified bounds would be multiplied by the penalty parameter β and added to the potential energy. Consequently, the minimization of the objective function eliminates solutions in which the C' position falls outside the specified bounds.

4   Global Optimization Approach αBB

As formulated in the previous section, the minimization of the energy function E_{Total} corresponds to a nonconvex nonlinear optimization problem. Even after reducing this optimization problem to a function of internal variables (dihedral angles), the multidimensional surface that describes the energy function has a very large number of local minima. This has become known as the multiple-minima problem. Because the objective function has many local minima, local optimization techniques exhibit a dependence on initial points selection. Therefore, global optimization algorithms are needed to effectively locate the global minimum corresponding to the native state of the protein.

The global optimization approach αBB [1, 6, 2, 3, 4], has been extended in this work to identifying the structure of the unknown binding sites. The αBB is a branch-and-bound based deterministic global optimization framework. The algorithm is shown to guarantee convergence to the global minimum of nonlinear optimization problems with twice–differentiable functions [19]. The application of this algorithm to the minimization of potential energy functions was first introduced for microclusters [39, 40], and small acyclic molecules [41, 42]. The αBB approach has also been extended to constrained optimization problems [1, 6, 2, 3, 4]. In more recent works, the algorithm has been shown to be successful for isolated peptide systems using the realistic ECEPP/3 potential energy model [38, 6] and for the quantitative determination of the binding specificity of a class II HLA molecule, HLA-DRB1*0101 allele, interacting with different peptides.

The αBB global optimization algorithm effectively brackets the global minimum solution by developing converging lower and upper bounds. These bounds are refined by partitioning the original search domain into sub-domains, and the upper bounding and lower bounding sequences can be shown to converge within ϵ to the global solution in a finite number of steps [44]. Upper bounds are obtained by minimizing E_{Total} using either function evaluations or local optimization methods. Lower bounds are generated by constructing valid convex relaxations of E_{Total}. These are defined as the original potential energy E_{Total} plus the summation of separable quadratic terms, in the form of $\alpha(x^L - x)(x^U - x)$, for all the independent dihedral angles, translation variables and euler angles that E_{Total} depends upon, as follows:

$$L = E_{Total} + \alpha \{ \sum_{m=1}^{M} \left(\phi^{mL} - \phi^m \right) \left(\phi^{mU} - \phi^m \right) + \left(\psi^{mL} - \psi^m \right) \left(\psi^{mU} - \psi^m \right) + $$

$$\left(\omega^{mL} - \omega^m \right) \left(\omega^{mU} - \omega^m \right) + \sum_{k=1}^{K} \left(\chi_k^{mL} - \chi_k^m \right) \left(\chi_k^{mU} - \chi_k^m \right) + \left(N_x^{mL} - N_x^m \right) \left(N_x^{mU} - N_x^m \right) + $$

$$\left(N_y^{mL} - N_y^m \right) \left(N_y^{mU} - N_y^m \right) + \left(N_z^{mL} - N_z^m \right) \left(N_z^{mU} - N_z^m \right) + \left(\varepsilon_1^{mL} - \varepsilon_1^m \right) \left(\varepsilon_1^{mU} - \varepsilon_1^m \right) + $$

$$\left(\varepsilon_2^{mL} - \varepsilon_2^m \right) \left(\varepsilon_2^{mU} - \varepsilon_2^m \right) + \left(\varepsilon_3^{mL} - \varepsilon_3^m \right) \left(\varepsilon_3^{mU} - \varepsilon_3^m \right) \}$$

where α is a nonnegative parameter which must be greater or equal to the negative one half of the minimum eigenvalue of the Hessian of E_{Total} in the considered domain defined by the lower and upper bounds (i.e., $x^L = -\pi, x^U = \pi$) of the dihedral angles, translation variables and euler angles. This parameter can be rigorously calculated based on the techniques introduces by Adjiman and Floudas [1], and Adjiman et al. [3, 4]. The overall effect of these terms is to overpower the nonconvexities of the original nonconvex terms by adding the value of 2α to the eigenvalues of the Hessian of E. The convex lower bounding functions, L, possesses a number of important properties which guarantee global convergence [42]:

(i) L is a valid underestimator of E;

(ii) L matches E at all corner points of the box constraints;

(iii) L is convex in the current box constraints;

(iv) the maximum separation between L and E is bounded and proportional to α and to square of the diagonal of the current box constraints. This property ensures that an ϵ_f feasibility and ϵ_c convergence tolerances can be reached for a finite size partition element;

(v) the underestimators L constructed over supersets of the current set are always less tight than the underestimator constructed over the current box constraints for every point within the current box constraints.

Once solutions for the upper and lower bounding problems have been established, the next step is to modify these problems for the next iteration. This is accomplished by successively partitioning the initial domain into smaller subdomains. The default partitioning strategy used in the algorithm involves successive subdivision of the original rectangle into two sub–rectangles by halving on the midpoint of the longest side (bisection). In order to ensure non–decreasing lower bounds, the sub-rectangle to be bisected is chosen by selecting the region which contains the infimum of the minima of lower bounds. A non–increasing sequence for the upper bound is found by solving the nonconvex problem, E, locally and selecting it to be the minimum over all the previously recorded upper bounds. Obviously, if the single minimum of L for any sub–rectangle is greater than the current upper bound, this sub–rectangle can be discarded because the global minimum cannot be within this subdomain (fathoming step).

The computational requirement of the αBB algorithm is proportional to the number of variables (global) on which branching occurs. Therefore, these global variables need to be chosen carefully. Obviously, in a qualitative sense, the branching variables should correspond to those variables which substantially influence the nonconvexity of the surface and the location of the global minimum. With this in mind [2, 3, 4] have developed principles to help identify the important variables.

For the problem of determining the binding sites of the unknown HLA molecules, the global variable set includes the ϕ, ψ and χ_1 variables. However, it should be highlighted that all the dihedral angles of the substituted residues, as well as the translation vector and the euler angles are continuous variables in the problem, and are treated as local variables.

5   Outline of the Proposed Approach

A systematic approach is presented for the structure prediction of a foreign antigen binding site based on the crystallographic data of HLA-DR1 molecule [59]. The proposed approach examines each of the binding sites separately and involves the following steps:

(1) The binding sites of HLA-DR1 molecule are evaluated. All amino acids within a radius of $\mathcal{R}=5.0\overset{\circ}{A}$ of the atoms of the binding amino acid in the crystallographic studies [59] are identified as shown in Table 1. A Program for Pocket Definition, as described in [28], constructs these pockets through the selection of all residues that are within a radius $\mathcal{R}$ of the atoms of the crystallographic binder.

(2) The amino acid substitutions between HLA-DR1 and the HLA-II molecule (e.g., HLA-DR3, I-E^k) are identified and are shown in Table 2. Note that pocket 1 of HLA-DR1 requires only one substitution Gly$\rightarrow$ Val in position $\beta86$ so as to result in pocket 1 of HLA-DR3. Pockets 4, 6 and 7 involve three substitutions while pocket 9 features only one substitution in the representation of the corresponding pockets of HLA-DR3. Note also that all pockets of HLA-DR1 require three or four substitutions so as to represent the corresponding pockets of I-E^k.

(3) For each one of the substituted residues, the intra and inter-molecular energy interactions are modeled. Specifically, the electrostatic, nonbonded, torsional, and hydrogen bonding contributions [51], are considered for the substituted residue, as well as the interactions of the substituted residues with the rest of amino acids that constitute the examined binding site. The solvation energy is also considered through solvent accessible areas [53, 65] as explained in section 3.3. The dihedral angles that define the three-dimensional structure of the substituted residues are considered explicitly as variables. Note also that the relative position of each amino acid has to be determined. This is done through the determination of the translation vector and the euler angles. For the substituted amino acids lower and upper bounds are considered for their N and C' coordinates, based on the available crystallographic data [59, 22, 21].

(4) Having the mathematical model that includes the intra and inter energetic interactions, and the solvation energy and which has as variables the dihedral angles of the substituted amino-acids, the translation vector and the euler angles, we minimize the total potential energy by employing the α**BB** deterministic global optimization approach [1, 6, 2, 3, 4] as described in section 4.

(5) The resulting global minimum energy conformer provides information on the predicted (x,y,z) coordinates of the atoms of the substituted residues. Structure verification is made by superposition of all atoms of the predicted structure and the ones derived from crystallographic data. The superposition is based on the global minimization of the root mean square differences of the distances between all the atoms involved in the pocket as described in section 5.2.

In the sequel, the algorithmic details of the proposed approach are given.

5.1   Algorithmic Framework

The implementation of the proposed approach involves the connection of the conformation energy program *PACK* [58], that allows the evaluation of all energy interactions when more than one protein chain are involved in the system, with the deterministic global optimization framework, αBB. *PACK* evaluates all energy components through repeated calls to the

Pocket				
1	**4**	**6**	**7**	**9**
pheα24	glnα09	gluα11	valα65	asnα69
ileα31	gluα11	asnα62	asnα69	leuα70
pheα32	asnα62	valα65	gluβ28	ileα72
trpα43	pheβ13	aspα66	tyrβ47	metα73
alaα52	leuβ26	leuβ11	trpβ61	argα76
serα53	glnβ70	pheβ13	leuβ67	trpβ09
pheα54	argβ71	argβ71	argβ71	aspβ57
gluα55	alaβ74			tyrβ60
asnβ82	tyrβ78			trpβ61
valβ85				
glyβ86				
pheβ89				
thrβ90				

Table 1: HLA-DR1 Pocket Compositions for $\mathcal{R} = 5.0$ Å

Pocket	Substitutions for HLA-DR3	Substitutions for I-E^k
1	β86: Gly $\rightarrow$ Val	β85: Val $\rightarrow$ Ile β86: Gly $\rightarrow$ Phe β90: Thr $\rightarrow$ Leu
4	β13: Phe $\rightarrow$ Ser β26: Leu $\rightarrow$ Tyr β74: Ala $\rightarrow$ Arg	β13: Phe $\rightarrow$ Ser β74: Ala $\rightarrow$ Glu β78: Tyr $\rightarrow$ Val β:71 Arg $\rightarrow$ Lys
6	β11: Leu $\rightarrow$ Ser β13: Phe $\rightarrow$ Ser β71: Arg $\rightarrow$ Lys	β11: Leu $\rightarrow$ Ser β13: Phe $\rightarrow$ Cys β71: Arg $\rightarrow$ Lys
7	β28: Glu $\rightarrow$ Asp β47: Tyr $\rightarrow$ Phe β71: Arg $\rightarrow$ Lys	β28: Glu $\rightarrow$ Val β47: Tyr $\rightarrow$ Phe β67: Leu $\rightarrow$ Phe β71: Arg $\rightarrow$ Lys
9	β9: Trp $\rightarrow$ Glu	α72: Ile $\rightarrow$ Val β9: Trp $\rightarrow$ Glu β60: Tyr $\rightarrow$ Asn

Table 2: Substitutions for HLA-DR3 and I-E^k binding sites

ECEPP/3 potential function program. A local optimization solver *NPSOL* is used for the minimization of the overall potential energy provided by *PACK* and for the minimization of the convexified potential function (*L*) provided by αBB. *MSEED* [53], the program for the determination of solvation energy is also connected to allow the consideration of the solvation energy at the local minima. The algorithmic procedure is represented graphically in Figure 1.

In particular, as Figure 2 illustrates the implementation of the proposed approach involves the following steps:

(1) The program for Pocket Definition, (*PPD*), uses the input files *residue.pdb* and *pocket.pdb* to generate the file with the coordinates of the residues involved in the considered pocket.

(2) The program *ARAS* is used to determine the translation vectors, euler angles and dihedral angles of the residues in the pocket given their (x, y, z) coordinates. This information together with initial values for the translation vector, euler angles and dihedral angles of the substituted residues are incorporated within the input file *name.input*.

(3) The program *prePACK* utilizes the *residue.dat* file involving a set of initial atomic coordinates that are based on fixed bond lengths, fixed bond angles and each variable dihedral angle initially set to 180°, the *mol.in* file for each one of the amino acids involved in the pocket and the *pre.name.abb* file which specifies the fixed and substituted residues. It then creates a standard input for the potential function program, *PACK*, *name.date*.

(4) The global optimization program α*BB* requires the *name.abb* file that defines the optimization problem, including the variable bounds. α*BB* also uses the *name.input* file, and the *name.bounds* file that involves the C' bounds used to evaluate the coordinates of C' as a function of the independent variables.

(5) The program *PACK* requires the *name.date* file and is connected with *ECEPP/3* in order to evaluate the potential function, which is minimized by the local optimization solver *NPSOL*.

(6) The *MSEED* solvation energy program uses the *JRF.dat* file which defines the solvation parameters σ_i and evaluates the solvation energy at the current local minimum structure.

5.2   Comparison with Crystallographic Data

To accurately compare the predicted structure of the pockets with the crystallographic data, the best rotation and translation to relate the two different sets of atomic position must be obtained. Assuming two pieces of proteins A and B with N_{atom} atoms, the best superposition is the one that minimizes the sum of squared distances between each B atom and the corresponding A atom. Existing approaches to this problem are based on:

(i) iterative minimization using rotation angles [55, 54];

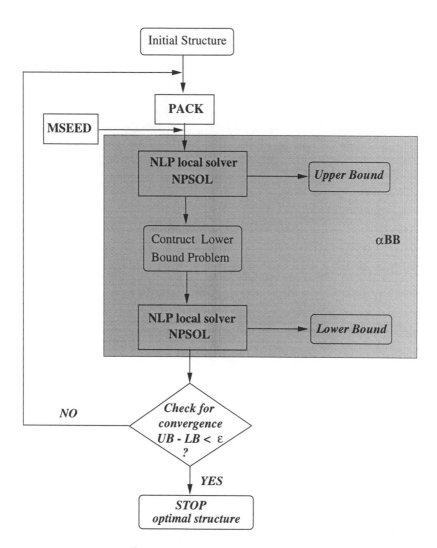

Figure 1: **Proposed algorithm**

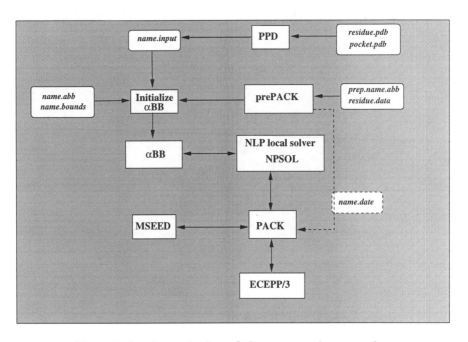

Figure 2: **Implementation of the proposed approach**

(ii) the use of decomposition approach, where the transformation matrix L is determined by calculating the best unrestricted linear transformation which converts A into B using least square matrix method [17]; or form a generalised inverse of the molecular structure [37], and then decompose L=RS where R is a rotation matrix and S is a symmetric distortion matrix;

(iii) the construction of a matrix U which yields an orthogonal rotation directly [30, 31, 45, 46].

As pointed out by McLachlan [46], the rotation angles method is very slow, where the rotation matrix methods depend on whether A is fitted to B or vice versa and does not minimize the r.m.s distance. McLachlan [46], proposed an approach to improve the speed and accuracy of determining the matrix U and moreover to cover all special cases which arise when U is degenerate or singular.

In this paper, the problem of obtaining the best fit of two protein structures is formulated and solved as a global optimization problem. The determination of the best rotation and translation matrix that minimize the "fitting distance" for the two protein structures are guaranteed to be found in all special cases without having to perform any additional tests and calculations.

Consider the two protein structures: A obtained from the crystallographic data and B determined from our proposed methodology. Both structures involve N_{atom} atoms with cartesian coordinates $(x_c(i), y_c(i), z_c(i))$, for the crystal and $(x_p(i), y_p(i), z_p(i))$ for the predicted structure. The mathematical formulation of the best fitting problem can then be posed as follows:

$$\min \quad RMS = (1/N_{atom})\sqrt{\sum_{i=1}^{N_{atom}} (x_c(i) - x'(i))^2 + (y_c(i) - y'(i))^2 + (z_c(i) - z'(i))^2}$$

s.t.

$$\begin{bmatrix} x'(i) \\ y'(i) \\ z'(i) \end{bmatrix} = R \begin{bmatrix} x_p(i) \\ y_p(i) \\ z_p(i) \end{bmatrix} + T$$

$$R = \begin{bmatrix} r_{11} & r_{12} & r_{13} \\ r_{22} & r_{22} & r_{22} \\ r_{31} & r_{32} & r_{33} \end{bmatrix}, \quad T = \begin{bmatrix} t_1 \\ t_2 \\ t_3 \end{bmatrix}$$

$$RR^\top = I \tag{12}$$

where R, T are the required rotation and translation vectors in order to translate the predicted binding sites which correspond to $(x_p(i), y_p(i), z_p(i))$ coordinates, to the cartesian system of the crystal $(x_c(i), y_c(i), z_c(i))$. The coordinates $(x'(i), y'(i), z'(i))$ correspond to the transformed system following the rotation and translation.

Problem (12) constitutes a special case of global optimization problems since it involves the minimization of a convex function subject to a set of linear equality and nonconvex equality constraints $RR^\top = I$. The deterministic global optimization algorithm αBB [1, 6, 2, 3, 4], presented briefly in section 4, is used for the solution of this global optimization problem. The results obtained for the superposition of the predicted HLA-DR3 and I-E^k binding sites with the crystallographic data are presented in the following sections. Four tests are performed in order to evaluate the prediction accuracy of the proposed methodology.

(i) for each predicted binding site the root-mean-square-deviations of Cartesian coordinates of all the atoms (cRMSD) and the C^α atoms are evaluated,

(ii) for each one of the substituted residues, the cRMSD is evaluated considering all the atoms,

(iii) for each one of the substituted residues, a relative-cRMSD is evaluated based on the following formula:

$$R-cRMSD = \frac{1}{N_{atom}} \sqrt{\sum_{i \in N_{atom}} \frac{1}{3} [\frac{(x_p(i) - x_c(i))^2}{x_c(i)} + \frac{(y_p(i) - y_c(i))^2}{y_c(i)} + \frac{(z_p(i) - z_c(i))^2}{z_c(i)}]}$$

to measure the relative predictive error of the proposed procedure,

(iv) computational binding studies are performed to compare the energetic-based rank ordering of the amino acids in the predicted binding site versus the rank ordering of the amino acids in the binding site based on the crystallographic data.

6   Prediction of HLA-DR3 Binding Sites

The proposed approach was applied for the prediction of the three dimensional structure of HLA-DR3 binding sites.

As presented in Table 2, by substituting Gly to Val in position $\beta 86$ in pocket 1 of HLA-DR1, the pocket 1 of HLA-DR3 is formulated. The predicted pocket of HLA-DR3 is shown in Figure 3 with the crystallographically obtained pocket superposition. The (cRMSD) difference between those two pockets is found to be 1.09$\overset{\circ}{A}$ based on the differences of the coordinates of all the atoms involved in the pocket. The relative cRMSD for the whole binding site is 0.0425 which corresponds to 4.25% difference of the predicted cartesian coordinates of the binding site and the crystallographic data. The (cRMSD) difference based on the α carbons is 0.55$\overset{\circ}{A}$. The (cRMSD) for the substituted residue (Val) is 1.584 $\overset{\circ}{A}$ and the relative-cRMSD is 0.04601 which indicates a 4.6% difference of the predicted Val versus the Val determined based on the crystallographic data of HLA-DR3 molecule [22].

To generate pockets 4 of HLA-DR3, three substitutions are made on the composition of the pockets of HLA-DR1 at the positions $\beta 13$: Phe $\rightarrow$ Ser, $\beta 26$: Leu $\rightarrow$ Tyr, and $\beta 74$: Ala $\rightarrow$ Arg. The predicted pocket is shown in Figure 4 together with the corresponding crystallographic data of HLA-DR3. The (cRMSD) difference for all the residues in the pocket is 1.11$\overset{\circ}{A}$ and the overall relative difference of the predicted pocket compared to the crystallographic data is 2.08%. The (cRMSD) difference based on the α carbons is 0.49 $\overset{\circ}{A}$. The (cRMSD) for each one of the substituted residues are 1.67$\overset{\circ}{A}$ for Ser, 0.83$\overset{\circ}{A}$ for Tyr and 1.46 $\overset{\circ}{A}$ for Arg and correspond to relative differences of 3.2%, 1.2% and 2.3%, respectively.

For pocket 6 of HLA-DR3, the substitutions are at positions $\beta 11$: Leu to Ser, $\beta 13$: Phe to Ser, and $\beta 71$: Arg to Lys. The predicted pocket is shown in Figure 5 together with the

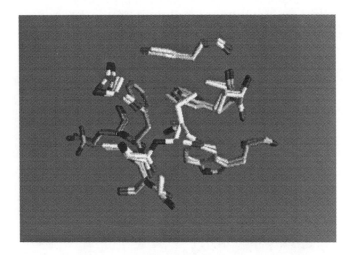

Figure 3: **Superposition of the predicted Pocket 1 of HLA-DR3 versus crystallographic data**

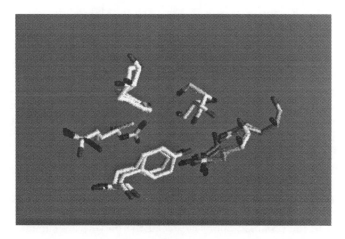

Figure 4: **Superposition of the predicted Pocket 4 of HLA-DR3 versus crystallographic data**

crystallographic data of pocket 6 of HLA-DR3. The (cRMSD) difference for this pocket is 1.22Å based on all atom deviations, which corresponds to a relative-cRMSD of 4.9%. The (cRMSD) difference based on the α carbons is 0.61 Å. The individual (cRMSD) for Serβ11 is 1.26 Å, for Serβ13 is 1.62 Å, and for Lys β71 1.82Å which correspond to relative predictive errors of 7.4%, 3.7% and 3.2%, respectively.

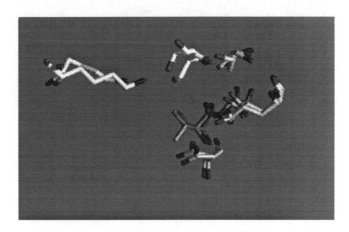

Figure 5: **Superposition of the predicted Pocket 6 of HLA-DR3 versus crystallographic data**

For pocket 7 of HLA-DR3 three substitutions are made at the positions β28: Glu to Asp, β47: Tyr to Phe, and β71: Arg to Lys. The (cRMSD) difference for this pocket is 1.94Å based on all atom deviations, which corresponds to a 4.69% deviation (see Figure 6). The (cRMSD) difference based on the α carbons is 0.71 Å. The (cRMSD) for each one of the substituted residues are 1.08Å for Phe, 3.08Å for Asp and 3.4 Å for Arg and correspond to relative differences of 1.4%, 5.1% and 4.7%, respectively.

Finally, for pocket 9 only one substitution is needed, namely Trp to Glu in position β9 to obtain pocket 9 of HLA-DR3 from pocket 9 of HLA-DR1. The resulting pocket is shown together with the one determined from the crystallographic data in Figure 7, and is found to have an (cRMSD) difference of 1.03 Å based on all atoms and 0.56Å based on the C^α atoms. The relative-cRMSD based on all atom deviations is 37.2%. Considering only the substituted residue, the (cRMSD) is 1.67 Å The large predictive deviation in this pocket is due to the large inherent deviation between the HLA-DR1 and the HLA-DR3 crystallographic data. As shown in Table 3, the (cRMSD) difference for pocket 9 is 1.05Å that corresponds to an inherent relative (cRMSD) of 20.7%.

The results of the proposed approach for all the pockets are summarized in Table 4. Note that the percentage predictive error is less than 5%, except for pocket 9 where the large inherent deviation between the two crystals prohibits a more accurate prediction.

As mentioned in section 3, the coordinates of N and C' are variables in the proposed formulation with bounded ranges for their values around the corresponding atoms in HLA-DR1. Based on the differences observed in the N and C' (x,y,z) coordinates of the HLA-

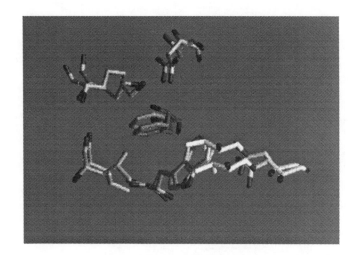

Figure 6: **Superposition of the predicted Pocket 7 of HLA-DR3 versus crystallographic data**

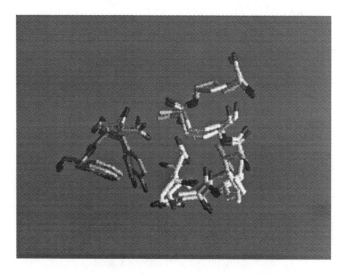

Figure 7: **Superposition of the predicted Pocket 9 of HLA-DR3 versus crystallographic data**

Pocket	HLA-DR1 vs HLA-DR3 Crystals - All atoms (cRMSD) (Å)	HLA-DR1 vs I-E^k-HB Crystals - All atoms (cRMSD) (Å)
1	1.03	1.24
4	0.84	1.23
6	0.84	0.84
7	0.996	0.997
9	1.05	1.092

Table 3: (cRMSD) differences between HLA-DR1, HLA-DR3 and I-E^k crystals

Pocket	Pocket		Substituted Residues	
	(cRMSD) (Å)		(Relative-cRMSD) (%)	(cRMSD) (Å)
	All atoms	C^α	All atoms	
1	1.09	0.55	4.6	Val: 1.58
4	1.11	0.49	2.1	Ser: 1.67 Tyr: 0.83 Arg: 1.46
6	1.22	0.61	4.9	Ser: 1.26 Ser: 1.62 Lys: 1.82
7	1.94	0.71	4.7	Asp: 3.08 Phe: 1.08 Lys: 3.40
9	1.32	0.56	37.2	Glu: 1.67

Table 4: Results for HLA-DR3 prediction

DR1, HLA-DR3 and I-E^k crystals [59, 22, 21] after superposition, tight bounds in the range of [0.3-1.0] suffice. To study further the effect of the bounds we considered bound variations of $(\pm 0.5), (\pm 0.7)$ and (± 1.0). The predicted structures of pocket 1 exhibit small (cRMSD) differences of 1.18, 1.11 and 1.09 Å, respectively, calculated based on all atoms.

The proposed approach considers the simultaneous substitution of all amino acids responsible for the differences of MHC class II molecules. The required substitutions are usually 2, 3 or 4 residues and give rise to a global optimization problem that include the dihedral angles of every residue and the translation vector and the euler angles defining the relative position of the residue. In order to reduce the size of the resulting global optimization problem, the following two simplifying alternative procedures were also explored: The first approach is sequential in nature. Instead of considering all amino acids substitutions simultaneously they are considered sequentially. In particular, the conformation of the first considered changed amino acid is determined by minimizing the intra and inter molecular interactions between the specific amino acid and the rest of the residues constituting HLA-DR1 binding site. Then, this residue is considered as part of the pocket and the structure of the second substituted residue is determined. In the second alternative approach each of the substituted amino acids are considered independently and their conformation is determined based on the minimum energy interactions with the rest of amino acids involved in the pocket of HLA-DR1 molecule. The results obtained for the case of pocket 1 of HLA-DR3 are better than that of the sequential approach having an (cRMSD) of 2.17 Å compared to 2.51 Å of the sequential procedure but worse than that of the simultaneous approach ((cRMSD)=1.09Å). The reason is that in the sequential approach the error from the first determined amino acid conformation is accumulated as its conformation affects greatly the conformation of the other sequentially considered amino acids.

7 Prediction of I-E^k Binding Sites

Pocket 1 of I-E^k, requires three substitutions, that is, $\beta 85$: Val → Ile, $\beta 86$: Gly → Phe, and $\beta 90$: Thr → Leu. The predicted pocket is illustrated in Figure 8 together with the crystallographic data of I-E^k [22]. The (cRMSD) difference based on all atoms deviations is 1.67 and corresponds to 9.2% relative predictive error. The (cRMSD) differences for the individual substituted residues are 2.45, 3.36, and 1.76 Å, for Ile, Phe and Leu, respectively.

For pocket 4 of I-E^k there are four substitutions needed as shown in Table 2 ($\beta 13$: Phe to Ser, $\beta 74$: Ala to Glu, $\beta 78$: Tyr to Val, and β:71 Arg to Lys). The predicted pocket is illustrated in Figure 9 superpositioned with the crystallographic data of pocket 4 of I-E^k. The (cRMSD) difference is 1.58Å, which corresponds to 3.49% predictive error. For the individual substituted residues the (cRMSD) differences are 0.78, 1.35, 2.88, and 1.61 Å, for Ser, Glu, Val, and Lys, respectively, and correspond to relative predictive errors of 1.59%, 2.16%, 4.48%, and 2.03%.

For pocket 6 of I-E^k three substitutions are required at the positions $\beta 11$: Leu → Ser, $\beta 13$: Phe → Cys, and $\beta 71$: Arg → Lys. The pocket predicted by the proposed methodology, together with the crystallographic data for pocket 6 of I-E^k, is shown in Figure 10. The (cRMSD) difference is 1.28 Å based on all atoms, which corresponds to 5.19% relative predictive error. For the individual substituted residues, the (cRMSD) differences are 1.89, 2.67, and 1.64 for Ser, Cys, and Lys, respectively. These differences correspond to 4.41%,

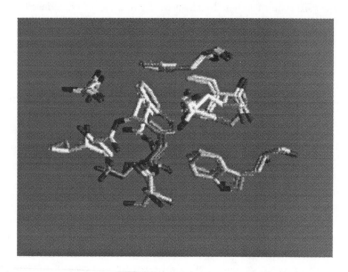

Figure 8: **Superposition of the predicted Pocket 1 of I-E^k versus crystallographic data**

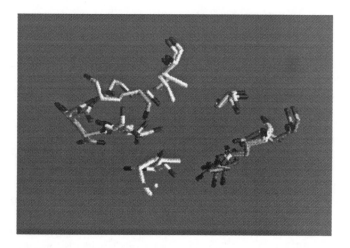

Figure 9: **Superposition of the predicted Pocket 4 of I-E^k versus crystallographic data**

14.06% and 2.82% relative predictive error.

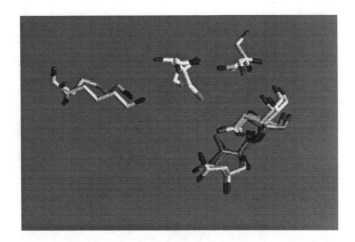

Figure 10: **Superposition of the predicted Pocket 6 of I-E^k versus crystallographic data**

Pocket 7 of I-E^k requires four substitutions as shown in Table 2 (β28: Glu to Val, β47: Tyr to Phe, β67: Leu to Phe, and β71: Arg to Lys). Figure 11 shows the predicted pocket 7, along with the crystallographic data for this pocket. The (cRMSD) difference is 2.03 Å and corresponds to 4.33% relative predictive deviation. For the individual residues the (cRMSD) differences are 2.89, 2.15, 2.20 and 3.23 Å for Val, Pheβ47, Pheβ67, and Lys, respectively, and correspond to 3.95%, 3.1%, 5.28% and 4.41% relative predictive deviation.

Finally, pocket 9 of I-E^k feature three substitutions at positions α72: Ile to Val, β9: Trp to Glu, and β60: Tyr to Asn. The resulting pocket 9 together with the crystallographic data of I-E^k is shown in Figure 12. The (cRMSD) difference is 1.35 Å that corresponds to 23.3% relative predictive deviation. For the individual residues the (cRMSD) differences are 1.56, 2.46, and 1.56 Å for Val, Glu, and Asn, respectively. Note, that the larger relative predictive deviation for this pocket is mainly due to the large relative error for Val at position α72, and the large deviation between the crystals HLA-DR1 and HLA-DR3 which gives a cRMSD of 1.09Å and a 21.4% relative deviation. The results for all the pockets are summarized in Table 5.

In order to study the effect of considering different bounds on N and C' coordinates, the proposed approach was applied to all the pockets for ±0.5 and ±0.3 Å bounds around the coordinates of the corresponding atoms of HLA-DR1 molecule. The results are shown in Table 6.

Note that, as was found from the crystallographic data of the I-E^k molecule binding with different peptides (i.e., a peptide derived from murine hemoglobin Hb(64-76), or a peptide from murine heat shock protein 70 Hsp(236-248)), there is some inherent variability in the range of 0.01-0.4 Å (for N and C' coordinates). These differences correspond to pocket flexibility to accommodate different peptides.

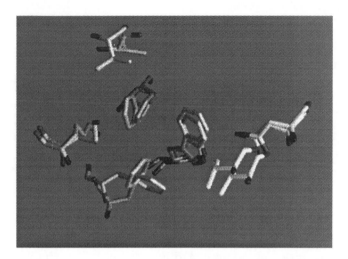

Figure 11: **Superposition of the predicted Pocket 7 of I-E^k versus crystallographic data**

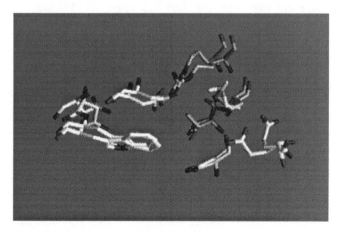

Figure 12: **Superposition of the predicted Pocket 9 of I-E^k versus crystallographic data**

Pocket	Pocket		Substituted Residues		
	(cRMSD) (Å)		(Relative-cRMSD) (%)	(cRMSD) (Å)	
	All atoms	C^α	All atoms		
1	1.67	0.47	9.2	Ile: 2.45 Phe: 3.36 Leu: 1.76	
4	1.58	0.83	3.5	Ser: 0.78 Glu: 1.35 Val: 2.88 Lys: 1.61	
6	1.28	0.65	5.2	Ser: 1.89 Cys: 2.67 Lys: 1.64	
7	2.03	0.93	4.3	Val: 2.89 Phe: 2.15 Phe: 2.20 Lys: 3.23	
9	1.35	0.63	23.3	Val: 1.56 Glu: 2.46 Asn: 1.56	

Table 5: Results for I-E^k prediction

Pocket	Bounds	(cRMSD) (Å)
1	± 0.5	2.26
	± 0.3	1.67
4	± 0.5	1.81
	± 0.3	1.58
6	± 0.5	1.28
	± 0.3	1.44
7	± 0.5	3.17
	± 0.3	2.41
9	± 0.5	1.84
	± 0.3	1.77

Table 6: Effect of different bounds on N and C' coordinates (I-E^k)

The obtained (cRMSD) for all predicted pockets illustrate the good agreement with the crystallographic data considering that there is an inherent difference between the crystals, as shown in Table 3. The (cRMSD) differences shown in Table 3 represent the differences in the common atoms of the pockets of HLA-DR1 and HLA-DR3 crystals, as well as the differences between HLA-DR1 and I-E^k crystals. These (cRMSD) differences serve as a reference point in the evaluation of the predicted pockets. For instance, for pocket 1 of HLA-DR3, the predicted structure via the proposed approach has an (cRMSD) difference of 1.09 Å while the crystallographic data of pocket 1 for HLA-DR1 and pocket 1 of HLA-DR3 exhibit an (cRMSD) of 1.03 Å among their common atoms. Comparing the results shown in Tables 3 and 4, 5, it is evident that the predicted structures are close to their reference points, even for pocket 9.

8   Implications for Binding Studies

8.1   Binding affinity evaluation

The structure prediction of the binding sites of MHC class II molecules has significant implications for the evaluation of peptide binding to the HLA type II molecules. The methodology proposed by Androulakis et al. [28] for the evaluation of the binding affinity of pocket 1 of HLA-DRB1 molecule is applied to the predicted pockets and compared to the results for the pockets obtained from the crystallographic data.

The basic idea of the method proposed by Androulakis et al. [28] is the determination of the conformation of the binding complex that corresponds to the global minimum of the interaction energy. An energetic-based criterion was introduced for the evaluation of the energy interaction between a given pocket and each naturally occurring amino acid. This measure, which is denoted as ΔE, corresponds to the difference between (i) the global minimum total potential energy that considers all the energetic atom-to-atom interactions, classified as inter-interactions between the atoms of the residues that define the pocket of HLA-DR1 protein and the atoms of the considered naturally occurring amino acid, and intra-interactions between the atoms of the considered naturally occurring amino acid, and (ii) the global minimum potential energy of the considered naturally occurring amino acid when it is far away from the pocket:

$$\Delta E = E^o_{Total} - E^o_{Res} \qquad (13)$$

where E^0_{Total} is the global minimum of the potential energy of the complex of the pocket and the binding peptide, and E^o_{Res} is the global minimum of the potential energy of the peptide away from the pocket. Note that ΔE does not represent a difference in the free energies of the complex and isolated aminoacids. Instead, it denotes the difference between potential energy plus solvation for the complex and the isolated aminoacids.

8.2   Binding studies for the HLA-DR3 molecule

This methodology was applied to the predicted pocket 1 of HLA-DR3 and to the pocket obtained from the crystallographic data [59] for the amino acids Phe, Ile, and Met. Based on the energy differences it is found that Phe is a better binder than Met by 1.1 kcal/mol which is better than Ile by 3.9 kcal/mol for the predicted pocket. For the pocket 1 based on the crystallographic data, the binding studies [28] determine the same sequence (i.e., Phe followed by Met and by Ile) with corresponding differences of 2.37 kcal/mol for Phe to Met and 2.06 for Met to Ile.

Application of the predictive binding approach [28] to the predicted, as well as to the crystallographicaly obtained, pocket 4 of HLA-DR3 for the amino-acids Asp, Glu, Ile and Phe shows that the negatively charged Asp and Glu are very strong binders. In contrast, Ile and Phe are weaker binders than Asp and Glu.

8.3 Binding studies for the I-E^k molecule

The aforementioned predictive binding approach was also applied to the predicted pocket 1 of I-E^k for the amino acids Ile, Val, and Phe. The results show that Phe is a better binder than Ile with an energy difference of 6.1 Kcal/mol, and Ile binds better than Val, with an energy difference of 2.8 kcal/mol. Similar results are obtained from the crystallographic data, and indicate that Phe is a better binder than Ile, and Ile is a a better binder than Val with energy differences of 4.4 and 0.7 kcal/mol, respectively.

8.4 Binding studies for the HLA-DR1 molecule

In this section, in order to further verify the correct prediction of the binding sites of HLA type II molecules, the crystal of HLA-DR3 molecule is used [22] for the prediction of the pocket 1 of HLA-DR1 molecule. The results obtained utilizing the predicted pocket were then compared to those found from the crystallographicaly obtained pocket [59]. As shown in Table 7, the binding studies using the predicted pocket illustrate the same trends as the binding studies of the crystallographic pocket.

The results for both the predicted and crystallographic pocket 1 with R=5.0 Å illustrate that Tyr, Phe and Trp are ranked at the top tier. At lower positions there are the Leu, Ile and Val, whereas at the bottom of the list finally are the negative charged residues Glu- and Asp-. Note that the differences in the energy interactions are very small ranging between [0,1.58].

Therefore, the proposed approach in this paper not only predicts the structure of the binding sites of HLA-II molecules, but also provides consistent results between the binding studies of individual amino acids and binding studies with sites based on the crystallographic data.

9 Conclusions

A novel and powerful theoretical approach is proposed for the structure prediction of the binding sites of MHC class II molecules based on the crystallographic data of HLA-DR1 [59]. This approach couples the modeling of energetic interactions and deterministic global optimization approaches and can predict the pockets of HLA-DR3 and I-E^k with small (rms) differences. Furthermore, application of our recently proposed predictive approach for binding studies [28] to pocket 1 of HLA-DR3, pocket 1 of I-E^k, and pocket 4 of HLA-DR3 demonstrates that the predicted rank ordered list of binders, for both the predicted

Residue	Δ E Crystal (kcal/mol)	Δ E Prediction (kcal/mol)	Difference (kcal/mol)	Difference %
Tyr	-20.000	-18.850	-1.15	5.75
Phe	-19.625	-18.040	-1.58	2.95
Trp	-16.950	-17.754	0.80	4.72
Gln	-15.396	-15.916	0.52	3.37
Met	-13.943	-13.928	-0.02	0.14
Asn	-13.784	-14.644	0.86	6.24
Thr	-13.297	-13.297	0.00	0.00
Leu	-12.481	-12.399	-0.08	0.64
Ile	-12.465	-12.486	0.02	0.16
Ser	-11.557	-11.187	-0.37	3.20
Cys	-11.280	-11.087	-0.19	1.68
Val	-11.209	-11.324	0.12	1.07
Ala	-10.355	-10.338	-0.02	0.19
Gly	-10.091	-9.996	-0.09	0.89
Glu-	-7.744	-6.891	-0.85	10.97
Asp-	-2.431	-2.594	0.16	6.58

Table 7: Comparison of Predicted vs Crystallographic Binding Studies in Pocket 1 of HLA-DR1 molecule ($\mathcal{R} = 5.0$ Å)

binding sites, using the approach proposed in this work, and the binding sites based on crystallographic data are in very good agreement.

Acknowledgments
C.A.F. gratefully acknowledges support from the National Science Foundation, and the National Institutes of Health (R01 GM52032). D.S.M. gratefully acknowledges support from the American Diabetes Association.

References

[1] C. S. Adjiman, and C. A. Floudas. Rigorous Convex Underestimators for General Twice-Differentiable Problems. *Jl. Global Opt.*, 9, 23-40, 1996.

[2] C. S. Adjiman, I. P. Androulakis, C. D. Maranas, and C. A. Floudas. A global optimization method, αbb, for process design. *Comp. Chem. Engng.*, 20, 419-418, 1996.

[3] C. S. Adjiman, S. Dallwig, C. A. Floudas and A. Neumaier. Global Optimization Method, αbb, for Twice-Differentiable Constrained NLPs - I Theoretical Advances *Comp. Chem. Engng.*, 22, 1137-1158, 1998.

[4] C. S. Adjiman, I. P. Androulakis, and C. A. Floudas. Global Optimization Method, αbb, for Twice-Differentiable Constrained NLPs - II Implementation and Computational Results *Comp. Chem. Engng.*, 22, 1159-418, 1998.

[5] N.L. Allinger, Y.H. Yuh, and J.-H. Liu. Molecular mechanics. the mm3 force field for hydrocarbons. *J. Am. Chem. Soc.*, 111:8551, 1989.

[6] I. P. Androulakis, C. D. Maranas, and C. A. Floudas. αbb : A global optimization method for general constrained nonconvex problems. *Journal of Global Optimization*, 7:337–363, 1995.

[7] I.P. Androulakis, C.D. Maranas, and C.A. Floudas. Prediction of oligopeptide conformations via deterministic global optimization. *Journal of Global Optimization*, 11:1–34, 1997.

[8] D.J. Bacon and J. Moult. Docking by least-square fitting of molecular surface patterns. *Jl. Mol. Biol.*, 225:849–858, 1992.

[9] T.L. Blundell, B.L. Sibanda, M.J.E. Sternberg, and J.M. Thornton. Knowledge-based prediction of protein structures and the design of novel molecules. *Nature*, 326:347, 1987.

[10] B. Brooks, R. Bruccoleri, B. Olafson, D. States, S. Swaminathan, and M. Karplus. Charm: A program for macromolecular energy minimization and dynamics calculation. *J. Comp. Chem*, 8:132, 1983.

[11] A. Calfisch, P. Niederer, and M. Anliker. Monte carlo docking of oligopeptides to proteins. *Proteins*, 13:223–230, 1992.

[12] R. Chandrasekaran and G.N. Ramachandran. Studies on the conformation of amino acids, xi. analysis of the observed side group conformations in proteins. *Int. J. Protein Res.*, 2:223, 1970.

[13] S.Y. Chung and S. Subbiah. A structural explanation for the twilight zone of protein sequence homology. *Structure*, 4:1123, 1996.

[14] M. L. Connolly. Analytical molecular surface calculations. *J. Appl. Cryst.*, 16:548–558, 1983.

[15] M.L. Connolly. Solvent-accessible surfaces of proteins and nucleic acids. *Science*, 221:709, 1983.

[16] P. Dauber-Osguthorpe, V.A. Roberts, D.J. Osguthorpe, J. Wolff, M. Genest, and A.T. Hagler. Structure and energetics of ligand binding to peptides: Escherichia coli dihydrofolate reductase–trimethoprim, a drug receptor system. *Proteins*, 4:31, 1988.

[17] R. Diamond. On the comparison of conformations using linear and quadratic transformations. *ACTA Cryst.*, 1, 1976.

[18] R.L. Dunbrack and M. Karplus. Backbone-dependent rotamer library for proteins: Application to side-chain prediction. *J. Mol. Biol.*, 230:543, 1993.

[19] C.A. Floudas. Deterministic global optimization in design, control, and computational chemistry. In L.T. Biegler, T.F. Coleman, A.R. Conn, and F.N. Santosa, editors, *Large Scale Optimization with Applications, Part II: Optimal Design and Control*, volume 93, pages 129–184. IMA Volumes in Mathematics and its Applications, Springer–Verlag, 1997.

[20] C.A. Floudas, P.M. Pardalos, C.S. Adjiman, W.R. Esposito, Z. Gumus, S.T Harding, J.L. Klepeis, C.A. Meyer and C.A. Schweiger. Handbook of Test Problems for Local and Global Optimization. Kluwer Academic Publishers, (1999).

[21] D. H. Fremont, W.A. Hendrickson, P. Marrack, and J. Kappler. Structures of an mhc class ii molecule with covalently bound single peptides. *Science*, 272:1001–1004, 1996.

[22] P. Ghosh, M. Amaya, E. Mellins, and D.C. Wiley. The structure of an intermediate in class ii mhc maturation: Clip bound to hla-dr3. *Nature*, 378:457–462, 1995.

[23] D.S. Goodsell and A.J. Olson. Automated docking of substrates to proteins by simulated annealing. *Proteins*, 8:195–202, 1990.

[24] T.N. Hart and R.J. Read. A multiple-start monte-carlo docking method. *Proteins*, 13:206–222, 1992.

[25] L. Holm and C. Sander. Fast and simple monte-carlo algorithm for side-chain optimization in proteins: application to model building by homology. *Proteins: Sruct. Funct. Genet.*, 14:213, 1994.

[26] J.K. Hwang and W.F. Liao. Side-chain prediction by neural networks and simulated annealing optimization. *Protein Eng.*, 8:363, 1995.

[27] I.D.Kuntz, J.M. Blaney, S.J. Oatley, R. Langridge, and T.E. Ferrin. A geometric approach to macromolecule-ligand interactions. *Jl. Mol. Biol.*, 161:269–288, 1982.

[28] I.P.Androulakis, N.N.Nayak, M.G.Ierapetritou, D.S. Monos, and C.A. Floudas. A predictive method for the evaluation of peptide binding in pocket 1 of hla-drb1 via global minimization of energy interactions. *Proteins*, 29:87–102, 1997.

[29] F. Jiang and S.H.Kim. Soft docking: Matching of molecular surface cubes. *Jl. Mol. Biol.*, 219:79–102, 1991.

[30] W. Kabsh. A solution for the best rotation to relate two sets of vectors. *ACTA Cryst.*, page 922, 1976.

[31] W. Kabsh. A discussion of the solution for the best rotation to relate two sets of vectors. *ACTA Cryst.*, page 827, 1978.

[32] J.L. Klepeis, I. P. Androulakis, M. G. Ierapetritou, and C. A. Floudas. Predicting solvated peptide conformations via global minimization of energetic atom-to atom interactions. *Comp. Chem. Engng.*, 22, 765-788, 1998.

[33] J.L. Klepeis, and C. A. Floudas. Free Energy Calculations for Peptides via Deterministic Global Optimization. *Jl. Chem. Phys.*, 110:7491-7512, 1999.

[34] P. Koehl and M. Delarue. Application of a self-consistent mean field theory to predict protein side-chains conformation and estimate their conformational entropy. *J. Mol. Biol.*, 239:249, 1994.

[35] B. Lee and F.M. Richards. The interpretation of protein structures: Estimation of static accessibility. *Jl. Mol. Biol.*, 55:379–400, 1971.

[36] M. Levitt. Protein folding by restrained energy minimization and molecular dynamics. *J. Mol. Biol.*, 170:723, 1983.

[37] A. L. Mackay. The generalized inverse and inverse structure. *ACTA Cryst.*, page 212, 1977.

[38] C. D. Maranas, I. P. Androulakis, and C. A. Floudas. A deterministic global optimization approach for the protein folding problem. In *DIMACS Series in Discrete Mathematics and Theoretical Computer Science*, volume 23, pages 133–150. American Mathematical Society, 1996.

[39] C. D. Maranas and C. A. Floudas. A global optimization approach for lennard-jones microclusters. *J. Chem. Phys.*, 97(10):7667–7677, 1992.

[40] C. D. Maranas and C. A. Floudas. Global optimization for molecular conformation problems. *Annals of Operations Research*, 42:85–117, 1993.

[41] C. D. Maranas and C. A. Floudas. A deterministic global optimization approach for molecular structure determination. *J. Chem. Phys.*, 100(2):1247–1261, 1994.

[42] C. D. Maranas and C. A. Floudas. Global minimum potential energy conformations of small molecules. *Journal of Global Optimization*, 4:135–170, 1994.

[43] C.D. Maranas, I.P. Androulakis, and C.A. Floudas. A deterministic global optimization approach for the protein folding problem. In P.M. Pardalos, D. Shalloway, and G. Xue, editors, *DIMACS Series in Discrete Mathematics and Theoretical Computer Science*, volume 23, pages 133–150. American Mathematical Society, 1995.

[44] C.D. Maranas and C.A. Floudas. Global minimum potential energy conformations of small molecules. *Journal of Global Optimization*, 4:135–170, 1994.

[45] A. D. McLachlan. A mathematical procedure for superimposing atomic coordinates of proteins. *ACTA Cryst.*, page 656, 1972.

[46] A. D. McLachlan. Gene duplications in the structural evolution of chymotrypsin. *J. Mol. Biol.*, 128:49, 1979.

[47] F. A. Momany, L. M. Carruthers, R. F. McGuire, and H. A. Scheraga. Intermolecular potential from crystal data. iii. *J. Phys. Chem.*, 78:1595–1620, 1974.

[48] F. A. Momany, L. M. Carruthers, and H. A. Scheraga. Intermolecular potential from crystal data. iv. *J. Phys. Chem.*, 78:1621–1630, 1974.

[49] F.A. Momany, L.M. Carruthers, R.F. McGuire, and H.A. Scheraga. Energy parameters in polypeptides. vii. geometric parameters, partial atomic charges, nonbonded interactions, hydrogen bond interactions, and intrinsic torsional potentials for the naturally occurring amino acids. *J. Phys. Chem.*, 79:2361, 1975.

[50] D. Monos, A. Soulika, E. Argyris, J. Corga, L. Stern, V. Magafa, P. Cordopatis, I.P. Androulakis, and C.A. Floudas. HLA–Peptide Interactions: Theoretical and Experimental Approaches. *Proceedings of the 12th International Histocompatibility Conference*, Vol 12, 1996.

[51] G. Némethy, K. D. Gibson, K. A. Palmer, C. N. Yoon, G. Paterlini, A. Zagari, S. Rumsey, and H. A. Scheraga. Energy parameters in polypeptides. 10. *J. Phys. Chem.*, 96:6472–6484, 1992.

[52] G. Némethy, M.S. Pottle, and H.A. Scheraga. Energy parameters in polypeptides. 9. updating of geometrical parameters, nonbinded interaction and hydrogen bond interactions for the naturally occurring amino acids. *J. Phys. Chem.*, 89:1883, 1983.

[53] G. Perrot, B. Cheng, K. D. Gibson, K. A. Palmer J. Vila, A. Nayeem, B. Maigret, and H. A. Scheraga. Mseed: A program for the rapid analytical determination of accessible surface areas and their derivatives. *J. Comp. Chem*, 13:1–11, 1992.

[54] S. T. Rao and M.G. Rossmann. Comparison of Super-Secondary Structures in Proteins. *J. Mol. Biol.*, 76:241, 1973.

[55] S.J. Remington and B.W. Matthews. General Method to assess similarity of protein structures, with applications to T4-Bacteriophage Lysozyme *Proc. Nat. Acad. Sci. USA*, 75:2180, 1978.

[56] H. Schauber, F. Eisenhaber, and P. Argos. Rotamers: to be or not to be? an analysis of amino acid side-chain conformations in globular proteins. *J. Mol. Biol.*, 230:592, 1993.

[57] H.A. Scheraga. *ECEPP/3 USER GUIDE*. Cornell University Department of Chemistry, January 1993.

[58] H.A. Scheraga. *PACK: Programs for Packing Polypeptide Chains*, 1996. online documentation.

[59] L. Stern, J. Brown, T. Jardetzky, J. Gorga, R. Urban, L. Strominger, and D. Wiley. Crystal structure of the human class ii mhc protein hla-dr1 complexes with an influenza virus peptide. *Nature*, 368:215–221, 1994.

[60] M.J. Sutcliffe, I. Haneef, D. Carney, and T.L. Blundell. Knowledge-based modeling of homologous proteins, part i: three dimensional frameworks derived from the simultaneous superposition of multiple structures. *Protein Eng.*, 1:377, 1987.

[61] P. Tuffery, C. Etchebest, S. Hazout, and R. Lavery. A new approach to the rabid determination of protein side-chain conformations. *J. Biomol. Struct. Dynam.*, 8:1267, 1991.

[62] W. F. van Gunsteren and H. J. C. Berendsen. *GROMOS*. Groningen Molecular Simulation, Groningen, The Netherlands, 1987.

[63] M. Vasquez. An evaluation of discrete and continuous search techniques for conformational analysis of side-chains in proteins. *Biopolymers*, 36:53, 1995.

[64] M. Vásquez, G. Némethy, and H. A. Scheraga. Conformational energy calculations on polypeptides and proteins. *Chemical Reviews*, 94:2183–2239, 1994.

[65] J. Vila, R.L. Williams, M. Vasquez, and H.A. Scheraga. Empirical solvation models can be used to differentiate native from non-native conformations of bovine pancreatic trypsin inhibitor. *Proteins*, pages 199–218, 1991.

[66] S. Weiner, P. Kollmann, D.A. Case, U.C. Singh, C. Ghio, G. Alagona, S. Profeta, and P. Weiner. A new force field for molecular mechanical simulation of nucleic acids and proteins. *J. Am. Chem. Soc.*, 106:765, 1984.

[67] S. Weiner, P. Kollmann, D. Nguyen, and D. Case. An all atom force field for simulations of proteins and nucleic acids. *J. Comp. Chem.*, 7:230, 1986.

Optimization in Computational Chemistry and Molecular Biology, pp. 191-207
C. A. Floudas and P. M. Pardalos, Editors
©2000 Kluwer Academic Publishers

A Coupled Scanning and Optimization Scheme for Analyzing Molecular Interactions

Julie C. Mitchell
San Diego Supercomputer Center
University of California, San Diego
9500 Gilman Dr.
San Diego, CA 92093-0537 USA
mitchell@sdsc.edu

Andrew T. Phillips
Department of Computer Science
131 Phillips Science Hall
University of Wisconsin, Eau Claire
Eau Claire, WI 54702-4004 USA
phillips@cs.uwec.edu

J. Ben Rosen
Department of Computer Science and Engineering
University of California, San Diego
9500 Gilman Dr.
San Diego, CA 92093-0114 USA
jbrosen@ucsd.edu

Lynn F. Ten Eyck
San Diego Supercomputer Center
University of California, San Diego
9500 Gilman Dr.
San Diego, CA 92093-0505 USA
lteneyck@sdsc.edu

The authors wish to thank the National Science Foundation (NSF DBI 9616115 & 9996165), the Department of Energy (DOE/OER DE-FG03-096ER62262), Proctor & Gamble, the La Jolla Interfaces in Science and Burroughs Wellcome Fund for helping to support our research.

Abstract

The past decade has brought major advances in the quality and variety of methods for computerized drug design and molecular docking, making the area ripe for the implementation of hybrid algorithms. Hybrid methods create improved algorithms from existing ones by mixing techniques in a way that maximizes advantages and minimizes disadvantages. Here, we outline a hybrid method for molecular docking which couples the rapid-scanning algorithm DOT with the global optimization algorithm CGU.

Keywords: Molecular docking, hybrid algorithm, global optimization, local optimization, fast Fourier transforms.

1   Introduction

Computer technology has revolutionized the way drugs are designed and analyzed. Not only do reliable methods for docking prediction exist, but there are a multitude of good techniques available (cf: [1, 6, 7, 8, 11, 13, 14, 17, 19, 20, 22, 26, 29, 35, 37, 38].) The CASP2 docking assessment compared a number of distinct techniques for studying protein interactions [4]. The quality of submitted structures was quite good on the whole, and no method stood out as being exceptional in comparison with the others.

Accuracy and speed are both important measures of quality in computer algorithms. Each can usually be given an objective formulation, but determining the "best" method is largely subjective. One often finds speed and accuracy to be a tradeoff in algorithms, and while some will wait years for the best answer possible, others are happier with a less optimal solution obtained quickly. Those methods expected to be fastest or most accurate can also vary greatly according to the docking problem, and no well-established set of benchmarks exists for making general comparisons.

The ideal computerized docking program would incorporate all the strengths of existing algorithms and avoid their weaknesses. While it might seem a fantasy, in fact this goal is quite realistic. Moreover, it can be attained quickly and easily from suitable combinations of existing techniques. When different methods are used cooperatively, the result can be an improvement over any used individually. The following gives an outline for a hybrid technique which couples the rapid scanning algorithm DOT with the global optimization method CGU. A general description of the method and its application to the docking of acetylcholine into the acetylcholinesterase-fasciculin complex may be found in [23]. Here, we have attempted to give a more thorough account of the underlying methodology.

2   Hybridization Techniques

A hybrid algorithm attempts to combine existing approaches into a new technique. By balancing the advantages and disadvantages of individual methods, a hybridization can reassemble them in a way that is more optimal. It is not surprising that this cross-fertilization of code has been put to good used in genetic and evolutionary algorithms [15, 18]. The basic underlying principle is, of course, much older than the computer and has a variety of creative uses. The following table illustrates several strengths and weaknesses for the two classes of algorithm used in our scanning and optimization hybrid. A review of the two columns suggests a strong complementarity. That is, the weaknesses of each approach are approximately balanced by strengths in the other.

	Rapid Scanning	**Global Optimization**
Strengths	Fast energy evaluation	Able to test any configuration
	Can use grid-based potentials	Energy to arbitrary precision
	Detailed energy profile	Answer is precise
Weaknesses	Grid-based restrictions	Expensive energy evaluation
	Energy is approximate	Requires a smooth potential
	Many results to analyze	Frugal landscape analysis

Rapid scanning algorithms are notable for their ability to quickly produce a comprehensive sketch of the energy landscape. They are also able to model electrostatic interactions using numerically-generated potentials, such as solutions to the Poisson-Boltzmann equation. However, their fast computational methods have the effect of introducing small errors in energy computations as well as grid-based restrictions on docked complexes. As a result, these methods are able to return an excellent sketch of low-energy basins, but are less able to identify an optimal docking configuration. Global optimization is, in contrast, designed to produce a single, optimal result. Global optimizers are able to compute energy terms and configurations to whatever precision is allowed by the interaction model. However, this precision typically requires smooth, explicit formulas for potential functions and a detailed computation of energy terms. Many optimization techniques are able to return some information on the global energy profile in addition to an optimal solution. To be efficient, though, optimization methods must be adept at hunting for the optimum without producing a detailed analysis of the energy landscape in the process.

In balancing the relative strengths of rapid scanning and global optimization, we have developed a hybrid or "coupled" optimization technique for molecular docking. Coupled is synonymous with united and joined, which seemed a most apt description of this particular method. Our approach combines the rapid scanning algorithm DOT [21, 35, 36] with the global optimization technique CGU [3, 27, 30, 28]. The two algorithms have not been blended, but rather linked in a cooperative scheme that is mutually beneficial. Moreover, because CGU and DOT share the advantage of being fully parallelizable, this unification will prove increasingly advantageous in the age of teraflops computing.

3 Rapid Scanning by Convolution

The most basic model for molecular docking poses the problem in terms of six variables. If Molecule A is assumed to be stationary, the position and orientation of Molecule B are determined by three space variables and three angular variables. If $V(x)$ is the scalar electrostatic field generated by Molecule A and ρ is a sum of Dirac delta functions centered at atoms in Molecule B and weighted according to their charges, the electrostatic potential energy is given by

$$E = \int_{R^3} V(x) \cdot \rho(x) dx \qquad (1)$$

The distribution $\rho(x - v)$ represents a shift of Molecule B, which in the six-dimensional configuration space corresponds to varying the spatial degrees of freedom. To specify an arbitrary configuration, we use the notation $\rho_\alpha(x - v)$ where α is Euler angle triple. If α is held fixed and v allowed to vary, we see that the electrostatic potential energy $E_\alpha(v)$ can be expressed as the integral

$$E_\alpha(v) = \int_{R^3} V(x) \cdot \rho_\alpha(x - v)dx \tag{2}$$

This is a type of convolution, called a *correlation*, and the Convolution Theorem for Fourier transforms allows one to rewrite the equality as

$$\hat{E}_\alpha = \hat{V} \cdot \hat{\rho}_\alpha^* \tag{3}$$

where $\hat{E}_\alpha, \hat{V}, \hat{\rho}_\alpha$ denote the Fourier transforms of E_α, V, ρ_α, and $\hat{\rho}_\alpha^*$ is the complex conjugate of $\hat{\rho}_\alpha$. By fixing a rotation of Molecule B, the energy values over all space coordinates are determined by this formula. When V and ρ_α are discretized to lie on a grid, the result will also hold when using fast Fourier transforms.

Convolution docking algorithms are designed to conduct an exhaustive search of the six-dimensional configuration space described above. Computational efficiency is achieved with the use of fast Fourier transforms and the formula (3). By combining this grid-based formulation for the space coordinates with a good rotation sample, it is possible to construct a uniform sample of the entire six-dimensional configuration space. Fast transform methods to perform geometric fitting and potential energy computations were originally proposed in [13, 19]. Currently available programs for convolution docking include DOT [21, 35, 36] as well as GRAMM [38, 39] and FTDOCK [8]. DOT has been developed by author Ten Eyck and collaborators at the Computational Center for Macromolecular Structures (CCMS). Rapid scanning algorithms have been shown to perform well at identifying docked complexes which are closely matched to experiment. At the CASP2 docking assessment, Vakser produced the lowest RMSD docking configuration for the hemagglutinin-antibody complex using GRAMM [39]. This is a large protein-protein docking problem with conformational change, which makes it very expensive to analyze using most docking methods.

Fast transform methods for molecular docking employ similar computational techniques but vary in the energy or "scoring" functions used. The scoring function used by DOT incorporates both geometric fit and electrostatic energy terms. While any electrostatic field can be used with DOT, the program has been designed to work best with a Poisson-Boltzmann model. The methods have been tested with solutions generated by UHBD [2] and DelPhi [16], and both lead to better docked complexes than those obtained using a Coulombic potential [21]. In addition to scanning for configurations having low potential energies, DOT is able to accurately pinpoint regions of space in which the free energy is low. Such information is essential to solving the ultimate docking problem, which is that of determining the pathways and mechanisms by which two molecules come together.

One well-acknowledged disadvantage of convolution algorithms is that correctly docked complexes are returned in a list which includes many "false positive" results. Since these incorrectly docked structures may score higher than the correct ones, determining an optimal configuration often requires considerable analysis based on biochemical information. This phenomenon stems in part from the fact that atoms in one of the two molecules must be

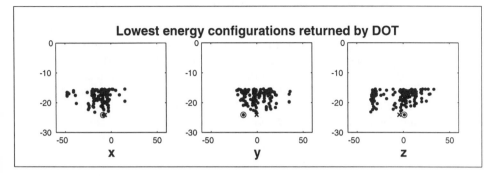

Figure 1: *This figure shows the x, y and z coordinates of DOT's favorable configurations plotted against their energy values. The angular variables of the six-dimensional configuration space have not been profiled, but minima shown have distinct angular components.*

rounded onto grid points in order to utilize fast transform methods. This has the effect of introducing small distortions in the shape fit and electrostatic energy computations. Figure 1 illustrates this by plotting the three spatial coordinates of configurations returned by DOT against their computed energies. The top-ranked solution returned by DOT has been circled, and a more biologically accurate answer is highlighted with an "x." The landscape returned by DOT resembles a bowl centered about the more biologically-precise solution, while the top-ranked answer appears inconsistent with the global energy profile.

Because "soft docking" programs such as DOT can allow close contact between two molecules, correctly approximating the electrostatic interactions can be tricky. One major improvement to DOT's scoring function has come very recently based on the observations of Victoria Roberts, one of the program's developers. She determined that the lower ranking of correctly docked structures was largely attributable to atoms being placed between the molecular boundary and solvent-accessible surface layer. Since the electrostatic field undergoes large changes in this region, computed energy values did not accurately model an induced fit between the molecules. Clamping the potential field to values found at the solvent-accessible surface layer appears to have largely eliminated this problem. A more thorough account of improvements in the quality of the rankings and free-energy computations can be found in [21].

Convolution docking is most efficient at lower resolution. Using a 64^3 grid and $1,800$ rotations, the program can complete its work in 30-45 minutes on a high-end workstation. A 128^3 grid and $1,800$ rotations consumes about 4-6 hours of computing time, and a large run with a 128^3 grid and $54,000$ rotations requires supercomputing machinery and over 100 processor hours. At higher resolution, the computations become too costly and produce far more information than is truly needed. High-resolution structures can be obtained more efficiently by refinement with a local optimization routine. This will be discussed in the next section, and this local refinement will later be used in conjunction with the CGU global optimization technique.

4   Fine-tuning Grid-based Results with Local Optimization

To relax grid-based approximations, a continuous refinement technique has been developed
in which favorable configurations returned by DOT are used as starting points for local
optimization. Refinement of rapidly-scanned results with local optimization has previously
been shown to give more biologically-accurate results [12]. However, DOT uses a numerical
approximation to the electrostatic potential of the stationary molecule, while a continuous
model is needed for local optimization. Explicit formulas for solutions to the Poisson-
Boltzmann equation are generally not known. Yukawa potentials have the form

$$V(r) = Q \cdot \frac{e^{-\lambda r}}{r} \tag{4}$$

and solve a linearized Poisson-Boltzmann equation with constant dielectric. These potential functions are sensitive to ionic concentration and temperature but do not incorporate
dielectric effects. In the region of solvent accessibility they are well-matched to potential
function values obtained from UHBD and DelPhi, and for rigid docking of small ligands
they appear to give highly comparable results.

Potential energy values were computed as the sum of pairwise interactions between
atoms in the stationary and moving molecules. For a pair of atoms separated by a distance
r_{ij} and having charges q_i and Q_j, the contribution to the electrostatic potential energy was
defined to be

$$V_{ij} = 330 \cdot (\ q_i \cdot Q_j \cdot \frac{e^{-\lambda r_{ij}}}{r_{ij}} \ ) \tag{5}$$

The constant 330 is a conversion factor for rewriting all atom potentials in units of kcal/mole,
which are units often used when solving Poisson-Boltzmann equation. The value of λ varies
according the temperature and ionic concentration, but $\lambda \approx 1$ is typical. This electrostatic
potential was combined with a Lennard-Jones potential

$$L_{ij} = \frac{1}{10} \cdot (\ (\frac{r_0}{r_{ij}})^{12} \ - \ 2 \cdot (\frac{r_0}{r_{ij}})^6 \ ) \tag{6}$$

to model steric repulsion and van der Waals interactions. A uniform value of $r_0 = 3.8$
was used regardless of atom types, but future implementations will likely use a more sophisticated interaction model. The constant $\frac{1}{10}$ determines the relative weighting of the
electrostatic and Lennard-Jones terms. This value is based on the typical depth of the van
der Waals attractive well. Numerous examples suggest this weighting gives optimal soft
docking results with DOT.

Local optimization was performed using NPSOL [9], which we have found to be computationally efficient and robust against poles in the potential energy. Because soft docking
simulates induced fit by allowing close contacts between atoms, the switch to continuum
methods requires a bit of care. Sequential Quadratic Programming algorithms, such as
NPSOL, are efficient because they can take larger, hence fewer, steps than gradient descent
techniques. However, large gradients at starting points can make it difficult to ensure that
local minima are close to the initial configurations. Bounding the size of initial minimization steps taken by NPSOL appears to solve this problem in the majority of cases without
sacrificing computational efficiency. The time required to refine solutions is, however, also

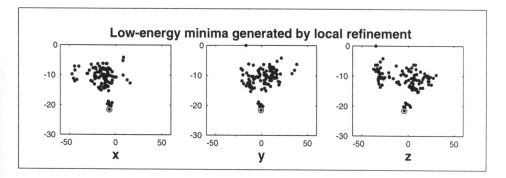

Figure 2: *This illustrates the result of refining DOT's rapid energy profile with local optimization. DOT has returned a softened landscape more likely to have a concentration of low-energy points near the optimal solution. Continuous refinement with "hard core" potentials eliminates false positive solutions and highlights the docking region.*

dependent on the energy model and the size of the molecules. Using the present all-atom model, one can refine about 100 solutions per hour for a small ligand (10-15 atoms) docked to a larger protein (500-600 residues).

Figure 2 illustrates the continuous refinement of favorable configurations returned by DOT. This refinement has the effect of relaxing grid-based restrictions and reranking solutions according to energy values at nearby local minima. While the fine-tuning of solutions does not radically change the docked complexes, its impact on the energy profile is much more dramatic. Landscape features which have been smoothed by soft docking appear more distinguished. The landscape now resembles a funnel rather than a bowl, and the docking site is highlighted by a cluster of low-energy minima. In cases where induced fit is expected between the molecules, the potentials can be softened using analytic techniques such as those implemented in [24, 34, 40]. Local refinement will continue to provide a precise fine-tuning of results, but it will impact the overall shape of the energy landscape less significantly. For rigid and soft docking problems, it is anticipated that refinement of DOT's rapidly-scanned results will often generate the global minimum of the potential energy function. Once the continuous model is revised to allow flexibility in the molecules, finding the global minimum ought to be less likely, although identifying a handful of low-lying local minima seems nearly ensured.

In many cases, the refined results will accurately pinpoint the location of the optimal configuration. However, in cases where a clear "winner" cannot be determined, this refinement can give the first step in global optimization using the CGU method. The next section provides a brief outline of CGU's underestimation technique, and this will be followed with a description of how mutual benefits are achieved by coupling the final stages of a DOT rapid scan and the initial stages of global optimization with CGU. By using DOT to conduct a fast global search and CGU to perform a detailed global optimization, the result is an enhancement compared to either algorithm alone. Moreover, the methods have been combined in a straightforward fashion that produces no wasted effort or unused results.

5   Convex Global Underestimation

Global optimization is a universal topic that has seen contributions from a multitude of distinct scientific fields such as engineering, finance and biophysics. Everyone needs to minimize something: cost, waste, effort, error, energy, etc. The needs of the business world accelerated the development of linear programming techniques, and the needs of bioscience continue to push boundaries in non-convex optimization. Convex Global Underestimation (CGU) is an optimization technique which arose from collaboration between researchers in computer science and biochemistry [3, 27, 30, 28]. Authors Rosen and Phillips developed CGU along with Ken Dill, and the method has been successfully tested on a variety of protein folding models with up to 70 degrees of freedom [3, 27, 30]. In addition to protein folding and docking problems, the algorithm has also been used to optimize parameters in potential energy functions [28]. While CGU has performed well on many types of functions, it was primarily designed to minimize potential energies encountered in protein folding applications. Two properties commonly found in such functions are the presence of many local minima and a single, well-defined primary basin.

The method employed by CGU attempts to trace out the underbelly of the potential energy landscape using local minima. It is an iterative procedure which at each stage generates a collection of local minima starting from random or user-specified seed points. For a problem with n degrees of freedom, the algorithm attempts to find at least $2n + 1$ distinct local minima. These minima are closely and rigorously underestimated by a convex quadratic function, and the search space is trimmed according to where this underestimating function lies below the smallest energy value known. In short, CGU uses local minima to trace out a bowl approximating the shape of the energy basin and then proceeds by redirecting its search near the bottom of this bowl. This trimming of the search space allows the algorithm to converge rapidly to the global minimum. The algorithm terminates when the global minimum of the underestimating function coincides with the position of the best local minimum it has found. This convergence typically occurs within 10 iterations of the underestimation technique.

Constructing the convex underestimator is straightforward. A compact representation for a quadratic function of n variables is given by

$$U(w) = w^t A w + B \cdot w + C \tag{7}$$

where A is an $n \times n$ matrix, B and w are a vectors of length n and C is a constant. When the matrix A has no negative eigenvalues, the function U is said to be *convex*. If $w_1, \ldots, w_k$ are local minimizers of the potential energy function $F(w)$, the function $U(w)$ underestimates all local minima provided

$$F(w_i) - U(w_i) \geq 0 \tag{8}$$

for $i = 1, \ldots, k$. To underestimate these minima as closely as possible, the parameters A, B, C are chosen in such a way that

$$\sum_{i=1}^{k} |F(w_i) - U(w_i)| \tag{9}$$

is minimized. If w_* is the lowest energy minimizer known, the new search space is defined to be the region in which

$$U(w) \leq F(w_*) \tag{10}$$

If two local minima have the same energy value, one is eliminated to ensure that mirror image conformations are not allowed for folding problems. Aside from this, no added checks on structural distinctness are needed by CGU. A formal outline of CGU methodology can be found in [27]. While this method is not guaranteed to find the global minimum, in practice it gives a rapid and accurate means of global optimization for many biomolecular applications.

The effectiveness of the CGU algorithm has been tested on hundreds of small protein folding problems [28] using a simplified H-P model due to Sun, Thomas and Dill [33]. A recent test of 20 problems produced results which are well-matched to general averages. These problems ranged in size from from $n = 4$ to $n = 70$ free variables and used 2 torsion angles per residue. The number of local minima expected for this model varies exponentially in the number of residues. Each problem was repeated 20 times using $8n + 4$ random seeds per iteration. For the above-mentioned model, experience indicates that this seeding rate (which is 4 times the minimum of $2n+1$) is sufficient to ensure a high rate of accuracy while minimizing computational costs. Since per residue potentials are likely to be smoother than all atom potentials, some adjustments may be necessary for optimal results in the latter case. For 13 of the 20 problems, the global minimum was found in 100% of the 20 trials. These problems included the 5 largest ($n = 34, 36, 38, 50, 70$). The remaining 7 problems had success rates of 75% or less. CGU generally returns a low "confidence factor" in cases where the true global minimum has not been found, and it rarely finds non-optimal solutions with any consistency. In only one of the 20 cases examined did the method produce results which were deceptive. For this example, the method returned the same answer in 19 of 20 trials. However, the remaining trial returned a lower energy value than the one found more frequently.

The success of CGU seems to depend almost entirely on the funneling properties of the landscape. When CGU fails, it is because the landscape lacks the qualities that the method was designed to exploit. If a single funnel exists, CGU is able to find global minima with consistency and unparalleled efficiency. However, caution is required in situations where multiple low-lying basins are likely. If the two lowest-energy minima known to CGU are well-separated but have similar energy values, the algorithm will be directed toward points which lie somewhere between them. This is clearly a *good idea* when this leads to a low-energy funnel and a *bad idea* anytime there is a high-energy barrier separating low-energy basins. To use CGU effectively, the domain would need to be separated into parts, each part containing just one distinguished basin. This would combine hierarchical strategies, such as those outlined in [10, 31, 32], with CGU's ability to tunnel to the bottom of bumpy, bowl-like landscapes.

The original local optimization routine used by CGU was recently replaced with the NPSOL optimization package mentioned earlier. This brand of local optimizer requires fewer function evaluations to converge, which in this context is a considerable savings. For protein folding problems, run times for the original version of CGU scaled as about $O(n^4)$. The new version runs faster and with a scaling of about $O(n^3)$ [30]. The speed of CGU at finding global minima depends on several things. For a six-dimensional docking problem,

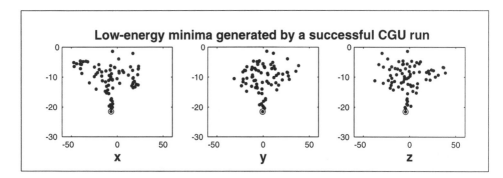

Figure 3: *This figure shows the energy profile for a successful CGU run. The CGU algorithm finds the global energy minimum through an iterative technique able to tunnel to the bottom of rough energy landscapes.*

the key issues are the speed of energy evaluation and local optimization. In protein folding and flexible docking models, the number of free variables is an additional consideration. Using simplified H-P models for folding protein backbones, the program can minimize a function with 16 variables in approximately one minute. For rigid and soft docking, only six variables are needed to model the system. The cost of optimization is then effectively proportional to the cost of energy evaluation, which scales according to the size of the molecules. The all-atom energy model used here is considerably more expensive than the per-residue potentials used in folding protein backbones. Our timings for smaller systems thus range from a few minutes to a few hours.

Since a randomly-seeded CGU run may terminate after 2 iterations in one run and 10 iterations in another, individual run times also vary according to the choice of seed points. We will now see why a seeding scheme which incorporates the results of a DOT scan can serve to minimize the number of iterations required by CGU to locate the global optimum. In addition to speeding the global optimization process, this addition is also likely to give improvements to CGU's success rate at finding optimal docking configurations.

6   Coupled Scanning and Global Optimization

It is clear that the complexity of optimization for a protein folding problem varies according to the size of the protein, and the number of free variables used to model a problem is typically a linear function of the protein's chain length. It is commonly thought that the number of energy minima in a folding problem is roughly proportional to the exponential of the protein's chain length, which suggests a similar relationship with the number of free variables used in the folding model. As mentioned earlier, CGU has shown a consistently high success rate at producing optimally-folded structures in this context. For protein folding problems with n free variables, the use of $8n$ seed points per CGU iteration gives consistent, reliable results. Since the number of local minima which must be found by CGU increases only *linearly* in the number of free variables while the total number of minima varies *exponentially*, the CGU algorithm gives an undeniably efficient means of folding proteins.

Energy landscapes for protein folding and docking problems share many commonalities, but they differ in one key respect. The basic interaction between a pair of molecules can be modeled using only six degrees of freedom, but the expected number of local minima remains dependent on the molecules. It is too early to determine how this will affect the success of CGU, but it seems logical that using more seed points per CGU iteration will be required in some cases. In a handful of preliminary trials which used 50-100 random seeds per CGU iteration, the algorithm was able to accurately locate global minima. The consistency in finding optimal solutions was good but slightly lower on the average than for protein folding problems. The lowest rate seen was 40% for docking acetylcholine into the acetylcholinesterase-fasciculin complex [23]. This case, however, is an exceptionally difficult one in which the toxin fasciculin inhibits the interaction between acetylcholine and acetylcholinesterase.

An alternative approach to using random seed points is to use *directed seeding* techniques able to produce *local minima of the greatest value* to CGU. Finding low-lying minima in the primary basin speeds the CGU optimization process, and these minima provide low energy landmarks which reduce the chance of eliminating the global optimum during CGU's successive trimming of the search space. Figures 3-4 illustrate the results of a successful and unsuccessful CGU run. The lowest-energy minima are concentrated in a very compact region of space, a situation which can occur when docking pathways are narrow.

When no active site is known, the space coordinates must encompass the stationary molecule and leave room for the docking one. The example displayed has a domain which is 128 Å on each side, while the region in which the lowest energy values are seen can be enclosed in a 10 Å box. By picking a point at random, one finds this region with a mere .0005 probability. Since a good initial choice for the angular variables is also required to generate the lowest energy minima, the chance of finding a good clue with 100 random seeds is grim. CGU will direct its search toward the low-energy region, increasing its odds of finding good local minima as it progresses. However, if the random sample taken in the initial iteration is too sparse, the first underestimator may lead CGU away from the global minimum instead of toward it. Later underestimators take into account all known minima and thus give a more comprehensive picture. While better samples could be generated with added random seeds, more benefit may be had by incorporating techniques that increase the odds of finding local minima in the primary basin.

The DOT results shown in Figure 1 are the top configurations taken from a scan of over 3.7 billion. Of these, 1.8 million had space coordinates in the low-energy region described above, although the vast majority of these were eliminated due to incorrect values for the angular variables. As may be seen in Figure 2, refinement of top DOT solutions produced a collection of eight minima with very low energies, one of which was globally optimal. A comparison of Figures 2-3 shows that the energy profile generated by refinement of top solutions from DOT bears a strong resemblance to the profile generated in a successful CGU run. Even if the global minimum had not been produced by refinement, the remaining local minima would have been enough to prevent CGU from avoiding the narrow docking pathway during the run shown in Figure 4. Moreover, using a healthy sample of low-lying minima in the first iteration is also apt to reduce the number of CGU iterations needed for convergence.

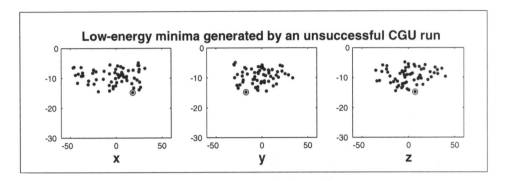

Figure 4: *This figure shows the profile of an unsuccessful CGU run. The program was able to sketch out the upper portion of the primary basin, but did not find the narrow region in which the lowest energy minima occur.*

The effects of directed seeding are clearly dependent on the seed points. If the collection of seeds produces a global minimum in the first iteration, the CGU algorithm will terminate at the second iteration with the correct answer. This would be the case when using the DOT-generated collection of seeds displayed in Figure 1. If the seed which gives the global minimum is removed, the algorithm consistently produces the global minimum in three iterations. For the given example, this behaviour appears to be typical provided at least four low-lying minima are present. With fewer low-lying minima, one sees continued improvements in success rate, but speed increases are less dramatic. Future work will examine the effect of coarsening the grid used in DOT's rapid scan. A coarse DOT run terminates very quickly, and it is apt to produce low-lying minima but not the global optimum. In general, it is expected that the size of DOT's scan and the length of CGU's optimization run can be balanced in an optimal way. We anticipate that a most efficient coupling of DOT and CGU should depend somewhat on molecule sizes and the complexity of the interaction model.

As noted in the section on local refinement, optimization times are highly-dependent on the expense of energy evaluation. Using an all-atom energy model, local optimization for docking small ligands to larger proteins can produce about 100 minima per hour. If 50 seed points are used for each iteration of CGU, then the cost per iteration is about 30 minutes. For the docking trials performed to date, the randomly-seeded CGU runs have required an average of six iterations, hence 3 hours of computing time. In contrast, runs coupled with DOT to perform smart seeding during the first iteration always converged in 2-3 iterations of the CGU method and had a nearly perfect rate of success. The two-way benefits of coupling DOT with CGU should now be evident. We see that DOT offers CGU increased speed and success at determining global energy minimizers, while CGU provides DOT with an enhancement to the precision and optimality of its docking results.

7 Concluding Remarks

The previous sections have served to illustrate the advantages of coupled optimization with CGU and DOT. Not only does this hybrid method offer advantages over either single method, but the technique has many extensions. Because CGU is not constrained to solving a six-dimensional problem, the optimization techniques have the ability to tackle flexible docking problems. Since the "softness" of DOT's docking model can be varied according to the problem, use of directed seeding should continue to give significant speed and accuracy improvements over random seeding schemes.

One important addition to future versions of this coupled optimization technique will be a more efficient and unified energy model. At present, the continuum computations use an expensive all atom model. This is acceptable for docking small ligands, but would be extremely slow for studying protein-protein interactions. Lennard-Jones potentials can be smoothed [24, 34, 40] to model induced fit in the continuum computations, while fast multipole [5, 41] or spline approximations [25] can be used for computational efficiency. The modeling of electrostatic interactions will be done using a new PDE solver able to approximate solutions to the non-linear Poisson-Boltzmann equation using local basis function expansions.

Finally, additional docking methods can be incorporated into this coupling scheme. The collection of results produced by CGU and DOT will contain far more than a single docked complex. DOT is capable of tracing out low-energy basins and pathways, while local refinement and global optimization give a collection of local minima leading to the optimal solution. Such a body of information is extremely valuable to dynamics-based studies. This suggests that the problem solving capabilities of this coupled optimization technique can be readily extended through the use of existing molecular dynamics software.

References

[1] Caflisch, A., Niederer, P. and Anliker, M. **(1992)**, "Monte Carlo docking of oligopeptides to proteins," *Proteins Vol. 13*, 223-30.

[2] Davis, M.E., Madura, J.D., Luty, B.A. and McCammon, J.A. **(1991)**, "Electrostatics and diffusion of molecules in solution: Simulations with the University of Houston Brownian Dynamics program," *Computer Physics Communications, Vol. 62*, 187-97.

[3] Dill, K.A., Phillips, A.T. and Rosen, J.B. **(1997)**, "Protein structure and energy landscape dependence on sequence using a continuous energy function," *Journal of Computational Biology, Vol. 3, No. 4.* 227-39.

[4] Dixon, J.S. **(1997)**, "Evaluation of the CASP2 docking section," *Proteins: Structure, Function and Genetics, Vol. 29, No. S1*, 198-204.

[5] Elliot, W.D. and Board, J.A. **(1994)**, *Fast multipole algorithm for the Lennard-Jones potential*, Technical Report 94-005: Department of Electrical Engineering, Duke University.

[6] Fischer, D., Lin, S.L., Wolfson, H.L. and Nussinov, R. **(1995)**, "A geometry-based suite of molecular docking processes," *Journal of Molecular Biology, Vol. 248, No. 2*, 459-77.

[7] Friedman, A.R., Roberts, V.A. and Tainer, J.A. **(1994)**, " Predicting molecular interactions and inducible complementarity: fragment docking of Fab-peptide complexes," *Proteins, Vol. 20, No. 1*, 15-24.

[8] Gabb, H.A., Jackson, R.M. and Sternberg, M.J. **(1997)**, "Modeling protein docking using shape complementarity, electrostatics and biochemical information," *Journal of Molecular Biology, Vol. 272*, 106-20.

[9] Gill, P.E., Murray, W., Saunders, M.A. and Wright, M.H. **(1986)**, *User's guide for NPSOL (Version 4.0): a Fortran package for nonlinear programming*, Technical Report: Department of Operations Research, Stanford University.

[10] Given, J.A. and Gilson, M.K. **(1998)**, "A hierarchical method for generating low-energy conformers of a protein-ligand complex," *Proteins: Structure, Function and Genetics, Vol. 33, No. 4*, 475-95.

[11] Goodsell, D.S. and Olson, A.J. **(1990)**, "Automated Docking of Substrates to Proteins by Simulated Annealing," *Proteins: Structure, Function and Genetics, Vol. 8, No. 3*, 195-202.

[12] Gschwend, D.A. and Kuntz, I.D. **(1996)**, "Orientational sampling and rigid-body minimization in molecular docking revisited: on-the-fly optimization and degeneracy removal," *Journal of Computer Aided Molecular Design, Vol. 10, No. 2*, 123-32.

[13] Harrison, R.W., Kourinov, I.V. and Andrews, L.C. **(1994)**, "The Fourier-Green's function and the rapid evaluation of molecular potentials," *Protein Engineering Vol. 7, No. 3*, 359-69.

[14] Hart, T.N. and Read, R.J. (1992), "A multiple-start Monte Carlo docking method," *Proteins, Vol. 13*, 206-22.

[15] Hart, W.E., Rosin, C., Belew, R.K. and Morris, G.M. (2000), "Improved evolutionary hybrids for flexible ligand docking in AutoDock," *Optimization in Computational Chemistry and Molecular Biology*. Floudas, C.A. and Pardalos, P.M., Eds. Kluwer

[16] Honig, B. and Nicholls, A. (1995), "Classical electrostatics in biology and chemistry," *Science, Vol. 268*, 1144-9.

[17] Jones, G., Willett, P., Glen, R.C., Leach, A.R. and Taylor, R. (1997), "Development and validation of a genetic algorithm for flexible docking," *Journal of Molecular Biology, Vol. 267, No. 3*, 727-48.

[18] Judson, R., Colvin, M., Meza, J., Huffer, A. and Gutierrez, D. (1992), "Do intelligent configuration search techniques outperform random search for large molecules?" *International Journal of Quantum Chemistry, Vol. 44, No. 2*, 277-90.

[19] Katchalski-Katzir, E., Shariv, I., Eisenstein, M., Friesem, A.A., Aflalo, C. and Vakser, I.A. (1992), "Molecular surface recognition: determination of geometric fit between proteins and their ligands by correlation techniques," *Proceedings of the National Academy of Sciences of the United States of America, Vol. 89, No. 6*, 2195-9.

[20] Klepeis, J.L., Ierapetritou, M.G. and Floudas, C.A. (1998), "Protein folding and peptide docking: A molecular modeling and global optimization approach," *Computers and Chemical Engineering, Vol. 22*, S3-S10.

[21] Mandell, J.G., Roberts, V.A., Pique, M.E., Kotlovyi, V., Nelson, E., Tsigelny, I. and Ten Eyck, L.F. (1999), "Rapid and Comprehensive Protein Docking by Fourier-Domain Methods," *In preparation. Publication information available at* http://www.sdsc.edu/CCMS/DOT.

[22] Meng, E.C., Shoichet, B.K., and Kuntz, I.D. (1992), "Automated docking with grid-based energy evaluation," *Journal of Computational Chemistry, Vol. 13*, 505-24.

[23] Mitchell, J.C., Phillips, A.T., Rosen, J.B., and Ten Eyck, L.F. (1999), "Coupled optimization in protein docking," *Proceedings of the Third Annual International Conference on Computational Molecular Biology (RECOMB99)*. ACM Press, New York.

[24] Moré, J. and Wu, Z. (1996), "Smoothing techniques for macromolecular global optimization," *Nonlinear Optimization and Applications*. Di Pillo, G. and Giannessi, F., Eds. Plenum Press, New York. pp. 297-312.

[25] Oberlin, D. and Scheraga, H. (1998), "B-spline method for energy minimization in grid-based molecular mechanics simulations," *Journal of Computational Chemistry, Vol. 19, No. 1*, 71-85.

[26] Oshiro, C.M., Kuntz, I.D. and Dixon, J.S. (1995), "Flexible ligand docking using a genetic algorithm," *Journal of Computer Aided Molecular Design, Vol. 9, No. 2*, 113-30.

[27] Phillips, A.T.,Rosen, J.B. and Walke, V.H. **(1995)**, "Molecular structure determination by convex global underestimation of local energy minima," *DIMACS Series in Discrete Mathematics and Theoretical Computer Science, Vol. 23, Global Minimization of Nonconvex Energy Functions: Molecular Conformation and Protein Folding.* Pardalos, P.M., Shalloway, D. and Xue, G., Eds.. American Mathematical Society, Providence, RI. pp. 181-98.

[28] Phillips, A.T., Rosen, J.B. and Dill, K.A. **(2000)**, "Energy landscape projections of molecular potential functions," *Optimization in Computational Chemistry and Molecular Biology.* Floudas, C.A. and Pardalos, P.M., Eds. Kluwer Academic Publishers B.V.

[29] Rarey, M., Kramer, B., Lengauer, T. and Klebe, G. **(1996)**, "A fast flexible docking method using an incremental construction algorithm," *Journal of Molecular Biology, Vol. 261, No. 3*, 470-89.

[30] Rosen, J.B., Phillips, A.T. and Dill, K.A. **(1999)**, "A method for parameter optimization in computational biology," *Preprint available at* http://cs.uwec.edu/~phillips.

[31] Shalloway, D. **(1992)**, "Application of the renormalization group to deterministic global minimization of molecular conformation energy functions," *Journal of Global Optimization, Vol. 2*, 281-311.

[32] Shalloway, D. **(1997)**, "Variable-scale coarse-graining in macromolecular global optimization," *In Large Scale Optimization with Applications to Inverse Problems, Optimal Control and Design, and Molecular and Structural Optimization.* Biegler, L.T., Coleman, T.F., Conn, A.R. and Santosa, F.N., Eds. Springer, New York. pp. 135-161

[33] Sun, S., Thomas, P.D., and Dill, K.A. **(1995)**, "A simple protein folding algorithm using binary code and secondary structure constraints," *Protein Engineering, Vol. 8, No. 8*, 769-778.

[34] Shao, C-S., Byrd, R.H., Eskow, E., and Schnabel, R.B. **(1997)**, "Global optimization for molecular clusters using a new smoothing approach," *IMA Volumes in Mathematics and Its Applications, Vol. 94.* Biegler, L.T., Coleman, T.F., Conn, A.R., and Santosa, F.N., Eds. Springer, New York. pp. 163-99.

[35] Ten Eyck, L.F., Mandell, J., Roberts, V.A. and Pique, M.E. **(1995)**, "Surveying molecular interactions with DOT," *Proceedings of the 1995 ACM/IEEE Supercomputing Conference.* Hayes, A. and Simmons, M., Eds. ACM Press, New York.

[36] Ten Eyck, L.F., Mandell, J., Kotlovyi, V. and Tsigelny, I. **(1999)**, "Fast Molecular Docking Methods," *Structure and Function of Cholinesterases and Related Proteins.* Doctor, B.P., Quinn, D.M., Rotundo, R.L. and Taylor, P. Eds. Plenum Press, New York

[37] Trosset, J-Y. and Scheraga, H. **(1999)**, "Flexible docking simulations: Scaled collective variable Monte Carlo minimization approach using Bezier splines, and comparison with a

standard Monte Carlo algorithm," *Journal of Computational Chemistry, Vol. 20, No. 2,* 244-52.

[38] Vakser, I.A. **(1995)**, "Protein docking for low-resolution structures," *Protein Engineering, Vol. 8, No. 4,* 371-7.

[39] Vakser, I.A. **(1997)**, "Evaluation of GRAMM low-resolution docking methodology on the hemagglutinin-antibody complex," *Proteins: Structure, Function and Genetics, Vol. 29, No. S1,* 226-30.

[40] Vieth, M., Hirst, J.D., Kolinski, A. and Brooks, C.L. III **(1999)**, "Assessing energy functions for flexible docking," *Journal of Computational Chemistry, Vol. 19, No. 14,* 1612-22.

[41] Xue, G.L. **(1998)**, "An $O(n)$ time hierarchical tree algorithm for computing force field in N-body simulations," *Theoretical Computer Science, Vol. 197,* 157-69.

Optimization in Computational Chemistry and Molecular Biology, pp. 209-229
C. A. Floudas and P. M. Pardalos, Editors
©2000 Kluwer Academic Publishers

Improved Evolutionary Hybrids for Flexible Ligand Docking in AutoDock

William E. Hart
Sandia National Laboratories
P.O. Box 5800, MS 1110
Albuquerque, NM 87185 USA
wehart@sandia.gov
(619) 844-2217

Chris Rosin
Department of Molecular Biology, MB-5
The Scripps Research Institute
La Jolla, CA
crosin@scripps.edu

Richard K. Belew
Department of Computer Science and Engineering
University of California, San Diego
La Jolla, CA
rik@cs.ucsd.edu

Garrett M. Morris
Department of Molecular Biology, MB-5
The Scripps Research Institute
La Jolla, CA
garrett@scripps.edu

Abstract

In this paper we evaluate the design of the hybrid EAs that are currently used to perform flexible ligand binding in the AutoDock docking software. Hybrid evolutionary algorithms (EAs) incorporate specialized operators that exploit domain-specific features to accelerate an EA's search. We consider hybrid EAs that use an integrated local search operator to refine individuals within each iteration of the search. We evaluate several factors that impact the efficacy of a hybrid EA, and we propose new hybrid EAs that provide more robust convergence to low-energy docking configurations than the methods currently available in AutoDock.

Keywords: drug docking, ligand binding, hybrid evolutionary algorithms, global optimization, local optimization.

1   Introduction

Computational methods for molecular docking are valuable tools for structure-based drug discovery. Methods for automated docking fall into two broad categories: matching methods and conformational search methods. Matching methods attempt to find a good docking based on the geometry of a rigid docking molecule and receptor site. The DOCK program [13] was one of the first matching methods developed, and current versions of it are still used. Conformational search methods typically model the ligand in greater detail, and they often allow conformational flexibility in either the ligand or receptor site, or both. These methods employ a simulation or optimization method to search through the space of ligand-receptor configurations.

AutoDock [7, 16] is an example of this approach to molecular docking. It uses a physically detailed model that allows for a fixed receptor site and flexible ligand. AutoDock employs a rapid grid-based method for energy evaluation and precalculates ligand-protein pairwise interaction energies so that they may be used as a look-up table during the conformational search. AutoDock has been successfully applied to a variety of applications using a simulated annealing search method [6].

More recently, evolutionary algorithms (EAs) have been incorporated into AutoDock and applied to standard test problems [18, 15]. EAs have become a popular choice for heuristic search in docking applications [1, 26], and in our evaluation of EAs with AutoDock they consistently perform better than simulated annealing. The molecular docking problem solved by AutoDock is a challenging global optimization problem, and the EAs perform a better global search across the range of positional, orientational and conformational parameters for flexible ligands. Two forms of EAs can currently be used with AutoDock: a genetic algorithm [5] and a hybrid EA that uses local search. The hybrid EAs apply local search in each iteration to refine points. Rosin et al. [18] and Morris et al. [15] show that this local refinement can significantly improve the performance of the EA.

In this paper we reconsider the design of these hybrid EAs. Specifically, we evaluate several factors that may impact the efficacy of these methods. First, we describe a new local search method that has more robust convergence properties than the method previously used with AutoDock. Next we consider the duration of local search, which impacts the balance between global sampling and local refinement in a hybrid EA. Finally, we consider the initial step length used by the local search method, which can be dynamically initialized using population statistics from the EA. We empirically evaluate the effects of these factors on the performance of hybrid EAs using standard test problems. Our results indicate that hybrid EAs using the more robust local search are usually better, and that running the local search method longer improved the search. Initializing the local search step length automatically did not appear to be an important factor, although using this approach can avoid certain worst-case scenarios where the fixed initial step length is poorly initialized.

2   AutoDock

AutoDock docks small flexible molecules to large rigid macromolecules like proteins [16]. A candidate docking gives specific positions and orientations for the protein and a small molecule. AutoDock uses an approximate physical model to compute the energy of a candidate docking, and uses a heuristic search to minimize this energy. This method makes most sense when there is a single docked configuration that is at a much lower energy than other configurations, so that we expect this low-energy configuration to be the consistent result of physical interaction between the two molecules. If the prediction of this configuration is to be accurate, the energy function must have its global minimum at or near this physical configuration.

Heuristic search operates on the configuration of the small molecule, assuming (without loss of generality) a fixed position for the protein. The small molecule can take any position around the protein, and can have any orientation. Global orientation is expressed as a quaternion, which can be thought of as a vector giving an axis of rotation, along with an angle of rotation about this axis. The small molecule may also have several internal rotatable bonds so that its shape is somewhat flexible. The representation of a candidate docking consists of 3 coordinates giving the position of the small molecule, followed by the 4 components of the quaternion specifying the overall orientation of the small molecule, followed by one angle for each of the rotatable bonds.

The docking potential used in AutoDock 3.0 is an empirical free energy potential. This energy potential is composed of five terms (see Morris et al. [15] for further details). The first three are pairwise interatomic potentials that account for short-range electrostatic repulsive forces and long-range weak van der Waals attractive forces. The standard Lennard-Jones 12-6 potential is used for the van der Waals forces, and a 12-10 potential is used for hydrogen bonds. The next term measures the unfavorable entropy of a ligand binding due to the restriction of conformational degrees of freedom, using a measure that is proportional to the number of sp^3 bonds in the ligand. The last term uses a desolvation measure adapted from Stouten et al. [22] which works well with the precalculated grid formulation used by AutoDock.

To account for internal energy in a flexible small molecule with internal rotatable bonds, we calculate the same energy contributions summed over all pairs of atoms within the small molecule. This sum is added to the total energy evaluation. This penalizes conformations of the small molecule that are energetically unfavorable independent of their interaction with the macromolecule.

To save time when computing energy of interaction with the macromolecule, 3-D potential grids are computed for each atom type before optimization begins. Interaction energy is computed as described above at each point in the grid. Then, when calculating total energy during optimization, the energy contribution of an atom is obtained via trilinear interpolation of its position within the grid specific to its atom type, based on the values at the nearest 8 points in the grid. Calculation of the energy due to pairwise interactions within the small molecule does not make use of these grids.

Computation of the grids for energy evaluation requires knowledge of the (assumed fixed) 3-D positions of each atom in the protein; these positions are usually obtained by X-ray crystallography. We also require the structure of the small molecule, along with the locations of internal rotatable bonds. Small molecules tend to be chemically simple, so that we can determine their structure (at least up to the degrees of freedom represented by the rotatable bonds) from their chemical composition alone. Partial charges are required to calculate electrostatic interaction potentials, but these partial charges can be computed from the structure with molecular modelling software such as MOPAC. So, it is possible to use AutoDock to test many candidate small molecules against a single target protein, after obtaining the structure of this protein experimentally. This makes AutoDock an important computational tool in the initial stages of drug design.

3 Optimization Methods

Docking is a difficult global optimization problem, and a variety of different optimization strategies have been proposed to solve docking problems (e.g. see [1, 26]). Simulated annealing [12, 25] is the first optimization method that was used to perform docking with AutoDock [7, 16]. Simulated annealing borrows from the natural metaphor of cooling metal in an attempt to globally minimize functions. Simulated annealing operates much like a steepest descent algorithm, but where a steepest descent algorithm rejects all inferior points, simulated annealing may accept an inferior point with probability p. This probability p is based on the inferiority of the alternate point and a temperature parameter, T:

$$p = e^{-\frac{\Delta E}{kT}},$$

where ΔE is the size of the energy gain, and k is the Boltzmann constant. A cooling schedule lowers temperature during the course of optimization. Simulated annealing does a more global search in early iterations, when high temperature allows transitions over energy barriers from one valley to another. In later iterations, the temperature becomes low, which places more focus on a local optimization within the current basin of attraction.

Subsequently, genetic algorithms [2, 5] have been used by Hart [8, 10], Rosin et al. [18] and Morris et al. [15] to perform docking with AutoDock. Genetic algorithms are evolutionary algorithms (EAs), which perform a multi-point search based on the mechanisms of natural evolution. Specifically, EAs utilize stochastic competition and multi-point recombination, which reflect the mechanisms of natural selection and sexual recombination. Figure 1 provides a basic overview of the main steps of an EA. In each iteration, an EA uses stochastic competition to select a subset of points from its current *population* of points. This subset is used to generate new points using evolutionary operators like recombination and mutation. Recombination generates a new point from two points, often forming a point that represents a convex combination of these two points. Mutation generates a point by varying a subset of the point's parameters. The set of new points generated by the evolutionary operators are typically used to form the population for the next iteration of EA. Although the stochastic competition tends to focus an EA's search, these methods often perform a robust global search because they sample across multiple points in a search domain.

Hart [8], Rosin et al. [18] and Morris et al. [15] also consider hybrid EAs that use local search. These EAs apply local search to a subset of the points generated by the evolutionary operators in an EA in each iteration. The motivation for these hybrids is that these methods could decompose the search by allowing the EA to globally sample across the range of possible docking configurations while the local search method quickly minimizes points to find locally optimal configurations. Hart, Kammeyer and Belew [8, 11] argue that these types of hybrid EAs are better global optimizers than either EAs or local search separately, and Törn and Žilinskas [24] note that most successful global optimization methods also apply the same principle of distinguishing the mechanisms for global and local search.

Uniformly generate P_0
For $t = 1, 2, \ldots$
  Stochastic competition generates $\overline{P}_t$ from P_{t-1}
  Apply recombination and mutation to points in $\overline{P}_t$ to generate P'_t
  Apply local search to points in P'_t to generate P''_t
  Compose P_t from P''_t and P'_t
EndFor

Figure 1: High-level description of the major steps in an EA. Local search is not used in a canonical EA, but it is used in a hybrid EA.

Global-local search hybrids may be especially effective for docking. We believe that there are multiple locations on the surface of the macromolecule where the small molecule could dock, and multiple orientations of the small molecule that are energetically plausible. Local search can reveal which of these locations and orientations is best by fitting the small molecule as closely as possible to the macromolecule within a small local neighborhood of a coarse location and orientation. But we do not expect smooth hills in energy from one location and orientation to a very different one, so that global search is required to choose among these. A global-local search hybrid can effectively sample distant locations and orientations with global search, and get accurate evaluations of each using local search.

The particular hybrid EAs that we have applied to docking are *Lamarckian* hybrid EAs. In Lamarckian hybrid EAs, the points used to start a local search are replaced by the final point generated by the local search. This is in contrast to a *Darwinian* hybrid EA, which simply gives the starting point the value of the final point (which generally increases the probability of selecting the point in the subsequent stochastic competition). Although Darwinian hybrid EAs are more biologically plausible, our previous work [8, 11] leads us to use Lamarckian GA-LS hybrids here.

The method previously used to perform local search in the hybrid EAs is Algorithm 1 from Solis and Wets [20]. The Solis-Wets method is a direct search method that performs a randomized local minimization. Each step starts with a current point x. A deviate δ is chosen from a normal distribution whose standard deviation is given by a parameter Δ_t. If either $x + \delta$ or $x - \delta$ is better, a move is made to the better point and a "success" is recorded. Otherwise a "failure" is recorded. After several successes in a row, Δ_t is increased to move more quickly. After several failures in a row, Δ_t is decreased to focus the search. Additionally, a bias term is included to give the search momentum in directions that yield success. This method is typically terminated if Δ_t falls below a given threshold Δ_{lb}.

An important feature of this type of local search is that it doesn't rely on gradient information. This is particularly important for docking with AutoDock because the docking potential is not differentiable throughout the entire search domain. AutoDock's grid-based intermolecular energy has a gradient that is undefined whenever an atom is on a grid boundary, and has discontinuities as atoms move across grid boundaries. This would make gradient-based local search a poor choice for AutoDock.

4 Hybridization Issues

Hybrid EAs using local search have been successfully applied to a range of applications. These techniques have been called memetic algorithms, genetic local search, hybrid genetic algorithms and genetic hillclimbing (for an extensive bibliography of these hybrid evolutionary algorithms see `alife.ccp14.ac.uk/memetic/~moscato/memetic_home.html`). Despite their success, basic principles have not been formulated to guide the development of effective hybrids, particularly for hybrid EAs applied to continuous search domains like the docking problem in AutoDock. In this section we discuss three hybridization issues that may affect the performance of the hybrid EAs used in our docking experiments.

Local Search Robustness   Since direct search methods do not explicitly employ derivative information, it is not possible for a direct search method to terminate with a guarantee that the final point is near a stationary point (where the gradient is zero). However, a basic expectation of a local search method for a hybrid EA is that it robustly converges to a stationary point. Solis and Wets [20] note that their algorithm will converge to a global minimizer if the step scale parameter is not adapted. However, in Algorithm A the step scale is adapted, and they argue that "a proof of convergence is unlikely."

A class of direct search methods that do have robust convergence properties are pattern search methods. Pattern search methods have been analyzed by Torczon and Lewis [23, 14], and they provide a general framework for describing a wide variety of direct search methods. In a general sense, pattern search methods sample the objective function from a given *pattern* of points that represent offsets from the current best point. If there is a better point in this pattern, then it is accepted as the new iterate and the sampling is repeated about it. If not, then the scale of the pattern is reduced (e.g. by halving it), and the function is again sampled about the best point. Lewis and Torczon's analysis provides a set of conditions which, if satisfied, guarantee a weak stationary point convergence. For unconstrained pattern search methods, they show that

$$\liminf_{t \to \infty} \| \bigtriangledown f(x_t) \| = 0,$$

where $\bigtriangledown f$ is the gradient of f.

A formal description of the pattern search algorithm used in our experiments is given in Figure 2. The pattern used in this algorithm is a set of offsets that form a positive basis [14]. Specifically, this method uses offsets generated by the $n + 1$ directions from the centroid of a regular simplex to each corner of the simplex [21]. These offsets lie at the corners of a triangle in two dimensions, the corners of a tetrahedron in three dimensions, and so on. Our implementation of pattern search uses a **shuffle** method to randomly select the order in which these offsets are considered.

Local Search Length   On combinatorial domains, the local search in a hybrid EA is often run until a locally optimal point is found. Thus methods like genetic hillclimbing are searching through the space of local optima. On continuous domains, the local search in a hybrid EA is typically truncated before the termination criteria stop the local search (e.g. when the step length becomes too small). Consequently the hybrid EA is not searching

Given x_0 and Δ_0
Let $\{v_1, \ldots, v_{n+1}\}$ be the canonical minimal positive basis
For $i = 1, \ldots, n+1$
   $r_i = i$
For $t = 1, 2, \ldots$
   shuffle(r)
   $s_t = \{0\}^n$
   For $i = 1, \ldots, n+1$
     $s = v_{r_i}$
     If$(f(x_t + \Delta_t s) < f(x_t))$
       $s_t = \Delta_t s$
       Break
     EndIf
   EndFor
   $x_{t+1} = x_t + s_t$
   If $(f(x_t) == f(x_t + s_t))$
     $\Delta_{t+1} = \Delta_t/2$
   Else
     $\Delta_{t+1} = 2\Delta_t$
   EndIf
   If $(\Delta_{t+1} < \Delta_{lb})$
     Break
EndFor

Figure 2: Pseudo-code for the pattern search algorithm.

with local optima, and the maximum duration of local search is a parameter that can fundamentally impact the dynamics of the hybrid EA. A hybrid EA with long local searches will execute fewer iterations of the EA than an EA hybrid with short local searches (if both terminate after the same number of function evaluations). Consequently, the duration of local search affects the balance between the amount of global sampling and local refinement performed by the hybrid EA.

In previous work, we have used both short and long local search durations for docking in AutoDock [18, 15]. We previously experimented with very short and very long local search durations in AutoDock, and found them to yield similar performance [18]. However subsequent experiments with the new energy potential in AutoDock suggest that this may be an important factor, particularly if the local search is only terminated when it exceeds the local search duration.

Initial Search Scale A fundamental feature of optimization problems on continuous domains is that the scale of changes in the objective function can vary dramatically in different regions of the domain. Consequently, a basic requirement of an optimization method is that it dynamically adapt the scale of its search in order to match the scale of changes in the objective function. This implies that hybrid EAs using local search need to dynamically adapt the initial scale of the local search method. If the initial scale of the local search is not adapted, then the scale of the entire hybrid is limited by how quickly the local search method adapts its scale. This limits the utility of local search, and in cases where the local search is truncated it can even prevent the local search method from productively refining a point.

Thus hybrid EAs that use direct search methods for local search need to adapt the initial local search scale to reflect the characteristics of the current search. To our knowledge, Mühlenbein, Schomisch and Born [17] are the only authors to have reported an initialization strategy for hybrid EAs using a direct search method. Their initialization strategy uses a statistic of the population spread to initialize all local searches with the same value. The spread of the population reflects the degree to which the EA's search has focused on a particular region of the search domain. If we assume that the local search should be as focused as the EA's search, then this represents a natural parameter for the initial local search step scale. Their method initializes Δ_0 to

$$\Delta_0 = \frac{|x_1 - x_m|}{\sqrt{n}},$$

where x_1 is the point with the best fitness in the population and x_m is the point with the median fitness.

5 Methods and Experiments

5.1 Search Algorithms

Our experiments compare the empirical performance of hybrid EAs on a set of standard docking problems. The hybrid EAs that we evaluate are the hybrid GA using SW that is provided in AutoDock [15] and hybrid GAs using the Solis-Wets method (SW) or pattern search method (PS) that are provided by the SGOPT optimization library [9], which was integrated into AutoDock 3.0 for this study. In all cases, the experimental parameters for the hybrid GAs were the same as those used by Morris et al. [15]. The GAs used a population of 50 points, applied a two-point crossover with a probability of 80% and applied a Cauchy mutation operator with a probability of 2%. When the Cauchy mutation operator is applied to a dimension of a point, it adds a Cauchy random variable with parameters $\alpha = 0$ and $\beta = 1$. The hybrid EAs provided by SGOPT scale the mutation steps to 0.01 times the range of the dimension. Stochastic competition was performed using proportional selection, where the baseline for computing the proportions was the worst point in the last 10 iterations [15]. Elitism was also used to keep the best point found so far.

Local search was performed on randomly chosen points in each iteration with a probability of 6%. Using local search infrequently is motivated by our preliminary work with this application where we varied the probability of local search [8]. The local search operator was either SW or PS. The SW method used in AutoDock implicitly scales the search in each dimension to a value that is approximately 0.01 times the range of the dimension. The lower bound on Δ_t is effectively zero for this method due to an incorrect implementation of SW in AutoDock; SGOPT correctly implements this lower bound. SW performs contraction after four consecutive failures and it performs expansion after four consecutive successes. The SW and PS methods provided by SGOPT also scale the search in each dimension. The initial search scale is 0.1 times the range of each dimension, and the SW and PS methods were terminated when the search scale fell below 0.001.

In each experiment, 20 trials were done with different random seeds. The hybrid EAs were terminated after 1.5 million function evaluations; this enables a comparison with previous work [18, 15]. Using a fixed number of function evaluations provides a reasonable basis of comparison for this problem because the calculation of the docking potential is the most expensive step in this optimization problem.

For notational convenience, we refer to the different hybrid EAs using the notation [ea]-[ls]. The values for **ea** are AD, the GA provided by AutoDock, GA0, the GA provided by SGOPT with fixed initial step length and GA1, the GA provided by SGOPT using the initialization of Mühlenbein et al. [17]. The value of **ls** is SW300, SW3000, PS300 or PS3000, which refer to the choice of local search method and the maximum duration of local search.

5.2   Experiments

A test suite of six cases was used in all of the experiments. Each test case consists of a macromolecule and a small substrate or inhibitor molecule. The salient features of the six test cases are summarized in Table 1. The different test cases were selected to test various aspects of the energy function [16].

Ligand/Protein Complex	PDB Code	Number of Torsions	Number of Dimensions
β-Trypsin/Benzamidine	3ptb	0	7
Cytochrome P-450cam/Camphor	2cpp	0	7
McPC-603/Phosphocholine	2mcp	4	11
Streptavidin/Biotin	1stp	5	12
HIV-1 protease/XK263	1hvr	10	17
Influenza Hemagglutinin/sialic acid	4hmg	11	18

Table 1: Summary of test cases. PDB codes taken from the Protein Data Bank.

The number of torsion angles is an important feature of these test cases because it determines the dimensionality of the search space. The representation used in each experiment consisted of a triple of Cartesian coordinates, a four dimensional quaternion, and the torsion angles. Thus, the dimensionality of the search space is 7+(number of torsion angles). The range of the coordinates defines a cube that is 23 angstroms long in each dimension. The quarternion parameters lie within $[-1, 1]^3 \times [-\pi, \pi]$, and each torsion angle lies within $[-\pi, \pi]$; the points in the initial population have each parameter generated randomly in its range.

For each method on each test case, we consider the minimum energy produced by the search. Because we have crystallographic structures of the true docked complex for each test case, we also measure the absolute accuracy of the final docked configuration. This is done by taking the square root of the average squared deviation of the spatial separation of corresponding atoms in the predicted configuration from the crystallographic configuration.

6 Results

Figures 3, 4 and 5 show boxplots of the final docking energies for each of the optimization experiments for each test case. Boxplots are a convenient method of summarizing data that provide a visual indication of the spread and skewness of the data. The dark bar in a boxplot shows the range between the first and third quartile; one quarter of the data is below the first quartile, and three quarters of the data is below the third quartile. The white line inside the dark bar represents the median. The whiskers at the top and bottom of each boxplot indicate the spread of the data up to 1.5 times the range of the first and third quartile.

We applied the nonparametric Kruskal-Wallis test [4] to identify significant differences between the hybrid EAs (at the 5% level). This test is appropriate because the data does not appear to be normally distributed, and because we are making multiple pairwise comparisons among more than 2 samples. These statistics are discussed in the following sections, which evaluate the effects of the hybridization issues that we discussed in Section 4.

6.1 AutoDock vs SGOPT

We can make a direct comparison between the hybrid EAs using SW provided by AutoDock and SGOPT. Figure 6 shows boxplots for the hybrid EAs using SW for all six test cases. For each test case, the trials for the hybrid EAs are ranked, and the boxplots show the distribution of ranks within each test case. This data has two distinct trends. First, the hybrid EAs from SGOPT find better solutions. Second, the hybrid EAs from AutoDock find better solutions when the local search is short, while the EAs from SGOPT find better solutions when the local search is long. Our statistical analysis shows that the AutoDock hybrid EAs are significantly different from almost all of the SGOPT hybrid EAs.

We believe that these differences may be explained by two factors. First, the initial step scale used by the AutoDock hybrid EAs is smaller than the initial step scale of the SGOPT hybrid EAs. Although a smaller initial step scale focuses the local search about the starting point, in early iterations of the hybrid EA the local searches are not simply refining to a local minimum, but they are also helping guide the EAs search (through the Lamarckian return of the final point into the EA's population).

A second factor concerns AutoDock's incorrect implementation of the lower bound on step scale. When the local search converges about a local minima, it will waste time refining the minima beyond the point where it is physically relevant. This effect will be particularly evident when the local search duration is long, which accounts for the worse performance for the AutoDock hybrid EAs with long local search.

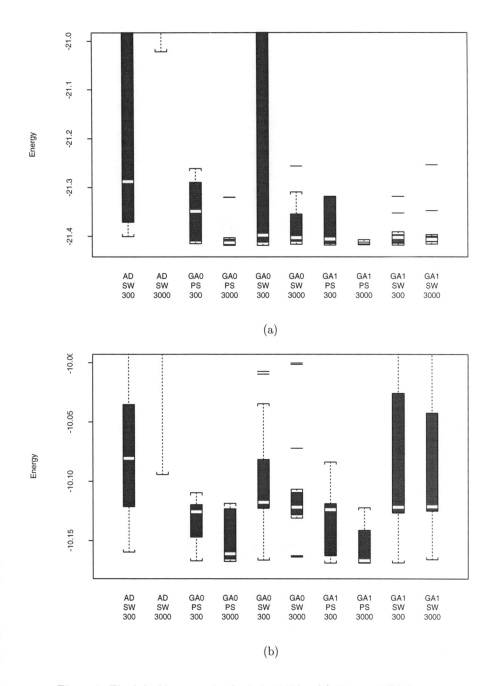

Figure 3: Final docking energies for hybrid EAs: (a) 1hvr and (b) 1stp.

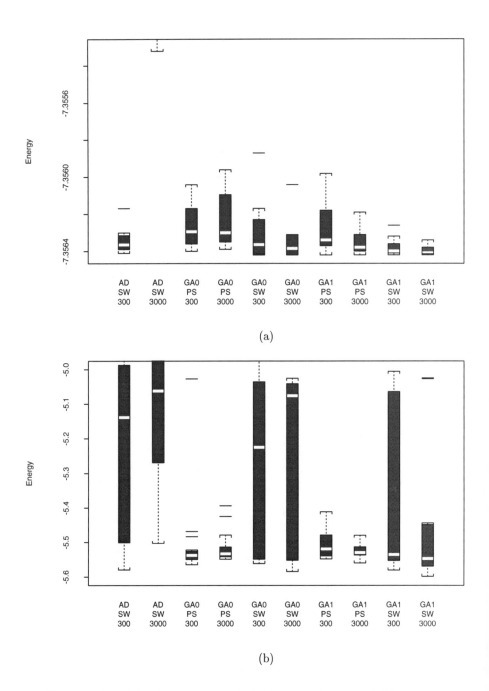

Figure 4: Final docking energies for hybrid EAs: (a) 2cpp and (b) 2mcp.

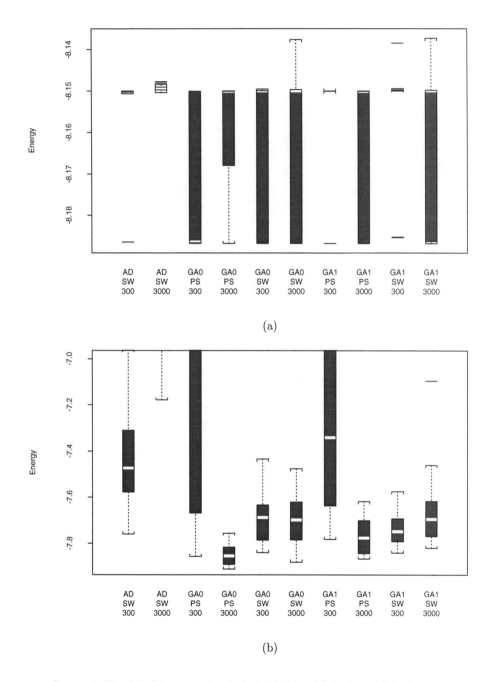

Figure 5: Final docking energies for hybrid EAs: (a) 3ptb and (b) 4hmg.

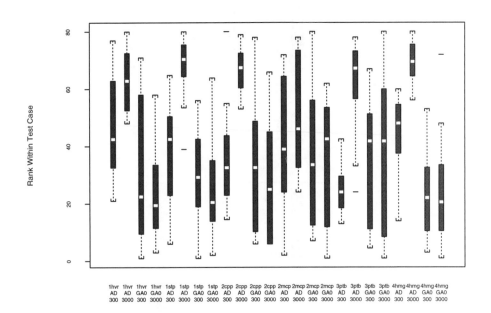

Figure 6: Comparison of rankings for AutoDock hybrid EAs and SGOPT hybrid EAs using SW.

6.2 SW vs PS

Our experiments generally support the use of the more robust PS method over SW for local search. Here, robustness takes into account both the ability to find the lowest energy conformations as well as the consistency at which the hybrid EA can find low energy conformations. The statistical analysis shows many significant differences between hybrid EAs using PS and SW local search, particularly when long local searches were used.

Figure 7 summarizes the relative ranks of each hybrid EA with SW and PS local search (using a fixed initial step length), grouped by the test case. The trend is particularly strong for 1hvr, 1stp and 3ptb. For 2cpp the EAs using PS are consistently better than the EAs using SW, but the EAs using SW can find better solutions often enough to balance the overall ranking. For 4hmg, the EAs using PS are better when long local searches are used, but worse when short local searches are used, which again balances the ranking. For 2mcp, the hybrid EAs using PS are generally worse than the EAs using SW. However, if the termination threshold for PS is reduced to 0.0001 then the hybrids using PS are generally better, which again provides evidence that longer local searches with PS are better.

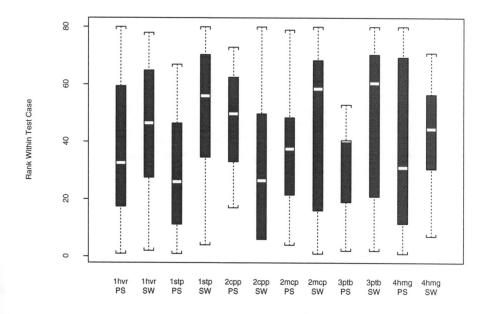

Figure 7: Comparison of rankings for SGOPT hybrid EAs using SW and PS local search.

6.3 Local Search Duration

In the significance test, the hybrid EAs using SW do not consistently exhibit significant differences. However, the significance test shows a consistent significant difference between the hybrid EAs using short PS and long PS for test cases 1stp, 1hvr and 4hmg. These are the most difficult test problems, which suggests that hybrid EAs using long local searches will be most effective for nontrivial problems.

6.4   Step Length Initialization

The experimental results indicate that adaptive methods for initializing the step length of local search do not have a substantial impact on the performance of the hybrid EAs. In fact, there are many cases where initializing the local search duration leads to worse performance. The adaptive hybrids have better performance with SW local search in several of the test cases, but in others they have worse performance. The adaptive hybrids with PS are generally indistinguishable from the nonadaptive hybrids. We also tried an alternative initialization strategy for local search, but it had similar results.

We investigated possible causes of this, and we discovered that the initialization methods often generate initial step lengths that are larger than the fixed value used by the GA0 hybrids. Also, we noted that the initial step lengths do not seem to converge to zero as we had anticipated. Instead, the variations in the population remain large enough to keep the adaptively determined initial step lengths rather large. This suggests that the dynamics of the EA have a significant impact on the utility of these adaptive mechanisms. For example, the linear rank selection used by Mühlenbein et al. [17] may be an important feature of the GA for the success of the adaptive methods.

Finally, it is possible that we selected a good default value for the initial step lengths. If the initial step length is too large then the local search effort will simply be spent on reducing it to a reasonable value. Similarly, if the initial step length is too small then the local search method will not make signficant progress until it can increase the step length to a reasonable value. In preliminary experiments, we confirmed that as the initial step lengths are raised the performance of the hybrid EAs using the fixed initial step length degrades. This was particularly true for hybrid EAs using short LS, which is expected since they have fewer iterations to adapt their search scale.

6.5   Length of Optimization

Although our experiments were run for 1.5 million function evaluations to enable direct comparisons with prior work, this number of function evaluations is relatively high for problems with a small number of dimensions. To evaluate how long optimization was productive, we examined the optimization trace and found the last iteration where an improvement of energies than 0.0001 was found in the energy. This gives a sense of whether the search has stalled. Figures 8 and 9 show the distribution of stall times for the different methods on the six test functions. In two problems (2cpp and 3ptb) the optimizers stall after as few as 100,000 function evaluations. In the remaining four problems, the optimizers typically require well over a million function evaluations.

The salient difference between these two classes of problems is that 2cpp and 3ptb have no rotatable torsion angles while the remaining problems are flexible ligands. The introduction of rotatable torsion angles makes docking problems significantly more difficult. Problems with rotatable torsion angles probably also increases the number of local optima, thereby making incremental progress of the objective function possible in a long optimization process like the one that we have applied.

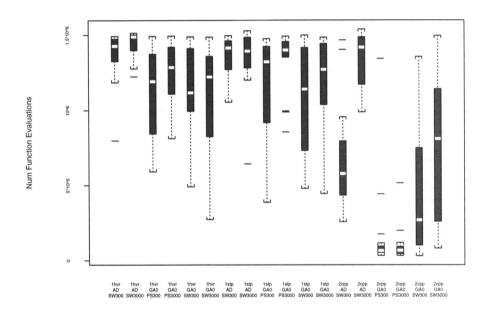

Figure 8: Distribution of stall times for AutoDock hybrid EAs and SGOPT hybrid EAs: 1hvr, 1stp and 2cpp

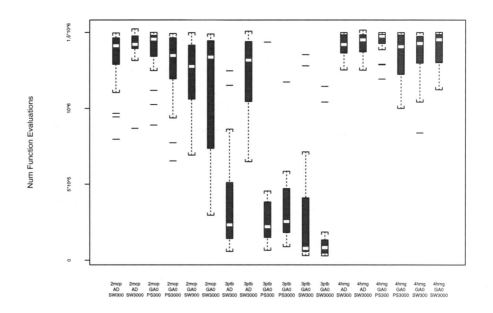

Figure 9: Distribution of stall times for AutoDock hybrid EAs and SGOPT hybrid EAs: 2mcp, 3ptb, 4hmg

7 Conclusions

Our experimental results demonstrate that the new hybrid EAs we have defined can perform significantly better than the methods currently available in AutoDock. In particular, the hybrids using the PS method with long local search were significantly better than other hybrids in most cases. The dynamic initialization of the local search step was not a signficant factor in our experiments. However, it is clear that this is an important algorithmic factor that needs to be developed to make hybrid EAs with direct search methods more robust.

Table 2 summarizes the performance of the GA0-PS3000 methods for the six test cases. In five of the six test cases, the best value found by the hybrid EA is better than the best value found by all of the methods reported by Morris et al. [15]. Furthermore, the mean rmsd of the solutions found is lower for all test cases than the results in Morris et al. In all cases the hybrid EA found solutions below or about 1 angstrom, which is very reasonable. This confirms that these hybrid EAs not only perform a more effective search, but they are also finding better solutions to the docking problem.

	Energy		RMSD	
	Minimum	Mean	Minimum	Mean
1hvr	-21.42	-21.40	0.65	0.70
1stp	-10.17	-10.15	0.49	0.51
2cpp	-7.36	-7.36	0.90	0.91
2mcp	-5.55	-5.52	0.89	0.98
3ptb	-8.19	-8.16	0.29	0.29
4hmg	-7.91	-7.85	0.97	1.03

Table 2: Performance statistics for GA0-PS3000: final energy and rmsd from known crystal structure.

Our experiments in this paper have focused on the factors that affect the utility of local search in a hybrid EA. Consequently, we have used the same problem formulation and type of GA that we have applied in previous studies with AutoDock [18, 15]. However, there are several ways that these methods could probably be improved. For continuous domains, the scale of mutation is often adapted [19] and recombination operators can be applied that do not impose a coordinate bias to the search [3]. Finally, the boundary constraints and the equality constraint on the quaternion's direction of rotation should be handled explicitly. Explicitly considering these factors will likely improve the global search performed by the EA.

Another factor that affects the global search concerns the total run time of the EA. Our experiments show that running the hybrid EAs for 1.5 million function evaluations often provide continued improvements in the solution. However, an alternative approach that needs to be considered is multiple executions of hybrid EAs that are shorter. This approach is closer to a multistart local search approach, but uses the EA to bias the selection of initial starting points. Although previous studies with this drug docking application suggest that multistart local search is less effective than hybrid EAs [8], this multistart hybrid EA may be more effective than running a single hybrid EA for a long time.

Acknowledgements

We thank Mark Land, Art Olson and David Goodsell for many useful discussions concerning drug docking and hybrid EAs. This work was performed in part at Sandia National Laboratories. Sandia is a multiprogram laboratory operated by Sandia corporation, a Lockheed Martin Company, for the United States Department of Energy under Contract DE-AC04-94AL85000.

References

[1] D. E. Clark and D. R. Westhead. Evolutionary algorithms in computer-aided molecular design. *J Computer-Aided Molecular Design*, 10:337–358, 1996.

[2] L. Davis, editor. *Handbook of Genetic Algorithms*. Van Nostrand Reinhold, 1991.

[3] L. J. Eshelman and J. D. Schaffer. Real-coded genetic algorithms and interval schemata. In L. D. Whitley, editor, *Foundations of Genetic Algorithms 2*, pages 187–202. Morgan-Kauffmann, San Mateo, CA, 1993.

[4] R. J. Freund and W. J. Wilson. *Statistical Methods*. Academic Press, 1997.

[5] D. E. Goldberg. *Genetic Algorithms in Search, Optimization, and Machine Learning*. Addison-Wesley Publishing Co., Inc., 1989.

[6] D. S. Goodsell, G. M. Morris, and A. J. Olson. *J Mol Recog*, 9(1), 1996.

[7] D. S. Goodsell and A. J. Olson. Automated docking of substrates to protiens by simulated annealing. *Proteins: Structure, Function and Genetics*, 8:195–202, 1990.

[8] W. E. Hart. *Adaptive Global Optimization with Local Search*. PhD thesis, University of California, San Diego, May 1994. ftp://ftp.cs.sandia.gov/pub/papers/wehart/thesis.ps.gz.

[9] W. E. Hart. SGOPT: A library for stochastic global optimization. (in preparation), 1998.

[10] W. E. Hart. Comparing evolutionary programs and evolutionary pattern search algorithms: A drug docking application. In *Proc. Genetic and Evolutionary Computation Conf*, 1999. (to appear).

[11] W. E. Hart, T. E. Kammeyer, and R. K. Belew. The role of development in genetic algorithms. In L. D. Whitley and M. D. Vose, editors, *Foundations of Genetic Algorithms 3*, pages 315–332, San Fransico, CA, 1995. Morgan Kaufmann Publishers, Inc.

[12] S. Kirkpatrick. Optimization by simulated annealing: Quantitative studies. *J. Stat. Phys.*, 34(5/6):975–987, 1984.

[13] I. D. Kuntz, J. M. Blaney, S. J. Oatley, R. Langridge, and T. E. Ferrin. A geometric approach to macromolecular-ligand interactions. *J Mol Bio*, 161:269–288, 1982.

[14] M. Lewis and V. Torczon. Rank ordering and positive bases in pattern search algorithms. *Mathematical Programming*, 1998. (submitted).

[15] G. M. Morris, D. S. Goodsell, R. S. Halliday, R. Huey, W. E. Hart, R. K. Belew, and A. J. Olson. Automated docking using a Lamarkian genetic algorithm and an empirical binding free energy function. *J Comp Chem*, 19(14):1639–1662, 1998.

[16] G. M. Morris, D. S. Goodsell, R. Huey, and A. J. Olson. Distributed automated docking of flexible ligands to proteins: Parallel applications of Autodock 2.4. *J. Comp.-Aid. Mol. Des.*, 10:293–304, 1996.

[17] H. Mühlenbein, M. Schomisch, and J. Born. The parallel genetic algorithm as function optimizer. In R. K. Belew and L. B. Booker, editors, *Proc of the Fourth Intl Conf on Genetic Algorithms*, pages 271–278, San Mateo, CA, 1991. Morgan-Kaufmann.

[18] C. D. Rosin, S. Halliday, W. E. Hart, and R. K. Belew. A comparison of global and local search methods in drug docking. In T. Baeck, editor, *Proc 7th Intl Conf on Genetic Algorithms*, pages 221–228, San Francisco, CA, 1997. Morgan Kaufmann.

[19] N. Saravanan, D. B. Fogel, and K. M. Nelson. A comparison of methods for self-adaptation in evolutionary algorithms. *BioSystems*, 36:157–166, 1995.

[20] F. Solis and R.-B. Wets. Minimization by random search techniques. *Mathematical Operations Research*, 6:19–30, 1981.

[21] W. Spendley, G. R. Hext, and F. R. Himsworth. Sequential application of simplex designs in optimisation and evolutionary operation. *Technometrics*, 4(4):441–461, Nov 1962.

[22] P. F. W. Stouten, C. Frömmel, H. Nakamura, and C. Sander. *Mol. Simul.*, 10, 1993.

[23] V. Torczon. On the convergence of pattern search methods. *SIAM J Optimization*, 7(1):1–25, Feb 1997.

[24] A. Törn and A. Žilinskas. *Global Optimization*, volume 350 of *Lecture Notes in Computer Science*. Springer-Verlag, 1989.

[25] P. van Laarhoven and E. Aarts. *Simulated Annealing: Theory and Applications*. Reidel, 1987.

[26] M. Vieth, J. D. Hirst, B. N. Dominy, H. Daigler, and C. L. Brooks III. Assessing search strategies for flexible docking. *J Comp Chem*, 19(14):1623–1631, 1998.

Optimization in Computational Chemistry and Molecular Biology, pp. 231-242
C. A. Floudas and P. M. Pardalos, Editors
©2000 Kluwer Academic Publishers

Electrostatic Optimization in Ligand Complementarity and Design

Erik Kangas
Departments of Chemistry and Physics
Massachusetts Institute of Technology
Cambridge, MA 02139 USA

Bruce Tidor
Department of Chemistry
Massachusetts Institute of Technology
Cambridge, MA 02139 USA tidor@mit.edu

Abstract

Analytic and numerical methods now allow optimization of the electrostatic contribution to the free energy of association of two molecules in solution. Using a continuum electrostatic approximation based on the linearized Poisson–Boltzmann equation, the electrostatic free energy of rigid bimolecular association becomes a quadratic function of the reactant-charge distributions. By optimizing the charge distribution of one reactant, we find that the electrostatic free energy can be minimized, and made favorable in many cases. Furthermore, a rigorous method for visualizing the extent of electrostatic complementarity between two molecules has been developed. In this paper we review the framework and progress of charge optimization and discuss some of the implications emerging to date.

Keywords: Electrostatics, global optimization, ligand design.

1 Introduction

The *in vivo* association of molecules is important for the regulation and performance of countless biological processes. Understanding the properties of these natural molecules that influence their binding as well as designing new molecules that bind to selected targets to produce desired pharmacological effects are active and growing areas of research in biophysical chemistry. In addition to properties such as non-toxicity, chemical stability, and bioavailability, any potential drug molecule must be engineered to bind tightly to the desired target molecule. This usually involves a choice of molecular shape which is complementary of the target molecule's active site so as to produce attractive intermolecular van der Waals interactions. It also involves choosing appropriate chemical groups in the drug molecule to induce both binding specificity [1, 2, 3, 4] and binding affinity.

The contribution of electrostatics to binding is non-intuitive. While placing a polar or charged chemical group in the prospective drug molecule (ligand) may create favorable interactions with the target molecule (receptor), it also creates favorable interactions with the solvent and ion atmosphere which must be sacrificed upon binding. It is this tradeoff between favorable interaction and unfavorable desolvation free energies that is the concern of charge optimization. Optimizing the ligand-charge distribution ensures that electrostatics contributes as favorably as possible to the binding process; optimization can have a dramatic effect on computed binding affinities.

In this paper, we review the framework and progress of this electrostatic free energy optimization problem. Section 2 describes the conceptual framework leading to an expression for the objective function, the electrostatic contribution to the free energy of binding, in terms of the ligand-charge distribution. Section 3 discusses the process of optimization and properties of the optima, including the favorability of the resulting optimized electrostatic binding free energy contribution. Section 4 examines one particular implication of the optimization process, namely a method to visually and numerically examine the extent of electrostatic complementarity for a ligand–receptor pair without explicitly computing the actual optimum. Section 5 contains concluding remarks.

2 The Electrostatic Free Energy of Binding

2.1 The Continuum Electrostatic Model

Fig. 1 depicts an example of a molecular association reaction in which two reactants, a ligand and receptor, associate rigidly to form a complex. The free energy change of the solution due to this binding can be separated into electrostatic and non-polar contributions [5, 6], where the non-polar contribution corresponds to the binding free energy of the reactants when their charge distributions are everywhere zero. The electrostatic component may often be accurately obtained through a continuum approximation [7, 8, 9, 10], wherein the solvent is treated as a dielectric continuum with free ions, the molecules as rigid dielectric cavities with embedded point charges, and the system obeys the Poisson–Boltzmann equation or its linearized form [6, 11]. The solute–solvent dielectric boundary is taken to be the solute molecular surface [12, 13], and an ion-excluding Stern layer [11, 14] is used to constrain the ion centers from inaccessible regions near the molecules.

For the sake of discussion, we make the following definitions. A generic molecule labeled (i) has n_i embedded point charges at positions $\mathbf{x}_{i,j}$. The values of the partial charges at these n_i positions are consolidated into a charge distribution vector $\mathbf{Q}_i$, such that the value of charge at position $\mathbf{x}_{i,j}$ is $Q_{i,j}$. We may specify receptor, ligand, or complex molecules by replacing (i) by (r), (l), or (c), respectively. The molecular dielectric constant is taken to be ϵ_m (typically of a value between 1 and 4) and the solvent dielectric constant, ϵ_s (typically about 80 for water at room or body temperature). The molecular dielectric constant is often chosen to be larger than one in order to approximately account, in a mean-field sense,

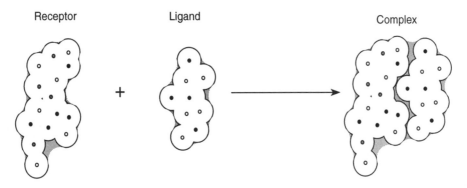

Figure 1: Illustration of a general rigid binding process wherein a receptor and ligand associate forming a complex. The small circles represent example locations of atomic centers shaded to indicate possible differences in charge. The solid line encloses the atomic volume of the molecules, which is essentially the union of the spherical volume occupied by the molecule's atoms. The shaded region outside the atomic volume represents the additional volume assigned to the molecule when the molecular surface is used to designate the molecule–solvent boundary. This excess volume arises because the polarizable solvent molecules cannot fit into all of the crevices left by the atomic volume.

for polarizability and internal degrees of freedom.

The electrostatic component, ΔG^{es}, of the rigid binding free energy is the difference in the total electrostatic free energy of the complex and that of the reactants at infinite dilution in solution [15],

$$\Delta G^{\text{es}} = G^{\text{es}}_{\text{c}} - G^{\text{es}}_{\text{l}} - G^{\text{es}}_{\text{r}}. \tag{1}$$

The total electrostatic free energy of any molecule consists of several components: the self-energy of each point charge, the coulombic intramolecular interaction, the interaction of the charges with the polarized solvent and perturbed ion atmosphere, and the enthalpic and entropic cost of polarizing the solvent and perturbing the ion atmosphere (compared to their hypothetical distributions if the molecule were completely hydrophobic).

The proper use of thermodynamic cycles allows these calculations to be carried out in cases where titration and conformational change accompany binding. Such cases are not discussed in the current overview; for the sake of clarity, the detailed methodology for treating these cases will not be presented here, but the reader is referred to [15].

The total electrostatic free energy of a rigid molecule in solution is given by [16, 17, 18]

$$G^{\text{es}}_i = \frac{1}{2} \sum_{j=1}^{n_i} Q_{i,j} \Phi_i(\mathbf{x}_{i,j}) \tag{2}$$

for systems obeying the linearized Poisson–Boltzmann equation [6, 16]

$$\nabla \cdot (\epsilon_i(\mathbf{x}) \nabla \Phi_i(\mathbf{x})) - \epsilon_s \kappa_i^2(\mathbf{x}) \Phi_i(\mathbf{x}) = -4\pi \sum_{j=1}^{n_i} Q_{i,j} \delta(\mathbf{x} - \mathbf{x}_{i,j}). \tag{3}$$

In this differential equation, $\Phi_i(\mathbf{x})$ is the total electrostatic potential, $\epsilon_i(\mathbf{x})$ is the dielectric constant of space (ϵ_m within the molecular volume and ϵ_s in the solvent), $\kappa_i(\mathbf{x})$ is the inverse Debye screening length of the solvent due to the ions, and $\delta(\mathbf{x}-\mathbf{x}_{i,j})$ is the three dimensional Dirac delta function. The linearized Poisson–Boltzmann equation is usually appropriate for protein–protein and protein–small molecule binding because the electrostatic potentials in the solvent are generally much smaller than kT/e and the energies and potentials produced by the linear and non-linear versions are very similar. The linearity is essential to the optimization as currently formulated. Note that Eq. (2) is really a free energy (including entropic terms for water and ion atmosphere rearrangement) and not just the total energy of the system [17, 19].

2.2 Quadratic Nature of the Electrostatic Binding Free Energy

Because the electrostatic potential in Eq. (3) is linearly related to the molecular charge distribution, $\mathbf{Q}_i$, the potential is a linear function of this charge distribution, i.e.

$$\Phi_i(\mathbf{x}) = \mathbf{A}_i(\mathbf{x}) \cdot \mathbf{Q}_i,$$

where $\mathbf{A}_i(\mathbf{x})$ is a position-dependent vector of proportionality constants (they are related to the Green's function for the linearized Poisson–Boltzmann equation [15]) with the coulombic singularities at $A_{i,j}(\mathbf{x}_{i,j})$ removed. Substituting this expression back into Eq. (2), we find that the total electrostatic energy of the molecule is given by

$$G_i^{es} = \sum_{j=1}^{n_i} \sum_{k=1}^{n_i} Q_{ij} \frac{A_{i,k}(\mathbf{x}_{i,j})}{2} Q_{ik}. \tag{4}$$

Rewritten as a matrix expression, Eq. (4) becomes

$$G_i^{es} = \mathbf{Q}_i^t S_i \mathbf{Q}_i, \tag{5}$$

with the matrix elements $S_{i,jk} = A_{i,k}(\mathbf{x}_{i,j})/2$.

Using Eq. (1) together with Eq. (5), the total electrostatic binding free energy takes the quadratic form

$$\Delta G^{es} = \mathbf{Q}_c^t S_c \mathbf{Q}_c - \mathbf{Q}_r^t S_r \mathbf{Q}_r - \mathbf{Q}_l^t S_l \mathbf{Q}_l. \tag{6}$$

If the receptor- and ligand-charge distributions are unaltered upon binding, then the complex-charge distribution vector is the union of the ligand- and receptor-charge vectors, $\mathbf{Q}_c = (\mathbf{Q}_r, \mathbf{Q}_l)$, the ligand–ligand and receptor–receptor intramolecular coulombic energies cancel, and this equation can be re-written as

$$\Delta G^{es} = \mathbf{Q}_r^t R \mathbf{Q}_r + \mathbf{Q}_r^t C \mathbf{Q}_l + \mathbf{Q}_l^t L \mathbf{Q}_l. \tag{7}$$

Here, inner products with the "desolvation matrices" L and R provide the change in solvation free energy of the ligand- and receptor-charge distributions, respectively, due to binding. The inner product with the matrix C provides the bound-state ligand–receptor screened coulombic interaction. If receptor- and ligand-charge distributions or conformations are altered upon binding, other expressions are obtained that can be treated in a similar manner [15].

3   Electrostatic Optimization

3.1   Optimization of the Electrostatic Binding Free Energy

Eq. (7) can be used to determine the electrostatic binding free energy of all potential ligands that have the same molecular shape but differ in their electrostatic charge distribution. Leaving aside initially the questions of chemical reasonableness and synthetic accessibility, examination of this physical space of charge distributions allows one to address questions about what physical properties are important for a good ligand. Moreover, optimization of the electrostatic binding free energy over all charge distributions can be performed, and the optima can be analyzed and used as templates in design studies. Because of the quadratic nature of Eq. (7), there will be extrema to the electrostatic binding free energy as a function of the ligand-charge distribution; because matrix L is non-negative definite when $\epsilon_s > \epsilon_m$ (the electrostatic desolvation penalty is unfavorable) [15, 20], all extrema are minima [21]. Therefore, there are "optimized" sets of charges $\mathbf{Q}_l^{opt}$ that minimize ΔG^{es}, producing the best possible electrostatic contribution to the total binding free energy. An optimized ligand-charge distribution is obtained by setting the gradient of the electrostatic binding free energy with respect to the ligand-charge distribution to zero (and using the reciprocity relation $L = L^t$)

$$\mathbf{Q}_r^t C + 2\mathbf{Q}_l^t L = 0 \qquad (8)$$

and then solving for the optimized charges [15, 20]

$$\mathbf{Q}_l^{opt} = -\frac{1}{2}L^{-1}C^t\mathbf{Q}_r. \qquad (9)$$

This optimization process is illustrated with a one-dimensional example in Fig. 2. As the ligand-charge distribution is varied, the receptor desolvation penalty remains constant, α, because the shape of the ligand is assumed constant (see below). The receptor–ligand interaction free energy varies linearly, βq, because the receptor-charge distribution is fixed. The ligand desolvation penalty varies quadratically, γq^2, as can be inferred from Eq. (7), because the charges interact with the polarization and ionic screening potentials, whose magnitudes are proportional to the inducing charge. The combination of the favorable linear contribution of the interaction energy and the unfavorable quadratic contribution of the ligand-desolvation penalty make the net electrostatic binding free energy quadratic, $\alpha + \beta q + \gamma q^2$, with a minimum at a non-zero value of ligand charge, $q^{opt} = -\beta/(2\gamma)$. Unless the receptor is entirely hydrophobic (all partial atomic charges are equal to zero), it is always better for the ligand to be charged or polar rather than hydrophobic.

The actual value of the optimized electrostatic binding free energy necessarily depends upon the choice of ligand point-charge locations, also known as the basis set. Additional locations increase the number of degrees of freedom in the optimization and thus may lower the optimized electrostatic binding free energy. The limiting value of the electrostatic binding free energy obtained by optimizing on a "complete set" of charge locations (i.e., no set of additional point charge locations can improve the optimal binding free energy appreciably) is denoted the optimal electrostatic binding free energy and the respective charge distribution is termed optimal [15]. Other basis sets, such as point multipoles, can also be used with this method [15, 20, 22].

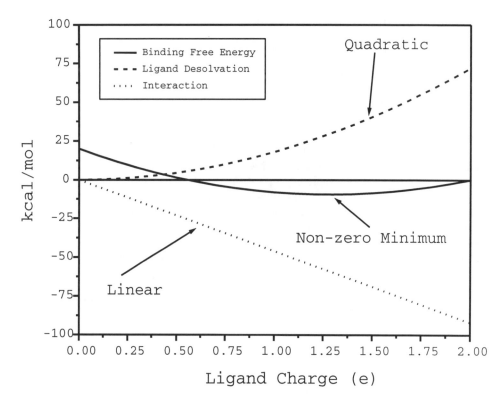

Figure 2: This graph depicts the how the linear ligand–receptor interaction energy and the quadratic ligand desolvation penalty (as a function of ligand charge), along with the constant receptor desolvation penalty, combine to make a quadratic total electrostatic binding free energy which has a minimum at a non-zero value of ligand charge. In this case, the minimum value of the total electrostatic binding free energy is negative.

3.2   Favorability of the Optimal Electrostatic Binding Free Energy

Fig. 2 also illustrates the possibility that the minimum of the electrostatic binding free energy may actually be favorable (negative). If this is the case, then there are also a whole range of additional ligand-charge distributions in the neighborhood of the optimum with a favorable electrostatic binding contribution. This result is perhaps unexpected because corresponding computations for docking *natural* complexes (generally carried out with fixed conformation and titration state) routinely find that the electrostatic contribution is unfavorable [9, 23, 24, 25, 26, 27]. However, calculation [15, 22, 28] and theory [29] have shown that the electrostatic contribution of optimal ligand-charge distributions to the total binding free energy can be favorable in many situations of physical interest.

    The favorability of the optimal electrostatic contribution has been proven [29] under the conditions of rigid binding, zero ionic strength ($\kappa = 0$), high-dielectric solvent ($\epsilon_s > \epsilon_m$), conservation of the total receptor plus ligand molecular volume (defined as the interior of the solvent–solute dielectric boundary) upon complex formation, and the absence of buried regions of disordered solvent in at least one of the reactants [29]. However, in practice many of these constraints may be relaxed or removed with the optimal electrostatic binding free energy remaining favorable.

3.3 Practical Considerations

Due to simplifications in the theory, a number of questions arise when considering its application to ligand design problems. The ligand-charge distribution obtained is optimal assuming a fixed ligand shape and ligand–complex geometry. It should be noted that the optimization is valid even in cases of receptor conformational change on binding because the receptor desolvation penalty is a constant that does not affect the optimization. What if, in practice, the ligand–complex alters conformation from that used in the design? Thermodynamics guarantees that conformational relaxation only serves to *lower* the free energy, so in this case the designed ligand would bind more tightly than predicted. In the design of tight-binding ligands, such an effect is not undesirable. If the conformational change is hypothesized (from theory) or known (from experiment), the optimization can be re-run in the changed conformation to seek further improvement.

In general, what conformation of the receptor should be targeted through charge optimization? This is a difficult question whose answer may depend on details of the system under study. If one has high-resolution structural data of the bound and unbound states, there are competing considerations. Often, details of the active site are better resolved in the bound state and the structural ligand can serve as the basis for further design efforts. If the unbound state is targeted, however, theory nearly guarantees that optimal electrostatics will favor binding in many cases of interest. The question really amounts to discovering whether the free energy cost of distorting away from the unbound state can be more than recovered in additional binding enhancements. Not enough work has been carried out to do more than frame the question at this point.

3.4 Non-Uniqueness of the Optimized Ligand-Charge Distributions

Optimized ligand-charge distributions may be non-unique for four reasons: (1) because the optimization condition, Eq. (8), is a statement about potentials rather than charges directly, two different charge distributions that produce the same potentials will be equally optimal [29]; (2) the numerical methods employed can lead to a non-empty null-space, which is a space of vectors that can be added to the ligand-charge distribution without significantly changing the binding free energy [21]; (3) the use of different basis sets, as a practical matter, may lead to different representations of charge distributions that produce essentially the same potentials; and (4) from a chemical standpoint, all molecules within a few tenths of a kcal/mol of the optimum binding energy may span a significant amount of chemical space while having very nearly optimal binding free energies. These forms of non-uniqueness are likely to be helpful in finding molecules whose shape and charge distribution are similar to optima. In such cases, one may apply constraints to find ligands that bind well, but which may have other more desirable properties, such as specificity or chemical feasibility. In a model sense, a refinement process of this type was used in previous work on an idealized problem to construct a spherical ligand for a barnase-like receptor consisting of only 11 point charges of physical separations and magnitudes, with a total charge of –1, and with favorable binding free energy [22]. This example illustrates how one may reassert the constraints of chemical feasibility on the charge distribution and satisfy them while still obtaining ligands with electrostatically favorable binding free energies.

3.5 Analytic and Calculational Studies and Considerations

The elements of some or all of the electrostatic binding matrices, R, C, and L, may be analytically determined in several simplified situations. If the ligand is treated as a sphere which binds to a receptor forming a larger spherical complex molecule, then the C and L matrices may be analytically obtained for point-charge or multipolar ligand-charge distributions [20], even in the presence of aqueous ions [30]. This spherical geometry has been extensively used in the past, in Tanford–Kirkwood theory [31, 32], to represent protein molecules. In the absence of aqueous ions, these two matrices may also be analytically obtained when the ligand and complex are ellipsoidal [30]. All three binding matrices can be analytically determined if the molecules are treated as "slabs," finite width in the z-direction and of infinite extent in the x- and y-directions, in an ion-free solution [15]. This particular geometry is amenable to the description of planar molecules or membranes [33, 34].

We have performed analytical optimization calculations in both spherical and slab geometries, and spherical geometries with aqueous ions [15, 20, 22, 30]. In some instances, we have found a strong correlation between the receptor desolvation penalty and the magnitude of the optimized electrostatic binding free energy. This suggests that the optimized ligands mimic and improve upon the interactions of the displaced solvent molecules of the active site. We have also found computationally that all electrostatic binding free energies in these geometries were favorable, as required by theory [29], and some were very favorable — contributing better than –10 kcal/mol to the total binding energy. This is remarkable, especially considering the relatively small size of the receptors and ligands examined. For the association of much larger proteins, the optimal contributions may be exceedingly favorable, leaving much room for the application of physical design constraints.

When the molecular geometry does not allow analytic calculation of the binding matrices, the elements may still be obtained through a numerical solution of the linearized Poisson–Boltzmann equation using a computer program package such as DELPHI [10, 18, 35]. The computational expense of charge optimization is dominated by the calculation of the ligand desolvation matrix, which requires a pair of solutions to the Poisson–Boltzmann (one for the bound complex state and one for the unbound ligand state) for each column of the matrix.

4 Electrostatic Complementarity

In the process of obtaining an optimized ligand-charge distribution, one takes the gradient of Eq. (7) with respect to the ligand-charge distribution and sets it equal to zero to find the minimum. This process yields Eq. (8). Returning to the definitions of the matrices involved, we see that this implies that the bound-state solvent-screened coulombic potential of the receptor charges and the difference in the bound- and unbound-state solvent-screened coulombic potentials of the ligand charges at each of the ligand's point-charge locations are equal in magnitude and opposite in sign. For an optimal ligand, this corresponds to a cancellation of these potentials everywhere within the ligand and on its surface [15]. If the ligand is non-optimal (remember, it may be optimized in some limited basis without being actually optimal), then the left-hand side of Eq. (8), denoted the "residual potential," will not be zero in all of these regions. Furthermore, due to the harmonic nature of this potential [15], the largest deviations from zero will be on the ligand surface.

The residual potential has several practical applications. All arise from the necessary cancellation of the potentials

$$\mathbf{Q}_r^t C = -2\mathbf{Q}_l^t L \qquad (10)$$

everywhere within and on the surface of an optimal ligand. Because of this, an optimal ligand and its receptor are called electrostatically complementary. Deviations from optimal electrostatic complementarity appear as deviations from equality in Eq. (10). This implies the existence of one particularly useful graphical tool. Displaying the two contributing potentials on the ligand surface in adjacent images, or simply displaying their sum, the residual potential, is a rapid way to visualize regions of particularly good or poor electrostatic complementarity. We find it useful to examine natural complexes this way. Scripts to carry out this analysis with the GRASP [36] software package are available at http://mit.edu/tidor/ along with example images. This is a particularly effective method for locating buried but uncompensated groups at a binding interface. The new method should be contrasted with a standard method used by structural biophysicists for examining binding complementarity, which involves examining the surface electrostatic potential of the free ligand and free receptor. While such an approach can offer a qualitative picture of whether oppositely signed potentials are arranged across the binding interface, it does not produce a quantitative measure of complementarity. In particular, it neglects all consequences of desolvation that accompany binding and does not permit a comparison of the relative magnitudes of desolvation and interaction effects that are essential for optimal binding. In actual practice, we anticipate that this visualization technique can be used to identify sites of non-complementarity at molecular interfaces. Appropriate mutations (charged to polar or hydrophobic, for instance) might be introduced that improve the computed complementarity. Synthesis and analysis of mutants experimentally could then be undertaken to test the modeling.

The residual potential may also be used for basis set improvement. It is possible that with the addition of more ligand point-charge locations, the optimized binding free energy could improve — dramatically if the initial set of locations are poorly chosen. By comparing the receptor interaction potential (LHS of Eq. (10)) with the optimized ligand desolvation potential (RHS of Eq. (10)), the extent of complementarity can be judged. Regions of the ligand that are especially non-complementary can be enhanced with additional basis points and the optimization re-run [15]. While basis sets consisting of point charges at atomic centers are commonly used, we have found that basis sets based on the "inverse-image" positions of the receptor charges often produce very complementary ligands [15].

5   Conclusions

The problem of optimizing the electrostatic free energy of binding in the continuum model is solvable for many cases of interest. The resulting optimal charge distributions minimize the electrostatic contribution to the binding free energy. This contribution has been numerically and theoretically shown to favor binding in many situations of biophysical interest. The ability to obtain optimal charge distributions may provide useful templates for design and for searching molecular libraries.

The theory is generalizable to other linear response models of molecular interaction.

Current work extends these basics of charge optimization to the optimization of the balance between specificity and affinity. Research on molecular recognition problems and the problem of multiple receptor titration states is also in progress, as well as improvement of the numerical methodology for optimizing realistically-shaped molecules and converting these optima to physically realizable charge distributions.

References

[1] Z. S. Hendsch and B. Tidor. Do salt bridges stabilize proteins? A continuum electrostatic analysis. *Protein Sci.*, 3:211–226, 1994.

[2] C. Tanford, P. K. De, and V. G. Taggart. The role of the α-helix in the structure of proteins. Optical rotatory dispersion of β-lactoglobulin. *J. Am. Chem. Soc.*, 82:6028–6034, 1960.

[3] C. H. Paul. Building models of globular protein molecules from their amino acid sequences. I. Theory. *J. Mol. Biol.*, 155:53–62, 1982.

[4] C. V. Sindelar, Z. S. Hendsch, and B. Tidor. Effects of salt bridges on protein structure and design. *Protein Sci.*, 102:4404–4410, 1998.

[5] B. Honig, K. Sharp, and A.-S. Yang. Macroscopic models of aqueous solutions: Biological and chemical applications. *J. Phys. Chem.*, 97:1101–1109, 1993.

[6] B. Honig and A. Nicholls. Classical electrostatics in biology and chemistry. *Science (Washington, D.C.)*, 268:1144–1149, 1995.

[7] Kim Sharp, J. Jean-Charles, and B. Honig. *J. Phys. Chem.*, 96:3822–3828, 1992.

[8] Jian Shen and Florante A. Quiocho. Calculation of binding energy differences for receptor–ligand systems using the poisson–boltzmann method. *J. Comput. Chem.*, 16:445–448, 1995.

[9] J. Shen and J. Wendoloski. Electrostatic binding energy calculation using the finite difference solution to the linearized Poisson–Boltzmann equation: Assessment of its accuracy. *J. Comput. Chem.*, 17:350–357, 1996.

[10] M. K. Gilson, K. A. Sharp, and B. H. Honig. Calculating the electrostatic potential of molecules in solution: Method and error assessment. *J. Comput. Chem.*, 9:327–335, 1988.

[11] O'M. Bockris and A. K. N. Reddy. *Modern Electrochemistry.* Plenum, New York, 1973.

[12] F. M. Richards. Areas, volumes, packing, and protein structure. *Annu. Rev. Biophys. Bioeng.*, 6:151–176, 1977.

[13] M. L. Connolly. Analytical molecular surface calculation. *J. Appl. Cryst.*, 16:548–558, 1983.

[14] M. K. Gilson and B. H. Honig. *Nature (London)*, 330:84, 1987.

[15] Erik Kangas and Bruce Tidor. Optimizing electrostatic affinity in ligand–receptor binding: Theory, computation and ligand properties. *J. Chem. Phys.*, 109:7522–7545, 1998.

[16] K. A. Sharp and B. Honig. Calculating total electrostatic energies with the nonlinear Poisson–Boltzmann equation. *J. Phys. Chem.*, 94:7684–7692, 1990.

[17] J. Theodoor G. Overbeek. The role of energy and entropy in the electrical double layer. *Colloids and Surfaces*, 51:61–75, 1990.

[18] M. K. Gilson and B. Honig. Calculation of the total electrostatic energy of a macro-molecular system: Solvation energies, binding energies, and conformational analysis. *Proteins: Struct., Funct., Genet.*, 4:7–18, 1988.

[19] Wolfgang K. H. Panofsky and Melba Phillips. *Classical Electricity and Magnetism.* Addison–Wesley, Reading, Massachusetts, second edition, 1962.

[20] L.-P. Lee and B. Tidor. Optimization of electrostatic binding free energy. *J. Chem. Phys.*, 106:8681–8690, 1997.

[21] G. Strang. *Introduction to Linear Algebra.* Wellesley–Cambridge Press, Wellesley, Massachusetts, 1993.

[22] L. T. Chong, S. E. Dempster, Z. S. Hendsch, L.-P. Lee, and B. Tidor. Computation of electrostatic complements to proteins: A case of charge stabilized binding. *Protein Sci.*, 7:206–210, 1998.

[23] J. Novotny and K. Sharp. Electrostatic fields in antibodies and antibody/antigen complexes. *Prog. Biophys. Molec. Biol.*, 58:203–224, 1992.

[24] V. K. Misra, K. A. Sharp, R. A. Friedman, and B. Honig. Salt effects on ligand–DNA binding: Minor groove binding antibiotics. *J. Mol. Biol.*, 238:245–263, 1994.

[25] V. K. Misra, J. L. Hecht, K. A. Sharp, R. A. Friedman, and B. Honig. Salt effects on protein–DNA interactions: The λcI repressor and EcoRI endonuclease. *J. Mol. Biol.*, 238:264–280, 1994.

[26] K. A. Sharp. Electrostatic interactions in hirudin–thrombin binding. *Biophys. Chem.*, 61:37–49, 1996.

[27] J. Novotny, R. E. Bruccoleri, M. Davis, and K. A. Sharp. Empirical free energy calculations: A blind test and further improvements to the method. *J. Mol. Biol.*, 268:401–411, 1997.

[28] Results from our research group.

[29] Erik Kangas and Bruce Tidor. Charge optimization leads to favorable electrostatic binding free energy. *Phys. Rev. E*, 59:5958–5961, 1999.

[30] Erik Kangas and Bruce Tidor.

[31] J. G. Kirkwood. Theory of solutions of molecules containing widely separated charges with special application to zwitterions. *J. Chem. Phys.*, 2:351–361, 1934.

[32] Charles Tanford and John G. Kirkwood. Theory of protein titration curves. i. general equations for impenetrable spheres. *J. Am. Chem. Soc.*, 79:5333–5339, 1957.

[33] V. A. Parsegian. Ion–membrane interactions as structural forces. *Ann. N. Y. Acad. Sci.*, 264:161–174, 1975.

[34] E. von Kitzing and D. M. Soumpasis. Electrostatics of a simple membrane model using Green's functions formalism. *Biophysical J.*, 71:795–810, 1996.

[35] K. A. Sharp and B. Honig. Electrostatic interactions in macromolecules: Theory and applications. *Annu. Rev. Biophys. Biophys. Chem.*, 19:301–332, 1990.

[36] A. Nicholls, K. A. Sharp, and B. Honig. Protein folding and association: Insights from the interfacial and thermodynamic properties of hydrocarbons. *Proteins: Struct., Funct., Genet.*, 11:281–296, 1991.

Optimization in Computational Chemistry and Molecular Biology, pp. 243-261
C. A. Floudas and P. M. Pardalos, Editors

Exploring potential solvation sites of proteins by multistart local minimization

Sheldon Dennis, Carlos J. Camacho, and Sandor Vajda
Department of Biomedical Engineering
Boston University
44 Cummington Street, Boston MA 02215
vajda@bu.edu

Abstract

The thermodynamics of solvation is studied by exploring the local minima of a function that describes the free energy of water around a protein. In particular, we determine if the ordered water positions in the crystal become preferred solvent binding sites in solution. The free energy is obtained by determining the electrostatic field of the solvated protein from a continuum model, and then calculating the interactions between this field and a single water molecule. The local minima in the neighborhood of selected points are explored by two different approaches. The first is a simple mapping of the free energy on a grid. The resulting maps show that the "free energy pockets" around crystallographic water sites are clusters of local minima. The second approach is based on the classical simplex algorithm which is used in two different implementations, one with a penalty function and the other modified for constrained minimization, called the complex method. Both the simplex and the complex methods are much faster than mapping the free energy surface. The calculations are applied to T4 lysozyme with data available on the conservation of solvent binding sites in 18 crystallographycally independent molecules. Results show that almost all conserved sites and the majority of non-conserved sites are within 1.3 Å of local free energy minima. This is in sharp contrast to the behavior of randomly placed water molecules in the boundary layer which, on the average, must travel more than 3 Å to the nearest free energy minimum. Potential solvation sites, not filled by a water in the x-ray structure, were studied by local free energy minimizations, started from random points in the first water layer.

Keywords: Solvation free energy, local minimization, clusters of local minima, simplex method, complex method, test problem

1   Introduction

Proteins neither fold nor function without bound water molecules, and understanding solvation is clearly a central problem in the biophysics of macromolecules. The experimental techniques that provide most information on the solvent structure around proteins are x-ray crystallography and NMR spectroscopy [1, 2]. In x-ray crystallography regions of extra electron density are interpreted as ordered water sites, resulting in about 200 water molecules in a typical high resolution protein structure [3]. NMR data reveal that in solution the water molecules around the protein are in rapid motion, including those that appear to be fixed in the x-ray structure. The exchange times are less than 500 ps with the exception of a few buried waters that may have residence times of up to 0.01 s [4, 5].

Since the electron density map derived from X-ray diffraction is averaged over a time scale measured in hours, discrete water density found in a crystal structure clearly does not indicate an actual water molecule at that position. It implies, however, that the potential of mean force has a local minimum; that is, the free energy of water must be lower than at all closely neighboring regions. As pointed out by Levitt and Park [3], if the free energy did not have a local minimum at that point, high electron density would not be found since, on average, water would be located relatively uniformly in that region.

The main computational tools used to study water behavior around proteins have been molecular dynamics and Monte Carlo simulations [6, 7, 8]. Such calculations confirmed that water molecules near the protein surface remain very mobile, with a diffusion coefficient decreased two- to four-fold relative to that of bulk water [8]. However, MD studies substantially differed in their evaluation of the influence of apolar, polar, and charged surface atoms on the mobility of the surrounding solvent [1, 7].

In this paper we search for preferred solvation sites by exploring local minima of a function that describes the free energy of water around the protein. The relationships between these preferred sites and the ordered water positions observed in the x-ray structure are also studied. We use a continuum model of electrostatics interactions [9, 10] in which the protein is represented by a low dielectric region containing discrete atomic charges at fixed positions, surrounded by a high dielectric medium representing the solvent, and calculate the electrostatic field of the solvated protein by solving the linearized Poisson-Boltzmann equation [9, 10]. The free energy of water is then calculated by translating and rotating a "probe" water molecule in the precalculated field. This approach removes the need for estimating the free energy change by averaging over a large number of trajectories as required in molecular dynamics, and reduces the problem to exploring the local minima of the free energy surface.

We focus on the following problems.

Problem 1: In order to determine if water positions in the x-ray structure are retained as preferred solvation sites in solution, we will explore the neighborhood of crystallographic water positions for local free energy minima.

Problem 2: In order to determine if there exist preferred solvation sites that are not in the vicinity of any crystallographic water site, we will start local minimization runs from a large number of points randomly placed in the first water layer.

In both problems we need robust local minimization algorithms that can explore the local minima in the vicinity of a starting point without jumping to other regions of the search space. Two approaches will be used. In the first, robustness is assured simply by calculating the free energy on a grid in the plane of the two Euler angles describing the position of the water molecule, thus mapping the free energy surface in a neighborhood of selected points [11]. At each grid point minimization is still performed in the subspace of remaining variables. The resulting free energy maps show that the "free energy pockets" around crystallographic water sites are actually clusters of local minima, very close to each other and are separated by moderate free energy barriers.

While the free energy maps are informative, the method is far from efficient for finding local minima. Therefore we performed further calculations using the simplex algorithm [12], one of the most robust approaches to local minimization. Two versions of the method were used. In the first we employed a penalty function to represent excluded volume constraints, thereby converting the problem into an unconstrained one. The second version, referred to as the complex method, works directly with the constrained problem without a penalty function [13, 14] Both the simplex and the complex methods are much faster than mapping the free energy surface.

The calculations were applied to the x-ray structure of the T4 lysozyme that includes 137 ordered water sites. This system is particularly interesting because x-ray structures are available for 18 crystallographycally independent T4 lysozyme molecules, including the wildtype 4lzm and nine mutants [15]. The comparison of these structures provides valuable information on the conservation of water sites across different crystal forms of the same protein.

2 Methods

2.1 Conformational space

The free energy will be calculated for a "probe" water molecule translating and rotating around the protein. The position of each water molecule is given in a local coordinate system centered at some point $w_0 \in R^3$. In the free energy mapping we use spherical coordinates, i.e., the location of the water molecule is given in terms of the Euler angles ϕ and θ, and the radius r. Further three Euler angles, ϕ_w, θ_w, and ψ_w specify the orientation of the water. Thus, a point in this space is given by the vector $s = (r, \phi, \theta, \phi_w, \theta_w, \psi_w)^T$, and the search space is defined by $S = \{0 \leq r, 0^o \leq \phi \leq 360^o, 0^o \leq \theta \leq 180^o, 0^o \leq \phi_w, \psi_w \leq 360^o, 0^o \leq \theta_w \leq 180^o\}$. By contrast, the water position is given in Cartesian coordinates when the search is performed by the simplex or the complex method and a point is specified by the vector $s' = (\Delta x_w, \Delta y_w, \Delta z_w, \phi_w, \theta_w, \psi_w)^T$. In Problem 1, w_0 is a crystallographic water position, whereas in Problem 2 w_0 is a randomly selected point in the first water layer.

2.2 Free energy function

Let Φ denote the electrostatic field of a solvated protein. The free energy of a water molecule at position $s \in S$ is given by

$$\Delta G_{el}(s) = \sum_{i=1}^{3} \Phi(x_i) q_i \qquad (1)$$

where x_1, x_2 and x_3 denote the positions of the atoms O, H_1, and H_2 in the water molecule, and q_1, q_2, and q_3 denote the corresponding atom-centered partial charges. We use the

TIP3 water model [16] which has rigid geometry, and hence for a given w_0 the vector $s \in S$ determines the atomic positions x_1, x_2 and x_3.

The electrostatic free energy is subject to steric constraints of the form

$$D_j < d_j, \ j = 1, ..., n \tag{2}$$

where d_j is the distance between the center of the water and the jth protein atom. The lower bound D_j is the sum of van der Waals radii of the jth protein atom and that of the water molecule, i.e., $D_j = r_j + r_w$.

To calculate the electrostatic field Φ, the linearized Poisson-Boltzmann equation is solved by a finite difference method as implemented in CONGEN [17]. The algorithm features adjustable rectangular grids, a uniform charging scheme that decreases the unfavorable grid energies, and smoothing algorithms that alleviate problems associated with discretization. The calculations were carried out using a 0.8 Å grid, with uniform charging, anti-aliasing, and 15-point harmonic smoothing. A 8 Å grid margin was maintained around the molecule. The dielectric constants of the protein and the solvent were set to 2 and 78, respectively, and the ionic strength was 0.05 M. Since the electrostatic field Φ is obtained only at the grid points, we employed a linear interpolation formula when calculating ΔG_{el} by Equation 1. Thus, with a precalculated field, free energy evaluation is extremely simple.

Before the free energy calculation, the x-ray structure of the protein has been refined by 200 steps of energy minimization, with the ordered water molecules included. We used version 19 of the Charmm potential [18] with polar hydrogens, and 20 Kcal/mol/Å ^{2} harmonic constraints on the positions of non-hydrogen atoms. These calculations placed the polar hydrogens and created a plausible hydrogen-bonding network between protein and water. The RMS shift of the water molecules due to the minimization was 0.1 Å.

2.3 Free energy maps

In order to understand the properties of the function defined by Equation 1 we first mapped the free energy surface in the vicinity of crystallographic solvent sites (Problem 1). Let w_0 denote the position of an ordered site. In the spherical coordinate system placed at w_0 we consider the lines defined by Euler angles (ϕ_i, θ_i), where $\phi_i = 0^o, 18^o, ...342^o$, and $\theta_i = 0^o, 18^o, ..., 180^o$. The free energy is minimized along each of these lines in terms of the remaining variables $(r, \phi_w, \theta_w, \psi_w)$ subject to the steric constraints given by Equation 2.

The constrained minimization problem is solved using a penalty function approach, i.e., by the unconstrained minimization of the extended target function

$$Q = \Delta G_{el} + CV_{exc} \tag{3}$$

where V_{exc} is an excluded volume penalty function defined by

$$V_{exc} = \begin{cases} \sum_{j=1}^{n} (D_j - d_j)^2 & \text{if } d_j < D_j; \\ 0 & \text{otherwise,} \end{cases} \tag{4}$$

and C is a weighting coefficient. As we will show, using $C = 10^5$ the atomic overlaps after the convergence of the minimization are so small that the ΔG_{el} term is not significantly affected. Notice that using the extended potential any steric overlap will show up as a sudden increase on the free energy maps.

For each (ϕ_i, θ_i) our goal is to find the local free energy minimum closest to the original water position w_0 along the line defined by (ϕ_i, θ_i), i.e., the smallest displacement r at which a local minimum occurs. Therefore we select $r = 0$ as the initial displacement. However, we also need to choose starting values for the variables $(\phi_w, \theta_w, \psi_w)$ that describe the orientation of the "probe" water molecule. As may be expected in a nonlinear problem and is confirmed by our calculations, for a given (ϕ_i, θ_i) and initial point $r = 0$ it is possible that, for some initial values of $(\phi_w, \theta_w, \psi_w)$, the Powell method does not find the local minimum corresponding to the smallest displacement r. In some cases the search may terminate early, e.g., due to being restricted to a subspace of the four-dimensional search space; in others the procedure may jump to a more distant local minimum. These artifacts can be easily identified by repeating the minimization for each (ϕ_i, θ_i) 30 times with different random orientations of the probe as the starting state. Since early termination or jumping over a local minimum along the line defined by (ϕ_i, θ_i) are relatively rare, the majority of the 30 runs yields very similar displacement values, well distinguishable from the few runs that end up in substantially different points. For each (ϕ_i, θ_i) we consider only the highly populated cluster of final water displacements, and choose the lowest ΔG_{el} value within this cluster as the free energy minimum along the line defined by the Euler angles (ϕ_i, θ_i).

Restricting consideration to the above minima the free energy surface can be visualized using two maps. The first map shows the minimum free energy as a function of ϕ_i and θ_i. The second map shows the displacement r at which the minimum occurs. The two maps together show if there is a free energy pocket in the vicinity of the crystallographic water site, and how far this pocket extends along each direction before the free energy starts to increase. We note that the analysis of the final maps provides an independent method to test if a local minimum found by the Powell method is the closest one to the origin along the line (ϕ_i, θ_i). In fact, an early termination or jump shows up as a discontinuity at the corresponding grid point on the displacement map, and most frequently also on the free energy map. As we will further discuss, selecting the dominant cluster from the 30 runs for each (ϕ_i, θ_i) assured the continuity of both maps for almost all water molecules.

2.4 Minimization by the simplex method

The classical simplex method is for unconstrained minimization, and will be used with the extended function defined by Equation 3. The algorithm proceeds as follows [12].

1. Select a starting simplex represented by $k = n + 1$ vertices $x^{(1)}, x^{(2)}, \ldots, x^{(k)}$. The method actually starts with only one feasible point $x^{(1)}$, and the remaining $k - 1$ points are found one at a time by random selection in the neighborhood of $x^{(1)}$.

2. Evaluate the target function Q at the k vertices and select the worst point x^{max} such that $f(x^i) \leq f(x^{max})$ for $i = 1, 2, \ldots, k$, and the best point x^{min} such that $f(x^i) \geq f(x^{min})$ for $i = 1, 2, \ldots, k$,

3. Calculate the the centroid of the simplex by

$$\bar{x} = [\sum_{i=1}^{k} x^{(i)} - x^{max}]/n \qquad (5)$$

Notice that the centroid excludes the worst point x^{max}.

4. Calculate (1) the reflection point x^* with $x^* = (1 + \alpha)\bar{x} - \alpha x^{max}$, where $\alpha > 0$, (2) the expansion point x^{**} with $x^{**} = \gamma x^* + (1 - \gamma)\bar{x}$, where $\gamma > 1.0$, and (3) the contraction point x^{***} with $x^{***} = \beta x^{max} + (1 - \beta)\bar{x}$, where $0 \leq \beta \leq 1.0$

5. Evaluate the function Q at the reflection point. If the reflection point is better than the best point, evaluate Q at the expansion point x^{**}. If the expansion point is better than the reflection point, replace the worst point in the simplex by the expansion point, otherwise replace it by the reflection point.

6. If the reflection point is better than the worst point, replace the worst point by the reflection point in the simplex. Otherwise evaluate Q at the contraction point. If the contraction point is better than the worst point, then replace the worst point by the contraction point in the simplex; otherwise reduce the size of the simplex leaving only the best point in place [12].

7. Terminate the iteration if the norm in the correction of the centroid is smaller than a threshold ϵ.

The algorithm has great versatility in adopting the simplex to the local free energy landscape. It will elongate and take a larger step if it can do so, it will change direction on encountering free energy barriers at an angle, and it will contract in the neighborhood of a minimum. However, for our purposes the most important feature of the method is its robustness which can be further increased by defining an upper bound on any side of the simplex, which will assure that the method will explore nearby local minima and will not jump to a far point of the conformational space. Although we may need to evaluate the function at more points than for a method with a superlinear convergence rate, the steps made by the simplex generally provide useful information on the form of the surface.

The simplex algorithm obviously provides only a local minimization tool. As we will further discuss, in the present application we want to explore the local minima in the neighborhood of a starting point. This will be accomplished by performing 30 minimization runs with randomly generated simplexes placed around the given starting state. The points representing the results of the 30 minimizations are clustered in the Cartesian space, i.e., differences in the rotational coordinates ϕ_w, θ_w, and ψ_w are ignored. We selected a simple clustering algorithm (see http://mvhs1.mbhs.edu/mvhsproj/projects/clustering/algorithm2.html). The original method introduces an appropriate number of clusters such that the distances within each cluster are smaller than the half-distance between any two clusters, where the inter-cluster distance is defined as the distance between the hubs of the corresponding clusters. This algorithm was modified by introducing a lower bound L on the inter-cluster distances, and thus any two clusters with a distance below L are concatenated into a single cluster. As we will discuss, this lower bound accounts for the fact that water positions in two different crystal forms of the same protein have been clustered into a single solvation site if they were closer than 1.2 Å to each other [15, 20].

2.5   Minimization by the complex method

The complex method is a straightforward extension of the simplex method ([14], p. 292) for solving constrained problems without introducing a penalty term. The method can handle only inequality constraints. The complex method is essentially a simplex method with the additional condition that any vertex must be a feasible point. Thus, every time a trial point

$x^{(j)}$ is generated, we find whether it satisfies all the constraints. If it violates any of the constraints, the point is moved half way toward the centroid of the already accepted points. This reduction in step size is continued until a feasible point is found. We will ultimately be able to find a feasible point by this procedure provided the feasible region is convex.

For comparison we tried to keep decision making in simplex and complex algorithms as similar as possible, but there are some essential differences. In the simplex method we use $k = n + 1$ vertices as suggested by Nelder and Mead [12]. A larger k leads to early convergence and generally causes the method to miss the extreme point being searched for. By contrast, for the complex method Box [13] recommended a value $k \approx 2n$, and we used $k = 2n$. In fact, if k is not sufficiently large, the complex tends to collapse and flatten along the first constraint boundary encountered. The other difference is in the selection of the starting simplex. In the case of the simplex method the only requirement is to "fill" the entire space, i.e., to avoid the flattening of the simplex to a subspace. By contrast, the complex method requires points that are feasible. As we described, feasibility can be assured by reducing the simplex if the feasible region is convex and at least one feasible point is already available. The first feasible point is placed by a randomized trial-and-error procedure. As in the case of the simplex method, we perform 30 minimization runs with different initial simplexes placed in the vicinity of the given starting point, and cluster the solutions to obtain information on the nearby local minima.

2.6 Application

We use the 1.7 Å resolution x-ray structure of the T4 lysozyme (PDB code 4lzm) that includes 139 water molecules. We placed the polar hydrogens and removed potential steric overlaps by 200 steps of energy minimization using version 19 of the Charmm potential [18] and 20 Kcal/mol/Å^2 harmonic constraints on the positions of non-hydrogen atoms. The refinement of the crystal structure results in an RMS shift of 0.1 Å.

T4 lysozyme is studied because the x-ray structure is available for a number of mutants that differ from the wildtype only at one or two positions [15]. Although the mutations affect the structure in the immediate vicinity of the amino acid substitution, the rest of the protein remains essentially unchanged. In some cases some "hinge-bending" motion occurs, but it can be corrected by bringing the rigid body fragments into a reference frame. The overlap of 18 x-ray structures results in a total of 1675 water molecules. Zhang and Matthews [15] clustered these waters by taking each water in turn, and counting how many other water molecules occurred within 1.2 Å, forming a total of 139 clusters. The analysis of these clusters provided information on the conservation of solvation sites. In particular, 40 sites that were occupied in at least 7 of the 18 molecules have been defined as conserved water positions.

3 Results and Discussion

In this paper we focus on the simplex and complex methods. A summary of free energy mapping results is given, with details reported elsewhere [21]. As we will show, the maps help to understand the optimization problem and the behavior of the algorithms.

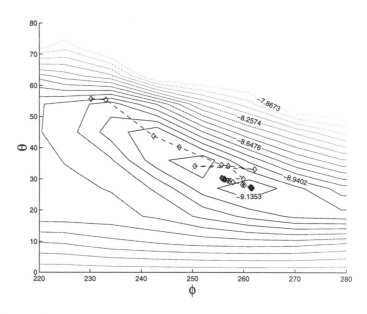

Figure 1: Contour lines of the minimum free energy surface in site 311 as a function of the Euler angles ϕ and θ. The dashed line shows the trajectory in a simplex minimization.

3.1 Free energy maps

Free energy and displacement maps have been constructed for the neighborhoods of crystallographic water sites in the x-ray structure 4lzm.

As an example, Figures 1 and 2 show such maps for water site No. 311. Figure 1 shows the minimum free energy as a function of the Euler angles ϕ and θ, i.e., the lowest free energy values attained along the lines defined by (ϕ, θ) as we move outward from the center of the local coordinate system. Figure 2 shows the displacement r at which this minimum is attained.

The free energy and displacement maps together provide substantial information on the local behavior of the free energy function. Based on the free energy maps we conclude that the free energy pockets discussed by Levitt and Park [3] are actually clusters of local minima, generally separated from other clusters by relatively high free energy barriers. Due to these energy barriers most clusters are well defined, and hence for each cluster we can identify the local minimum with the lowest free energy and determine its displacement r from the crystallographic water position. However, since there may exist several local minima with similar energies, we frequently study all minima within the cluster rather than focusing on the lowest energy one.

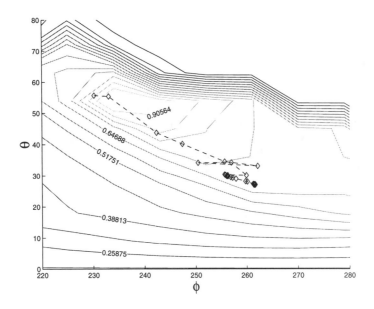

Figure 2: Displacement r at the free energy minima shown in Figure 1, including the pathway of the centroid in the simplex minimization.

Although the particular geometries differ, the site around water 311 is typical of most crystallographic water sites. Figure 1 shows only the region containing the two lowest energy minima, but the free energy pocket around this site includes two other minima that are beyond the boundaries of the plot. As the displacement map in Figure 2 shows, the cluster is well defined, outward moves from the crystallographic position may reduce the free energy for some distance, but after less then 1 Å displacement the free energy invariably starts to increase. In many cases the border of the pocket is very sharp because even a very small move along the line defined by the corresponding ϕ and θ pair steeply increases the free energy. Such steep increase is generally due to steric overlaps, indicating that the site is close to the protein surface. For water 311 the maximum displacement occurs at $\phi = 235^o$ and $\theta = 55^o$, and it is close to 1 Å. The lowest free energy value, -9.3 kcal/mol, is at $\phi = 261^o$ and $\theta = 29^o$ (Figure 1), and it is attained with slightly less than 0.8 Å displacement (Figure 2).

We attempted to construct free energy and displacement maps for all the 139 crystallographic water sites in 4lzm. However, in seven cases the maps showed sudden changes in the free energy at some points of the (ϕ, θ) plane.This may signal an inherent discontinuity of the free energy function due to the instability of the water molecule at the particular sites when the protein is in solution rather than in a crystal. However, as we described in the method, it is also possible that our procedure of performing 30 local minimizations at (ϕ_i, θ_i) and then selecting the most populated cluster is unable to identify the minimum closest to the original crystallographic site. Therefore we increased the number of mini-mization runs to 60, but were unable to remove the discontinuities. As we will discuss, the seven water sites were re-examined by the simplex method, and were shown to be inherent discontinuities.

Of the 132 remaining sites, 100 are within 1.3 Å of the closest local free energy mini-mum. In terms of displacement statistics, there is a slight difference between conserved and non-conserved water positions. The closest minimum is within 1.3 Å for the majority of conserved sites (31 out of the 36, or 86.1%). In the case of non-conserved sites, this fraction is 71.8%. However, in spite of the substantial shifts in the position of a few non-conserved waters, a comparison of displacement distributions for conserved and non-conserved sites does not show a significant difference. Similarly, there is no significant difference between the free energy distributions, although a few buried water molecules have very low free energies (≤ -12 kcal/mol).

The finding that most crystallographic sites are within 1.3 Å of a local minimum is in sharp contrast to the behavior of water molecules randomly placed in the first water layer. Such molecules, on the average, must move 3.2 Å to reach the nearest local minimum. Thus, we conclude that ordered water sites in the x-ray structure are at least partially due to favorable electrostatic interactions between the protein and the water, and the majority of such positions remain preferred solvation sites when the protein is in solution.

3.2   Search by simplex and complex methods: crystallographic sites

The free energy maps show the local minima in the neighborhood of 132 crystallographic water sites. To test the simplex and complex methods, we explored the same regions by performing 30 minimization runs with randomly selected initial simplexes around each site. As mentioned in Methods, the search is in the space defined by the vectors $s' = (\Delta x_w, \Delta y_w, \Delta z_w, \phi_w, \theta_w, \psi_w)^T$. In the translational subspace, the vertices of the initial simplex have been obtained by random displacements between -1 Å and 1 Å along each coordinate axis. The rotational coordinates ϕ_w and ψ_w of the vertices have been randomly selected in the 0° to 360° interval and θ_w from the 0° to 180° interval.

Figures 1 and 2 also show the trajectory of the centroid in a minimization by the simplex method, superimposed on the free energy and displacement maps, respectively. The trajectory shown contains 350 simplex iterations. As described in the method, for each water site we perform 30 minimization runs from different initial points, and cluster the solutions. For water 311, 24 of the 30 solutions form a single cluster (are within 1.2 Å to each other), and the remaining 6 points distribute among three other clusters, indicating three further local minima. Comparison with the free energy map shows that the most populated cluster is also the lowest free energy local minimum.

Figures 3 and 4 show the trajectory of the centroid in a minimization by the complex method. Since the algorithm includes a search for a feasible initial simplex in the $k = 2n = 12$-dimensional space, the initial position of the centroid generally differs from the one in the simplex method, even when the same seed is used in the random number generator. As shown by the relatively high energy, for the particular run in Figure 3 the initial simplex is in a region that is close to the protein but is still feasible (i.e., the extended target function is below -7 kcal/mol, indicating the lack of any steric overlap). Without a penalty term,

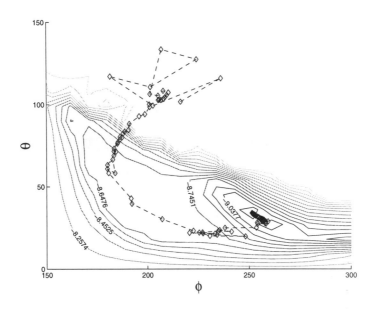

Figure 3: The pathway of the centroid of the simplex in a minimization by the complex method in site 311. The contour lines represent the minimum free energy surface shown in Figure 1.

the electrostatic component ΔG_{el} in this region changes slowly, and thus the simplex moves around before starting toward the minima in the free energy pocket. The trajectory of the centroid shown is actually the result of over 1,500 iterations.

Table 1 lists the average number of function evaluations and the average CPU time for the three different methods, the latter for runs on a single SiliconGraphics R10000 processor. For the grid-based approach both numbers are shown for the 30 minimization runs by the Powell method at a given (ϕ_i, θ_i) in the space of the four variables $(r, \phi_w, \theta_w, \psi_w)$. Notice that selecting the most populated cluster provides the free energy and displacement values at (ϕ_i, θ_i), i.e., a single grid point of the free energy and displacement maps. Since with the selected grid density each map consists of 220 grid points, the total average CPU time required for the construction of one map is 27,280 s, or about 7.6 hrs on a single R-10000 processor.

For the simplex and complex methods Table 1 shows the (average) total number of function evaluations and the (average) total CPU time in 30 minimization runs, starting with different initial simplexes. In contrast to the 30 minimizations in the grid-based approach that yield a single point of the free energy map, the 30 runs explore an entire cluster of local minima (see below). As shown in the Table, the general relationship between simplex and complex trajectories is well represented by Figures 1 and 3; i.e., the complex method, on the average, requires about five times as many iterations as the simplex method. In spite of this large difference in the number of function evaluations, the CPU times differ only by about 16%. This apparent contradiction is due to the extreme simplicity of function evaluation in this particular case, which is essentially interpolating in a table. Therefore, the CPU times are determined by the computational overhead rather than by the number of function evaluation, and the former seems to be very similar for simplex and complex methods. Both are much more efficient than mapping the free energy surface. For example, exploring a water site with the simplex method, on the average, requires 807 s and thus about 13 minutes instead of the 7.6 hrs CPU time for constructing a free energy map.

Table 1. *Performance characteristics of the three methods*

	Method		
	Grid-based	Simplex	Complex
Number of function evaluations	5153 ± 5112	6063 ± 360	28428 ± 2559
CPU time, s	124 ± 16	807 ± 109	944 ± 134

Why does the simplex method require much fewer function evaluations than the complex method? We recall that the simplex and complex methods work with $k = n + 1$ and $k = 2n$ vertices, respectively. We have evidence that using fewer vertices in the simplex method has a favorable effect, because increasing the number of vertices without any other change in the simplex algorithm increases the number of function evaluations. However, the better performance of the simplex method is mainly due to the use of the penalty function given by Equation 2 that facilitates the elimination of steric overlaps. Since the effect of the penalty term on the position of the minima can be neglected, particularly at the limited resolution of the grid, in the application considered here the simplex method with a penalty function is superior to the complex method that accounts for the constraints in a binary fashion, without any concept of direction.

For each water site, the results of the 30 runs are clustered as described in the Methods, using 1.3 Å as the lower bound L on the cluster size. Each cluster essentially represents a set of points that are within the region of attraction of a local minimum, and is labeled by the name of the atom closest to the hub of the cluster. By comparing the lists of local minima to the free energy maps we concluded the 30 runs for each site were sufficient to find the lowest minimum in each free energy pocket as one of the clusters, but not necessarily the lowest energy ones: for the 139 crystallographic sites, the most populated cluster is also the lowest energy one in 92 cases (66%). Comparison to the free energy maps shows that it is more meaningful to characterize the position of the free energy "pocket" (i.e., the cluster of local minima separated from other clusters by free energy barriers) by the position of the most populated cluster rather than by the position of the lowest energy cluster. The explanation is that in spite of the simplicity of the simplex method, some trajectories may end up in a different cluster, particularly if the free energy of the particular "pocket" is much

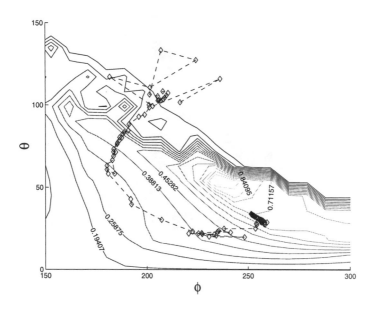

Figure 4: Displacement r at the free energy minima shown in Figure 3, including the pathway of the centroid in the complex minimization.

higher than the free energy of a neighboring pocket. However, the "jump" to a different cluster happens only in a small fraction of the trajectories, and hence the displacement of the most populated cluster is a more robust characterization of the distance between the original water position and the free energy pocket. Indeed, the most populated cluster is at the local minimum closest to the original site in 105 of the 139 crystallographic sites (75%). Furthermore, for the 36 conserved water sites that are generally better defined than the non-conserved sites, the most populated cluster is also the closest one in 29 cases (81%). In 4 of the remaining 7 sites the most populated cluster contacts the side chain that interacts with crystallographic water in the x-ray structure.

Using the simplex and complex methods we were able to explore the seven water sites for which no meaningful free energy maps have been obtained due to the sudden changes. It turns out that all corresponding free energy pockets are very shallow, resulting in five to eight clusters. In two cases (water sites 205 and 209) some trajectories end up close to the crystallographic water positions, but at very high energies (-1.95 kcal/mol and -2.07 kcal/mol, respectively), although for water 209 other water positions (230 and 337) are also obtained. For the other five sites (220, 233, 234, 255 and 260) none of the trajectories remains close to the crystallographic positions. Since all the seven sites discussed here are non-conserved, the weak energetics do not contradict the experimental evidence. However, it remains an open question as to why these water molecules are ordered in the x-ray structure.

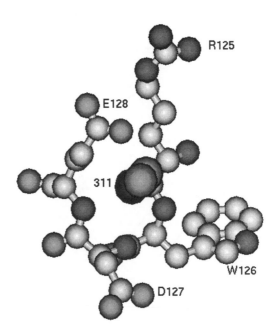

Figure 5: The two most populated local minima (large shaded spheres) in the simplex search from water position 311 (large dark sphere). Residues 125 through 128 of the T4 lysozyme are also shown using shades as follows: carbons lightly shaded, oxygens darker, and nitrogens dark.

For the 139 water sites the calculations yield 570 clusters (i.e., 4.1 clusters per site), 203 of which (40%) correspond to water molecules For 130 of the 139 sites, the atom closest to one of the clusters is an atom of the original water in its crystallographic position, with hydrogens placed by Charmm [18]. The corresponding clusters are generally well populated. This result is in complete agreement with the results of the free energy mapping, and confirms that the majority of crystallographic positions become preferred solvation sites in solution.

The 203 water clusters for the 139 sites means that for a number of water molecules some of the 30 trajectories end up in another water site, i.e., the crystallographic sites are not necessarily separated by high energy barriers. In addition, there are 367 non-water clusters. Most of these are close to side chains that interact with the particular water molecule in the x-ray structure. However, the additional minima are far enough from the crystallographic water sites to be regarded as separate clusters. In fact, high resolution protein structures reveal that a charged side chain, particularly Glu or Asp, can simultaneously contact up to five ordered water sites. The additional clusters correspond to such potential solvation sites that in the given x-ray structure are not filled with a well determined water. As we will see, most of these sites are on the "other side" of side chains that already contact at least one ordered water.

For example, Figure 5 shows the two most populated local minima (large gray spheres) when starting minimizations from water position 311 (large black sphere). Residues 125 through 128 of the T4 lysozyme are also shown. Notice that the site is surrounded by the charged side chains of Asp-127 and Glu-128, as well as by Arg 125 that is a bit farther from the water position. Rather than shifting to any of these side chains, water 311 is seen in a free energy pocket that is cooperatively determined by the three, and by additional charges on nearby backbone atoms. Free energy minimizations robustly identify this region of the free energy, but they also show that the region contains several local minima and thus several potential water positions, two of which are shown in Figure 5. Since these minima are close to each other and are separated by moderate free energy barriers, water molecules can shift from one to another.

As we have discussed, the most populated cluster is generally a useful characterization of the entire free energy pocket and hence we also collected statistics by restricting consideration to these clusters. The corresponding free energy was negative for all sites. The closest atom to the cluster was that of a water molecule for 103 of the 139 sites, but in two cases the trajectory actually ended close to a different water molecule, not the one in the original site. In 18 cases the populated cluster was close to a side chain that contacts a water molecule in the x-ray structure. This means that the minimized water orientation relative to the side chain differs from the one observed in the crystal. In 13 cases a backbone atom is the one closest to the populated minimum. The free energy of such water molecules ranges from -6.1 to -3.8 kcal/mol. Although these interactions are relatively weak (the free energy of some water molecules can be as low as -13 kcal/mol), the mean free energy for backbone-contacting water molecules is not significantly higher than the mean free energy for all ordered sites. In fact, some of the crystallographic water sites interact with the backbone.

There are 5 cases in which the minimized water contacts a side chain that has not been observed to contact an ordered water in the x-ray structure. Most frequently this situation occurs for charged or polar side chains that could interact favorably with water molecules, but are too ill-defined to be seen in the x-ray structure. In fact, Thanki et al. [22] observed that the highly mobile Lys side chains have no preferred orientation for water contacts. The same applies to most Tyr and Trp side chains.

We note that the initial simplex size, particularly in the translational subspace, is an important parameter that controls how well the local minima in a neighborhood are explored. If the size is too small, all runs remain in the same local minimum, whereas a simplex that is too large may move over to a different site. With the current selection at least 90% of trajectories remained in the free energy pocket, and essentially all local minima, observable on the free energy maps, have been found. We also found additional clusters corresponding to local minima that were not seen on the maps, most likely due to their limited resolution.

3.3 Search by simplex and complex methods: random positions

Although the results just described already provide some information on the potential water positions (i.e., free energy minima that do not correspond to crystallographic water sites), we randomly selected 100 water positions in the first solvation layer and explored the local minima in their neighborhoods by the simplex method. The hundred runs results in 414 clusters, i.e., the number of clusters per residue is 4.1, exactly the same as in the minimizations around crystallographic sites. Only 29 of these clusters are closest to a water site. This is 7% of the total, much less than around the ordered sites.

Restricting consideration to the most populated clusters we drop 8 minimization runs in which the final free energy is positive. 14 clusters are closest to an ordered water site. Although in the remaining 78 trials the most populated clusters are closer to protein atoms than to a crystallographic water, 36 of these interact with side chains that also contact at least one ordered water. There are 26 trials in which the dominant cluster moves to a backbone atom. For some of these sites the water-protein interaction is relatively weak, but the mean does not significantly differ from the mean calculated for all sites.

In the remaining 16 trials we find side chains that are not seen to contact an ordered water in the x-ray structure. These include four Lys residues, Lys-16, Lys-43 (twice), Lys-60, and Lys-162. Based on the thermal factors, all these side chains are very mobile in the crystal, and water molecules associated with them would not be seen. There are two Asn side chains, Asn-2 and Asn-81, that interact weakly with water. According to our results, the side chains of Asp 72 and Asp 92 must also be preferred solvation sites, and we do not fully understand why no ordered water is seen around them in the x-ray structure.

4   Conclusions

The structural and functional importance of water associated with proteins is well known [1, 2]. While x-ray crystallography and NMR techniques provide a wealth of structural data, they leave open important questions concerning the origin of preferred water sites, and the relationship between water positions in crystal structures and in solution. On the basis of the experimental data and the somewhat incomplete theoretical studies, Levitt and Park [3] formulated a number of assumptions. To explain the origin of an ordered water they concluded that the potential of mean force at that point must have a local minimum, i.e., the free energy of a water at closely neighboring regions (within say 2-3 A) must be relatively high, forming a free energy "pocket".

In this paper we studied the above free energy "pockets" by performing local minimization runs in the neighborhoods of interest, and then clustering the obtained local minima. The free energy was calculated for an explicit water molecule ("probe" molecule), interacting with the electrostatic field of the solvated protein. Using a precalculated field this approach results in a very simple free energy function. The computational problem is to find the local minima that are closest to a given point in space. However, accounting for both translation and rotation of the "probe" water, the search is in the space of six variables: the Euler angles ϕ and θ, the center-to-center distance r, and the Euler angles ϕ_w, θ_w, and ψ_w that specify the orientation of the water. The minimization is subject to steric constraints that prevent water-protein overlaps.

We used two different approaches to find the local minima in the neighborhood of selected points. The first employs a penalty functions and attempts the minimization by Powell's method. We have found that this method lacked the necessary robustness, i.e., depending on the initial values for the rotational variables ϕ_w, θ_w, and ψ_w, the algorithm frequently converged to distant local minima, ignoring others close to the initial point. We improved the performance by reducing the dimensionality of the problem, i.e., by converting it into a grid-based search for the variables ϕ and θ, and performing repeated local minimizations in the space of the remaining variables r, ϕ_w, θ_w, and ψ_w. Thus, in terms of ϕ and θ, this is a mapping of the free energy on a grid. The resulting maps showed that the free energy "pockets" around crystallographic water sites are actually clusters of local minima, and the different clusters are generally but not always separated by relatively high free energy barriers.

The second approach was based on the classical simplex algorithm which was used in two different implementations, one with a penalty function and the other modified for constrained minimization, called the complex method. Both the simplex and the complex methods were much faster than mapping the free energy surface. The methods have been used with multiple runs to explore the various local minima in the free energy "pockets". In particular, we have selected 30 random initial simplexes, performed the searches, and clustered the solutions. Each cluster essentially represented a set of points that were within the region of attraction of a local minimum. As we have shown, the positions of the most populated clusters provided the best key to describe the free energy pocket.

The calculations were applied to T4 lysozyme with data available on the conservation of solvent binding sites in 18 crystallographycally independent molecules. Results show that the majority of crystallographic water sites are within 1.3 Å of local free energy minima. This is in sharp contrast to the behavior of randomly placed water molecules in the boundary layer which, on the average, must travel more than 3 Å to the nearest free energy minimum. Potential solvation sites, not filled by a water in the x-ray structure, were studied by local free energy minimizations, started from random points in the first water layer, and generally found good agreement with the crystallographic site. There were some false positives, i.e., local minima that were not seen in the crystal structure as ordered water sites. However, the dominant majority of these false positives were not very wrong, because they were simply on the "other side" of side chains that contacted one or more water molecules anyway, but not in the same orientation.

Acknowledgments: This research has been supported by grant DBI-9630188 from the National Science Foundation and by grant DE-F602-96ER62263 from the Department of Energy.

References

[1] Karplus, P.A., Faerman, C.H. (1994), "Ordered water in macromolecular structure," *Curr. Opinion Struct. Biol. Vol. 4*, 770-776.

[2] Teeter, M.M. (1991), "Water-protein interactions: Theory and experiment," *Annu. Rev. Biophys. Biophys. Chem. Vol. 20*, 577-600.

[3] Levitt, M., Park, B. (1993), "Water: now you see it, now you don't," *Structure, Vol. 1*, 223-226.

[4] Otting, G., Liepinsh, E., Wuthrich, K. (1991), "Protein hydration in aqueous solution," *Science Vol. 254*, 974-980.

[5] Belton, P.S. (1994), "NMR studies of protein hydration," *Prog. Biophys. Mol. Biol. Vol. 61*, 61-79.

[6] Levitt, M., Sharon, R. (1988), "Accurate simulation of protein dynamics in solution," *Proc. Natl. Acad. Sci. USA. Vol. 85*, 7557-7561.

[7] Brunne, R.M., Liepinsh. E., Otting, G., Wuthrich, K,, van Gunsteren, W.F. (1993), "Hydration of proteins. A comparison of experimental residence times of water molecules solvating the bovine pancreatic trypsin inhibitor with theoretical model calculations," *J. Mol. Biol. Vol 231*, 1040-1048.

[8] Makarov, V.A., Feig, M., Andrews, B.K, Pettitt, B.M. (1998), "Diffusion of solvent around biomolecular solutes: a molecular dynamics simulation study," *Biophys. J. Vol. 75*, 150-158.

[9] Gilson, M.K., Honig, B. (1988), "Calculation of the total electrostatic energy of a macro-molecular system: solvation energies, binding energies, and conformational analysis," *Proteins Vol. 4*, 7-18.

[10] Honig, B., Nicholls, A. (1995), "Classical electrostatics in biology and chemistry," *Science Vol. 268*, 1144-1149.

[11] Camacho, C.J., Weng, Z., Vajda, S.,and DeLisi, C. (1999), "Free energy landscapes of encounter complexes in protein-protein association," *Biophys. J. Vol 76*, 1176-1178.

[12] Nelder, J.A., Mead, R. (1964) " A simplex method for function minimization," *Computer J. Vol. 7*, 308-313.

[13] Box, M.J. (1965) "A new method of constrained optimization and a comparison with other methods," *Computer J. Vol. 8*, 42-52.

[14] Rao S.S. (1978), *Optimization. Theory and Applications*, John Wiley and Sons, New York.

[15] Zhang, X.J., Matthews, J.W. (1994), "Conservation of solvent-binding sites in 10 crystals of T4 lysozyme," *Prot. Science Vol 3*, 1031-1039.

[16] Jorgensen, W.L, Chandrasekhar, J., Madura, J.D. (1983), "Comparison of simple potential functions for simulating liquid water," *J. Chem .Phys. Vol. 79*, 926-935.

[17] Bruccoleri, R.E. (1993), "Grid positioning independence and the reduction of self-energy in the solution of the Poisson–Boltzmann equation," *J. Comp. Chem. Vol. 14*, 1417-1422.

[18] Brooks, B.R., Bruccoleri, R.E., Olafson, B., States, D.J., Swaminathan, S., Karplus, M. (1983), "CHARMM: A program for macromolecular energy, minimization, and dynamics calculations," *J. Comp. Chem. Vol. 4*, 197-214.

[19] Press, W., Flannery, B.P., Teukolsky, S.A., Vetterling, W.T. (1990), *Numerical Recipes*, Cambridge University Press, Cambridge.

[20] Sanschagrin, P.C., Kuhn, L.A.(1998), "Cluster analysis of concensus water sites in thrombin and trypsin shows conservation between serine proteases and contributions to ligand specificity," *Prot. Science Vol. 7*, 2054-2064.

[21] Dennis, S., Camacho, C., Vajda, S. (1999), "A continuum electrostatic analysis of preferred solvation sites around proteins in solution, " *Proteins*, in press.

[22] Thanki, N., Thornton, J.M., Goodfellow, J.M. (1988), "Distribution of water around amino acid residues in proteins," *J. Mol. Biol. Vol. 202*, 637-657.

Optimization in Computational Chemistry and Molecular Biology, pp. 263-266
C. A. Floudas and P. M. Pardalos, Editors
©2000 Kluwer Academic Publishers

On relative position of two biopolymer molecules minimizing the weighted sum of interatomic distances squared

Andrei B.Bogatyrev
Institute for Numerical Math.
Gubkina, 8, Moscow GSP-1,
RUSSIA
gourmet@inm.ras.ru

Abstract

An exact analytical solution is given to a problem of relative arrangement of two molecules which minimizes the weighted sum of squared interatomic distances.

Keywords: Minimization on a Lie group, critical point, singular value, singular decomposition of a matrix.

In molecular genetics studies and drug design a problem of a search for similar portions in the chains of two distinct biopolymer molecules arises. This similarity is sometimes understood as the proximity of 3-D atomic configurations of the molecules. To estimate this value they consider the following extremum problem on the group of $\mathbf{R}^3$-isometries.

Given the positions of two sets of atoms $\{x_i\}_{i=1}^N$; $\{y_\alpha\}_{\alpha=1}^M$ $(x_i, y_\alpha \in \mathbf{R}^3)$ *and a set of nonnegative weights* $\{A_{i\alpha}\}$ *that measure the semblance of atoms* x_i *and* y_α *to obtain the minimum and the extremal arguments of the function*

$$J(\Omega, b) := \sum_{i=1}^{N} \sum_{\alpha=1}^{M} A_{i\alpha} |x_i - \Omega y_\alpha - b|^2, \tag{1}$$

here $\Omega \in SO(3)$ *is a rotation matrix and* $b \in \mathbf{R}^3$ *is a translation vector.*

This problem had in essence been solved in the series of papers [1, 2, 3], with the description of (all but one) degeneraces of this problem in the latter paper. Our approach to the solution differs from that of the quoted papers since we use the elementary differential geometry of Lie groups. This approach gives nothing new on the algorithm level, but it allows us to give a very detailed and complete classification of all degenerated cases for the above problem. The author is thankful to the reviewer for the references on the previous works in this field.

With the rotation matrix Ω being fixed minimum of J with respect to b is attained in the unique point $b^*(\Omega)$ where $\dfrac{\partial J}{\partial b}(\Omega, b^*) = 0$:

$$b^*(\Omega) = A^{-1}\left(\sum_{i\alpha} A_{i\alpha}(x_i - \Omega y_\alpha)\right), \tag{2}$$

where $A := \sum_{i\alpha} A_{i\alpha} > 0$. We arrived at the problem of minimization on group $SO(3)$ of a function

$$J(\Omega, b^*(\Omega)) = const(\{x_i\}, \{y_\alpha\}, \{A_{i\alpha}\}) + 2tr(X\Omega), \tag{3}$$

where the value of $const(\cdot, \cdot, \cdot)$ is independent of Ω and the matrix X is defined as:

$$X := A^{-1}\sum_{i\alpha j\beta} A_{i\alpha}A_{j\beta}y_\alpha x_j^t - \sum_{i\alpha} A_{i\alpha}y_\alpha x_i^t \tag{4}$$

The following two lemmas allow us to find and classify all critical points of

$$J_X(\Omega) := tr(X\Omega), \qquad \Omega \in O(3). \tag{5}$$

Lemma 1. *The point $\Omega \in O(3)$ is the critical one for J_X if and only if the matrix $H := X\Omega$ is symmetric*

Lemma 2. *Let $\Omega \in O(3)$ be a critical point of a function J_X; $\lambda_1, \lambda_2, \lambda_3$ are eigenvalues of $H := X\Omega$. Then the quadratic form $d^2 J_X(\Omega)$ defined on a tangent space $T_{O(3)}(\Omega)$ have the same number of positive and negative squares as the form determined by the matrix*

$$-\text{diag}\left((\lambda_1 + \lambda_2),\ (\lambda_2 + \lambda_3),\ (\lambda_3 + \lambda_1)\right).$$

Let us sketch the proofs of the stated above lemmas. The tangent space at $\Omega \in O(3)$ we identify [4] with the tangent space at the neutral element of the group $O(3)$ that is with the algebra $so(3)$ of real skew-symmetric matrices K of dimension 3.

$$0 = \langle dJ_X, K\rangle := \frac{d}{dt}tr(X\Omega\exp Kt)\Big|_{t=0} = tr(X\Omega K).$$

The last expression is equal to zero for all skew-symmetric matrices K if and only if the matrix $X\Omega$ is symmetric.

$$\langle d^2 J_X, K\rangle := \frac{d^2}{dt^2}tr(X\Omega\exp Kt)\Big|_{t=0} = tr(X\Omega K^2) =$$

$$= tr\left(\text{diag}(\lambda_1, \lambda_2, \lambda_3)(\text{Ad } U^t \cdot K)^2\right) =: tr(\text{diag}(\lambda_1, \lambda_2, \lambda_3)K_1^2),$$

where $U \in SO(3)$ diagonalizes H, $so(3) \ni K_1 := U^t K U$. If

$$K_1 = \left\| \begin{array}{ccc} 0 & a & b \\ -a & 0 & c \\ -b & -c & 0 \end{array} \right\|,$$

then the last expression in the previous chain of equalities has the appearance

$$-[(\lambda_1 + \lambda_2)a^2 + (\lambda_2 + \lambda_3)c^2 + (\lambda_3 + \lambda_1)b^2].$$

To simplify the usage of the lemmas 1, 2, we reduce minimization of J_X to minimization of $J_D(\Omega) := tr(D\Omega)$ with diagonal matrix $D := \text{diag}(\mu_1, \mu_2, \mu_3)$ where $\mu_1 \geq \mu_2 \geq \mu_3 \geq 0$ are the singular values [5] of X.

Lemma 3 *There exist orthogonal matrices Ξ_1, Ξ_2 composed of the right eigenvectors of respectively XX^t and X^tX for which holds the factorization*

$$X = \Xi_1 D \Xi_2^t. \tag{6}$$

Corollary

$$\min_{SO(3)} J_X = \begin{cases} \displaystyle\min_{SO(3)} J_D, & \text{if } \Xi_1\Xi_2 \in SO(3), \\ \displaystyle\min_{O(3)\backslash SO(3)} J_D, & \text{if } \Xi_1\Xi_2 \in O(3) \backslash SO(3). \end{cases}$$

Below are listed all the critical points of J_D on the group $O(3)$ and their classification.

Case 1. $\mu_1 > \mu_2 > \mu_3$.

	$SO(3)$	$O(3) \backslash SO(3)$	$d^2 J_D(\Omega)$ indices	point type
	$\text{diag}(1,1,1)$	$\text{diag}(1,1,-1)$	$(-,-,-)$	max
Ω	$\text{diag}(1,-1,-1)$	$\text{diag}(1,-1,1)$	$(-,-,+)$	saddle
	$\text{diag}(-1,1,-1)$	$\text{diag}(-1,1,1)$	$(-,+,+)$	saddle
	$\text{diag}(-1,-1,1)$	$\text{diag}(-1,-1,-1)$	$(+,+,+)$	min

Case 2. $\mu_1 = \mu_2 > \mu_3$.

	$SO(3)$	$O(3) \backslash SO(3)$	$d^2 J_D(\Omega)$ indices	point type
	$\text{diag}(1,1,1)$	$\text{diag}(1,1,-1)$	$(-,-,-)$	max.
Ω	$\Omega_-(e_2)$	$\Omega_+(e_2)$	$(-,0,+)$	saddle
	$-\text{diag}(1,1,-1)$	$-\text{diag}(1,1,1)$	$(+,+,+)$	min.

$$\Omega_{\pm}(e_2) = \left\| \begin{matrix} 2e_2 \cdot e_2^t - I_2 & \vdots & 0 \\ \cdots\cdots\cdots & \vdots & \cdots \\ 0 & \vdots & \pm 1 \end{matrix} \right\|, \quad e_2 \in \mathbf{R}^2 - \text{unit vector}.$$

Case 3. $\mu_1 > \mu_2 = \mu_3 > 0$.

	$SO(3)$	$d^2 J_D(\Omega)$ indices	$O(3) \backslash SO(3)$	$d^2 J_D(\Omega)$ indices	point type
	$\text{diag}(1,1,1)$	$(-,-,-)$	$\Omega^+(e_2)$	$(-,-,0)$	max.
Ω	$\text{diag}(1,-1,-1)$	$(-,-,+)$	$\text{diag}(-1,1,1)$	$(-,+,+)$	saddle
	$\Omega^-(e_2)$	$(+,+,0)$	$-\text{diag}(1,1,1)$	$(+,+,+)$	min.

$$\Omega^{\pm}(e_2) = \left\| \begin{matrix} \pm 1 & \vdots & \\ \cdots & \vdots & \cdots\cdots\cdots \\ & \vdots & 2e_2 \cdot e_2^t - I_2 \end{matrix} \right\|, \quad e_2 \in \mathbf{R}^2 - \text{unit vector}.$$

Case 4. $\mu_1 > \mu_2 = \mu_3 = 0$.

	$SO(3)$	$O(3) \setminus SO(3)$	$d^2 J_D(\Omega)$ indices	point type
	$\Omega_*^+(e_2)$	$\Omega^+(e_2)$	$(-,-,0)$	max.
Ω	$\Omega^-(e_2)$	$\Omega_*^-(e_2)$	$(+,+,0)$	min.

$\Omega_*^\pm(e_2) = \mathrm{diag}(1,1,-1)\Omega^\pm(e_2)$.

Case 5. $\mu_1 = \mu_2 = \mu_3 > 0$.

	$SO(3)$	$d^2 J_D(\Omega)$ indices	$O(3) \setminus SO(3)$	$d^2 J_D(\Omega)$ indices	point type
	$\mathrm{diag}(1,1,1)$	$(-,-,-)$	$-\Omega(e_3)$	$(-,0,0)$	max.
Ω	$\Omega(e_3)$	$(+,0,0)$	$-\mathrm{diag}(1,1,1)$	$(+,+,+)$	min.

$\Omega(e_3) = 2e_3 \cdot e_3^t - I_3$, $e_3 \in \mathbf{R}_3$ – unit vector.

Case 6. $\mu_1 = \mu_2 = \mu_3 = 0$.

$J_D = 0$ identically.

Summarizing cases 1-6 we arrive to the theorem 1:

Theorem 1. *The minimum of $J(\Omega, b)$ is attained at the point $(\Omega^*, b^*) \in SO(3) \times \mathbf{R}^3$ where b^* is from (2) and*

$$\Omega^* = -\Xi_2 \mathrm{diag}(1, 1, \pm 1)\Xi_1^t, \tag{7}$$

the sign "+" or "−" on the right-hand side is chosen so that $\Omega^* \in SO(3)$, Ξ_1 and Ξ_2 satisfy (6) with X from (4). If two smaller singular values of X coincide then the point of minimum may be not unique.

References

[1] Kabsh W., Acta cryst., 1976, p. 922.

[2] Kabsh W., Acta cryst., 1978, p. 827.

[3] McLachan A.D., J.Mol.Biol., 128 (1979), p. 49.

[4] Warner F., *The foundations of differentiable manifolds and Lie groups* – Springer, 1983.

[5] Voevodin V.V., Kuznetsov Yu.A. *Matrices and computations* – Moscow: Nauka, 1984.

Optimization in Computational Chemistry and Molecular Biology, pp. 267-286
C. A. Floudas and P. M. Pardalos, Editors
©2000 Kluwer Academic Publishers

Visualization of Chemical Databases Using the Singular Value Decomposition and Truncated-Newton Minimization

Dexuan Xie and Tamar Schlick
Departments of Chemistry and Mathematics
Courant Institute of Mathematical Sciences
New York University and The Howard Hughes Medical Institute
251 Mercer Street, New York, NY 10012
dexuan@cims.nyu.edu, schlick@nyu.edu

Abstract

We describe a rapid algorithm for visualizing large chemical databases in a low-dimensional space (2D or 3D) as a first step in chemical database analyses and drug design applications. The compounds in the database are described as vectors in the high-dimensional space of chemical descriptors. The algorithm is based on the singular value decomposition (SVD) combined with a minimization procedure implemented with the efficient truncated-Newton program package (TNPACK). Numerical experiments show that the algorithm achieves an accuracy in 2D for scaled datasets of around 30 to 46%, reflecting the percentage of pairwise distance segments that lie within 10% of the original distance values. The low percentages can be made close to 100% with projections onto a ten-dimensional space. The 2D and 3D projections, in particular, can be efficiently generated and easily visualized and analyzed with respect to clustering patterns of the compounds.

Keywords: chemical databases, clustering analysis, visualization, SVD, TNPACK, optimization, drug design.

1 Introduction

The field of combinatorial chemistry was recognized by *Science* as one of nine areas of study in 1997 that have great potential to benefit society [30]. The systematic assembly of chemical building blocks to form potential biologically-active compounds and their rapid testing for bioactivity has experienced a rapid growth in both experimental and theoretical approaches [4]. As experimental synthesis techniques are becoming cheaper and faster, huge chemical databases are becoming available for computer-aided design, and the development of reliable computational tools for their study is becoming more important than ever.

The specific computational problems involved in chemical libraries can be associated with certain mathematical disciplines. **Library characterization** involves the tools of *multivariate statistical analysis* and *numerical linear algebra* (see below for specific applications). The **similarity problem** in drug design involves finding from the database a drug that binds to a specific target or a drug that is similar to another drug with known bioactive properties. This search can be performed using 3D structural and energetic searches or using the concept of molecular descriptors introduced below. In either case, *multivariate nonlinear optimization* and optionally *configurational sampling* is involved. The **diversity problem** in drug design involves defining the most diverse subset of compounds within the given library. This problem is a *combinatorial optimization* task, and is known to have a non-polynomial time complexity [8, 24].

Typically, these combinatorial optimization problems are solved by stochastic and heuristic approaches [26]. These include genetic algorithms, simulated annealing, and tabu-search variants. As in other applications, the efficiency of simulated annealing is strongly dependent of the choice of cooling schedule and other parameters. In recent years, several potentially valuable annealing algorithms such as deterministic annealing, multiscale annealing, and adaptive simulated annealing have been extensively studied.

In special cases, combinatorial optimization problems can be formulated as integer programming and mixed-integer programming problems [8, 24, 17]. In this approach, linear programming techniques such as interior methods, can be applied to the solution of combinatorial optimization problems, leading to branch and bound algorithms, cutting plane algorithms, and dynamic programming techniques. Parallel implementation of combinatorial optimization algorithms is also important in practice to improve the performance [26].

One way to analyze a database of n potential *biologically active* compounds (drugs) is to characterize each compound in the database by a list of m *chemical descriptors*. These variables reflect atom connectivity, molecular topology, charge distribution, electrostatic properties, molecular volume, and so on. These descriptors can be generated from several commercial packages such as the popular Molconnx program [1]. Assigning associated *biological activity* for each compound (e.g., with respect to various ailments or targets, which may include categories like headache, diabetes, protease inhibitors) requires synthesis and biological testing. Hence, analyses of chemical databases (such as clustering, similarity, or dissimilarity sampling) can be performed on the collection of m-dimensional real vectors in the space R^m. However, due to large size of the dataset, some database-analysis tasks (say the diversity problem) are extremely challenging in practice because exhaustive procedures are not realistic. Any systematic schemes to reduce this computing time can be valuable.

In this paper we describe an algorithm that produces rapidly two-dimensional (2D) or 3D views of the compounds in a chemical database for clustering analysis. This visualization problem is often formulated as a distance-geometry problem: find n points in 2D (or 3D) so that their interpoint distances match the corresponding values from R^m as closely as possible. This approach was implemented by Sammon with the steepest descent (SD) minimization algorithm for clustering analysis in 1969 [28]. More recently, Sammon's method has been applied to the analysis and 2D projection mapping of molecular databases [3, 27]. The SD algorithm with a randomly chosen starting point generally suffers from slow convergence and may generate a 2D mapping that poorly approximates the original distances. As an alternative to the distance-geometry approach, a neural network procedure by Kohonen – the self-organizing map method [18] – has also been applied to the visualization of chemical databases [6, 11]. This method usually defines a mapping based on a 2D regular grid of nodes such that similar compounds are close to one other on the grid. Compared with a distance-geometry mapping method, however, it is unclear how the distance relationships of chemical database are preserved.

Our visualization algorithm consists of two parts. The first part defines a 2D projection mapping by the *singular value decomposition* (SVD) [15], a technique used for data compression in many practical applications like image processing. This factorization, in contrast to optimization, only requires the input (high-dimensional) data vectors; it has a complexity of order $O(n^2m)$ floating point operation and $O(nm)$ memory locations; no initial projection guess is needed. We find that the accuracy of the SVD mapping depends on the distribution of the singular-value magnitudes: if the first two singular values are much larger than the others, the 2D mapping has a high accuracy. This generalizes to mapping in higher dimensions as well; that is, if the first ten singular values can be largely separated from the rest, a 10D projection can be accurate. For scaled datasets as used in practice, however, two or three dominant singular values cannot generally be found.

The second part refines the SVD projection based on the distance geometry approach when the accuracy of the SVD projection is not satisfactory. Here, the SVD projection is used as a starting point for the truncated-Newton minimization iterative method. Determining a good initial guess for a minimization algorithm is an important and difficult objective in the distance-geometry approach. Our new distance error objective function is minimized with our efficient truncated-Newton program package, TNPACK [29, 32]. We call our algorithm the *SVD/TNPACK* method. This method is also described in [33], along with other applications. The applications in this paper all involve a natural scaling of the datasets rather than range-scaling as used in [33]. The projection analysis done here also illustrates an application to diversity and similarity sampling and presents 3D in addition to 2D projections.

We report numerical tests of the SVD/TNPACK procedure for two chemical datasets: an artificial dataset made of eight groups of compounds with different pharmacological activities (ARTF) and a dataset of monoaminooxidase inhibitors (MAO). ARTF and MAO contain 402 and 1623 compounds, respectively. All compounds in these datasets have been characterized with 312 topological descriptors. In addition, the MAO dataset has also been characterized by 153 binary descriptors (MAO_{01}). Since the various chemical descriptors vary drastically in their magnitudes as well as the variance within the dataset, scaling is important for proper assessment of distance relationship. Given no chemical/physical guidance, we consider a natural scaling procedure for ARTF and MAO, such that all scaled chemical descriptors have a mean of zero and a standard deviation of one.

For these scaled databases, SVD alone produced poor 2D projections (e.g., only about 0.004% of the distance segments are within 10% of the original distances for MAO_{01}), and the TNPACK minimizations that follow SVD become crucial (e.g., TNPACK increased this number 0.004% to 30% in less than one minute on an SGI R10000 processor). We also find that a larger number than three of the projection space is required to reach higher accuracy. Namely, the accuracy can be improved to 96% when the dimension number of the projection space is increased from two to ten for both scaled ARTF and MAO.

Numerical results also show that SVD is very fast: the computational time is one second for ARTF (402 compounds) and six seconds for MAO (1623 compounds) on an SGI R10000 processor; TNPACK is also very efficient (several minutes), and much more efficient than SD.

In section 2, we describe a mathematical framework for analysis of chemical datasets. Sections 3 describes the SVD/TNPACK method. Section 4 presents the numerical results and chemical structure analyses for the scaled datasets. Conclusions are summarized in Section 5.

2 Mathematical framework for analysis of chemical databases

We consider a database S of n potential *biologically active* compounds (drugs), where each compound is described by a list of m chemical descriptors. Thus, we can express the database S as a collection of n vectors

$$S = \{X_1, X_2, \ldots, X_n\},$$

where vector $X_i = (x_{i1}, x_{i2}, \ldots, x_{im})^T$ denotes the i-th compound in S, and the real numbers $\{x_{ik}\}$ are values of the associated chemical descriptors.

The database S can also written as a rectangular matrix X by listing, in rows, the m descriptors of the n compounds:

$$X = (X_1, X_2, \ldots, X_n)^T = \begin{bmatrix} x_{11} & x_{12} & \cdots & x_{1m} \\ x_{21} & x_{22} & \cdots & x_{2m} \\ \vdots & \vdots & \ddots & \vdots \\ x_{n1} & x_{n2} & \cdots & x_{nm} \end{bmatrix}. \tag{1}$$

This rectangular matrix typically has $n \gg m$ for large databases, where n may be of order million.

To measure the similarity or diversity for each pair of compounds X_i and X_j, we define distance quantities δ_{ij} on the m-dimensional vector space $\mathcal{R}^m$. The simplest one is the *Euclidean distance*:

$$\delta_{ij} = \|X_i - X_j\| = \sqrt{\sum_{k=1}^{m}(x_{ik} - x_{jk})^2}, \tag{2}$$

where $\| \cdot \|$ denotes the Euclidean norm. There are $n(n-1)/2$ distance segments $\{\delta_{ij}\}$ in S for pairs $i < j$.

Scaling may be important for proper assessment of distance quantities because the various chemical descriptors vary drastically in their magnitudes. Generally, scaled descriptors $\{\hat{x}_{ik}\}$ can be defined by the following formula: For $k = 1, 2, \ldots, m$,

$$\hat{x}_{ik} = a_k(x_{ik} - b_k), \qquad 1 \leq i \leq n, \tag{3}$$

where a_k and b_k are real numbers, and $a_k > 0$. They are called the scaling and displacement factors, respectively.

In practice, however, it is very difficult to determine the appropriate scaling and displacement factors for the specific application problem [34]. Given no chemical/physical guidance, the following two scaling procedures are often used. The first modifies each column of X by setting

$$b_k = \min_{1 \le i \le n} x_{ik} \quad \text{and} \quad a_k = 1/(\max_{1 \le i \le n} x_{ik} - b_k) \quad \text{for } k = 1, 2, \ldots, m. \tag{4}$$

This makes each column in the range $[0, 1]$. The second sets

$$b_k = \frac{1}{n} \sum_{i=1}^{n} x_{ik} \quad \text{and} \quad a_k = 1/\sqrt{\frac{1}{n} \sum_{i=1}^{n} (x_{ik} - b_k)^2} \quad \text{for } k = 1, 2, \ldots, m, \tag{5}$$

so that each scaled column of X has a mean of zero and a standard deviation of one. The scaling procedure with (4) is also referred to as a standardization of descriptors. Both scaling procedures (4) and (5) assume that no one descriptor dominates the overall distance measures.

The distances $\{\delta_{ij}\}$ can be used in performing similarity searches among the database compounds and between these compounds and a particular target. This task can be formulated as finding:

$$\min_{\substack{1 \le i \le n \\ i \ne j}} \delta_{ij},$$

where $\delta_{ij} = \|X_i - X_j\|$, and X_j is a given target. Note that each distance segment δ_{ij} requires $O(m)$ floating-point operations (flops) to compute, an exhaustive calculation over all n candidates requires a total of $O(nm)$ flops. An effective scheme is sought when n and m are large.

More difficult and computationally-demanding is the diversity problem. Namely, we seek to reduce the database of the n compounds by selecting a "representative subset" of the compounds contained in S, that is one that is "the most diverse" in terms of potential chemical activity. This problem naturally arises since pharmaceutical companies must scan huge databases each time they search for a specific pharmacological activity. This molecular diversity problem can be formulated as determining:

$$\max_{S_0 \subset S} \sum_{\substack{X_i, X_j \in S_0 \\ X_i \ne X_j}} \|X_i - X_j\|,$$

where each S_0 contains n_0 representative compounds ($n_0 \ll n$, a fixed integer number). This is a *combinatorial optimization* problem, an example of a very difficult computational task (NP-complete). An exhaustive search of the most diverse subset S_0 requires a total of $O(C_n^{n_0} n_0^2 m)$ flops because there are $C_n^{n_0}$ possible subsets of S of size n_0 and each subset takes $O(n_0^2 m)$ flops. Here $C_n^{n_0} = n(n-1)(n-2) \cdots (n - n_0 + 1)/n_0$.

As a first step in solving such similarity and diversity problems, methods that produce a *low*-dimensional projection view of the compounds can be used for clustering analysis. Assume we have a mapping from $\mathcal{R}^m$ to $\mathcal{R}^{low}$ that takes each point $X_i \in \mathcal{R}^m$ to $Y_i \in \mathcal{R}^{low}$, where $low \ll m$. Typically the integer low is 2 or 3 but we use $low > 3$ in some cases discussed below; the projection cannot be easily visualized for $low > 3$, but the compressed matrix from X ($n \times low$ instead of $n \times m$) can be useful in reducing computer time for database applications. The intercompound distances for the vectors Y_i and Y_j is denoted as $d(Y_i, Y_j)$. We define it as

$$d(Y_i, Y_j) = \sqrt{\sum_{k=1}^{low} (y_{ik} - y_{j_k})^2}.$$

An ideal projection mapping will generate points $\{Y_i\}$ such that their distance values match the original values, i.e., satisfy

$$d(Y_i, Y_j) = \delta_{ij} \tag{6}$$

for all $\{i, j\}$ pairs. However, no such a mapping exists in general because the problem is typically over-determined — finding $n \times low$ unknowns $\{y_{ik}\}$ satisfying $n(n-1)/2$ equations of form (6). An optimal approximate mapping is thus sought based on the distance geometry approach [25]. Specifically, an objective error function E to describe the discrepancy between $\{\delta_{ij}\}$ and $\{d(Y_i, Y_j)\}$ is constructed, and then we find a minimum point $Y^* = (Y_1^*, Y_2^*, \ldots, Y_n^*)$ with $Y_i^* \in R^{low}$ for $i = 1, 2, \ldots, n$ such that

$$E(Y_1^*, Y_2^*, \ldots, Y_n^*) = \min_{Y_i \in R^{low}, \, 1 \leq i \leq n} E(Y_1, Y_2, \ldots, Y_n), \tag{7}$$

where each $Y_i = (y_{i1}, y_{i2}, \ldots, y_{ilow})^T$. The objective function E can be formulated in many different ways [3, 25, 27]. Here we use the following expression:

$$E(Y_1, Y_2, \ldots, Y_n) = \frac{1}{4} \sum_{i=1}^{n-1} \sum_{j=i+1}^{n} \omega_{ij} \left(d(Y_i, Y_j)^2 - \delta_{ij}^2 \right)^2, \tag{8}$$

$$\omega_{ij} = \begin{cases} 1/\delta_{ij}^4 & \text{if } \delta_{ij}^4 \geq \eta, \\ 1 & \text{if } \delta_{ij}^4 < \eta, \end{cases}$$

where $\{\omega_{ij}\}$ denote weights, and the parameter η is a small positive number such as 10^{-12}. The first and second derivatives of E are well defined, and an efficient second-derivative method like Newton-type algorithms [12] can be applied.

Various error measures can be used to assess the agreement between the original and projected pairwise distances. Besides the value of the objective function E, we use the following percentage ρ to measure the quality of the approximation of $d(Y_i, Y_j)$ to δ_{ij} for all pairs $i < j$:

$$\rho = \frac{T_d}{n(n-1)/2} \cdot 100. \tag{9}$$

The variable T_d is the total number of the distance segments $d(Y_i, Y_j)$ satisfying

$$|d(Y_i, Y_j) - \delta_{ij}| \leq \epsilon \delta_{ij} \quad \text{when } \delta_{ij} > d_{min}, \tag{10}$$

or

$$d(Y_i, Y_j) \leq \tilde{\epsilon} \quad \text{when } \delta_{ij} \leq d_{min}, \tag{11}$$

where ϵ, $\tilde{\epsilon}$, and d_{min} are given small positive numbers less than one. For example, we set $\epsilon = 0.1$ to specify a 10% accuracy ($d_{min} = 10^{-12}$ and $\tilde{\epsilon} = 10^{-8}$). The second case above (very small original distance) may occur when two compounds in the datasets are similar highly. The greater the ρ values, the better the mapping and the more information can be inferred from the projected views of the complex data.

3 The SVD/TNPACK method

The SVD decomposition of the database rectangular matrix X (defined in (1)) as $U\Sigma V^T$ can be written as the sum of rank-1 matrices [15]:

$$X = \sum_{k=1}^{r} \sigma_k u_k v_k^T, \tag{12}$$

where r is the rank of matrix X ($r \leq m$), $u_k \in R^n$ and $v_k \in \mathcal{R}^m$, respectively, are left and right singular vectors, and σ_k is the singular value. All singular values are arranged in decreasing order:

$$\sigma_1 \geq \sigma_2 \geq \ldots \geq \sigma_r > 0 \quad \text{and} \quad \sigma_{r+1} = \ldots = \sigma_m = 0.$$

Let $u_k = (u_{1k}, u_{2k}, \ldots, u_{nk})^T$. Using (12), we can express each vector X_i as a linear combination of orthonormal basis vectors $\{v_k\}_{k=1}^{m}$ of $\mathcal{R}^m$:

$$X_i = \sum_{k=1}^{m} \sigma_k u_{ik} v_k = \sum_{k=1}^{r} \sigma_k u_{ik} v_k, \quad i = 1, 2, \ldots, n$$

since $\sigma_{r+1} = \ldots = \sigma_m = 0$. Hence, the compound vector X_i is expressed in terms of new coordinates

$$X_i = (\sigma_1 u_{i1}, \sigma_2 u_{i2}, \ldots, \sigma_r u_{ir}, 0, \ldots, 0)^T. \tag{13}$$

Based on (13), we define the *low* dimensional vector Y_i as the natural projection of X_i onto the subspace $\mathcal{R}^{low}$ of R^m:

$$Y_i = (\sigma_1 u_{i1}, \sigma_2 u_{i2}, \ldots, \sigma_{low} u_{ilow})^T, \quad i = 1, 2, \ldots, n. \tag{14}$$

When the percentage ρ defined in (9) is not large enough, we improve the accuracy of the SVD projection (14) by our truncated Newton program package, TNPACK [29, 32] by minimizing the objective error function E defined in (8).

The truncated Newton method [9] consists of both outer and inner loops. The outer loop defines a sequence of solution vectors $\{Y^k\}$ expressed in the form

$$Y^{k+1} = Y^k + \lambda_k P^k, \quad k = 0, 1, 2, \ldots,$$

where Y^k and P^k are vectors of $R^{low \cdot n}$, P^k is a descent direction, λ_k is the steplength, and Y^0 is an initial guess. The inner loop defines P^k by a "truncated" preconditioned conjugate gradient scheme. The steplength λ_k is generated by using a line search scheme ([21], for example).

TNPACK was first published in 1992 [29] and updated recently [32]. One of the features of TNPACK is an application-tailored preconditioner matrix (that approximates the Hessian of the objective function) used to accelerate convergence [31]. This novel preconditioner makes TNPACK an efficient tool for the minimization of molecular potential functions in comparison to other available minimizers [10, 31]. For the present applications, we used the new version of TNPACK [32] in combined with a simple preconditioner, namely, the

Table 1: Performance of the 2D SVD and SVD/TNPACK (TN) mappings. Here E is the minimization objective function defined in (8), and ρ the percentage defined in (9), which measures the quality of the approximation of the 2D mapping

Datasets	E		ρ		TN	CPU time	
	SVD	TN	SVD	TN	Itn.	SVD (sec.)	TN (min.)
ARTF	7.06×10^3	2.77×10^3	25.91	45.95	31	1.18	0.45
MAO	1.31×10^5	5.41×10^4	5.51	43.94	33	6.24	7.49
MAO_{01}	2.4×10^5	9.79×10^4	0.004	29.10	11	3.65	0.77

diagonal part of the Hessian, or terms $\partial^2 E(Y_1, Y_2, \ldots, Y_n)/\partial y_{ik}^2$ (for $i = 1, 2, \ldots, n$ and $k = 1, 2, \ldots, low$).

We use the SVD projection (14) as the starting point Y^0, and terminate TNPACK iteration at Y^k provided that it satisfies

$$\|g(Y^k)\| < \epsilon_g(1 + |E(Y^k)|), \tag{15}$$

where ϵ_g is a small positive number (we used 10^{-5}), and g is the gradient vector of E. Such an Y^k defines the SVD/TNPACK projection.

4   Numerical examples

Two datasets were used for testing our SVD/TNPACK scheme: MAO ($n = 1623$ and $m = 312$) and ARTF ($n = 402$ and $m = 312$). ARTF merges eight different groups of molecules with different types of pharmacological activities. Descriptors for ARTF and MAO were generated from the software package Molconn-X [1]. We scaled descriptors using scaling procedure (5), and deleted all zero columns of dataset matrix X. We thus obtained dense rectangular matrices with $m = 202$ for scaled ARTF and $m = 204$ for scaled MAO. We also considered MAO with binary descriptors, MAO_{01} ($m = 153$). The binary descriptors were generated from the software MACCS II [20].

We used the NAG library [2] to compute the SVD of each dataset. For simplicity, we used all default parameters of TNPACK [29, 32] for the minimization that follows the SVD projection. The target accuracy ϵ in (10) was set to 0.1. The termination rule for TNPACK is (15) with $\epsilon_g = 10^{-5}$. All computations were performed in double precision on a single R10000/195 MHZ processor of an SGI Power Challenge L computer at New York University.

Table 1 displays the performance of SVD and SVD/TNPACK in defining 2D mappings for these datasets. The accuracy of 2D mapping is indicated by the percentage ρ defined in (9) (i.e., the portion of the distance segments that are within 10% of the original distance values). From Table 1 we see that both SVD and TNPACK are efficient: computer CPU time ranges from one second to seven minutes. SVD alone yields poor accuracies in terms of distance preservation (ρ ranges from 0.004 to 25%). TNPACK greatly improves the SVD projection in this regard (ρ ranges from 30 to 46%).

To illustrate the reason why the 2D SVD mapping is poor for the scaled datasets, Figure 1 presents the distributions of the normalized singular values $\hat{\sigma}_i$ on seven intervals:

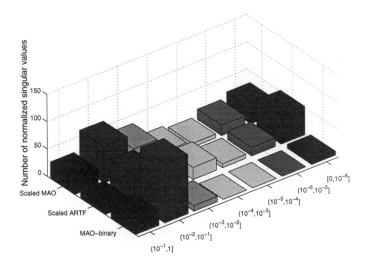

Figure 1: The distribution of the normalized singular values $\{\hat{\sigma}_i\}$

$(10^{-k}, 10^{-(k-1)}]$ for $k = 1$ to 6 and $[0, 10^{-6}]$. Here the normalized singular values are defined by

$$\hat{\sigma}_i = \sigma_i / \max_{1 \leq j \leq r} \sigma_j \quad \text{for } i = 1, 2, \ldots, r.$$

From Figure 1 we see that most normalized singular values are not small for the scaled datasets, implying that the first two singular values are not significantly larger than the others. Hence, the 2D mapping is poor for the scaled datasets.

Figure 2 shows that the accuracy (i.e., the percentage ρ defined in (9)) of the SVD and SVD/TNPACK projections for the scaled datasets can be improved sharply when the number of dimensions (low) of the projection space is increased from two to ten. We also found it useful to use higher-order SVD mappings for the purpose of selecting initial points for minimization refinement.

Table 2 compares the performance of TNPACK with that of the steepest descent (SD) method since SD has been used in similar applications [3, 27]. Here both TNPACK and SD used the same termination rule (15) and the same SVD starting point. Table 2 shows that TNPACK is more efficient (a factor of three) to find a minimum point. This efficiency will likely become more significant as the database size n increases.

Table 3 compares the performance of TNPACK using the SVD projection as a starting point with that using a randomly selected starting point. It shows that the SVD starting point helps accelerate the minimization process significantly, and generate better 2D mappings (smaller values of E). Again, the improvements are likely to be more more significant as n increases.

Figure 3 displays the 2D mappings of the scaled ARTF, the scaled MAO, and the binary MAO_{01}. These figures also compare the plots of the 2D mappings generated by SVD alone and SVD/TNPACK (blue vs. red symbols). The SVD plots have been significantly changed by TNPACK so as to improve the distance values in 2D with respect to the original values.

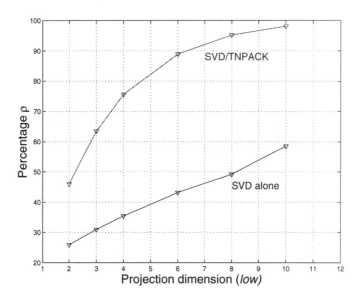

Figure 2: The percentage ρ defined in (9) increases with the number of dimensions of the projection space ($\eta = 0.1$ for ARTF)

Table 2: Comparison of TNPACK versus SD for minimizing E

Method	Final E	Final $\|g\|$	Iterations	CPU time (min.)
ARTF				
SD	2.77×10^3	2.77×10^{-3}	1375	1.17
TNPACK	2.77×10^3	1.05×10^{-4}	31	0.45
MAO				
SD	5.42×10^4	5.42×10^{-1}	1768	26.35
TNPACK	5.41×10^4	2.21×10^{-1}	33	7.49

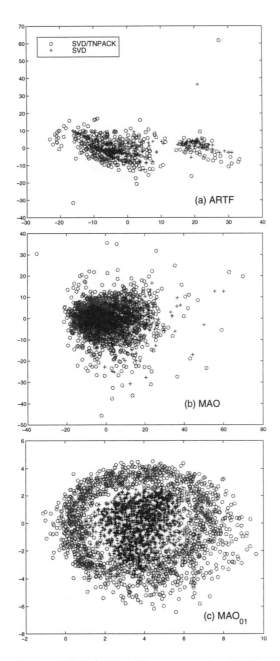

Figure 3: The 2D SVD and SVD/TNPACK mappings for ARTF, MAO, and MAO_{01}

Table 3: Comparison of TNPACK minimization using the SVD mapping as a starting point (SVD) versus a randomly selected starting point (RAN)

Starting point X^0	Final E	Final $\|g\|$	Iterations	CPU time (min.)
ARTF				
SVD	2.77×10^3	1.05×10^{-4}	31	0.45
RAN	2.87×10^3	1.44×10^{-2}	61	0.91
MAO				
SVD	5.41×10^4	2.21×10^{-1}	33	7.49
RAN	5.55×10^4	3.14×10^{-1}	133	25.83

Figure 4a displays the distribution of eight chemical/pharmacological classes of compounds in ARTF as a result of the 2D SVD/TNPACK mapping. The number of compounds in each class is indicated in the figure next the class name. One selected chemical structure for each class is marked by a black circle and shown in Figure 5.

Noting that the 2D mapping has several small subclusters and a few singletons, we selected six spatially distant points (marked as A1 to A6) from different pharmacological classes on Figure 4b. See Figure 6 for their chemical structures. This is an application of the projection to the diversity sampling problem. Note that even within one family the chemical structures may differ. As an application to the similarity problem, we also selected three spatially close points (B1 to B3) from the same H1 ligand class on Figure 4b. Their similar chemical structures are presented in Figure 6.

Finally, we generated the 3D SVD/TNPACK mapping for the scaled ARTF. As expected, the accuracy of the 3D mapping is higher than the 2D mapping ($\rho = 63.46\%$ for 3D while $\rho = 46$ for 2D with $\eta = 0.1$). Four different views of the 3D mapping are displayed in Figure 7; a single point corresponding to A1 in Figure 4b was removed for better resolution. From these figures we see that the 3D mapping is quite similar to the 2D mapping: the ecdysteroids (red spheres in 3D and red triangles in 2D) and the AChE inhibitors (green spheres in 3D and green squares in 2D) classes continue to appear separate from the rest and a strong overlap between D1 agonists, D1 antagonists, H1 ligands, and 5HT ligands persists.

5   Conclusions

We have presented a mathematical framework for analysis of chemical databases. Our SVD/TNPACK method is easy to implement and efficient to use in visualizing large chemical databases in a low-dimensional space (2D or 3D).

The scaled databases make it difficult to calculate 2D/3D projections that approximate well the original distance distributions. This is because all scaled descriptors lie within the same range and there are in general no dominant singular values. However, we showed that higher-accuracy projections can be obtained for these scaled datasets when the projection dimension is increased from two to ten or so. Though these higher-dimensional projections

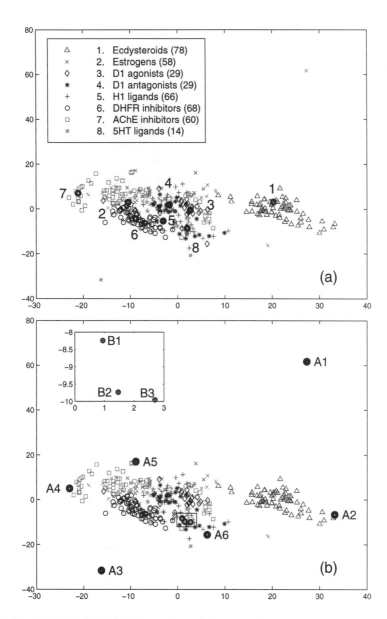

Figure 4: The 2D SVD/TNPACK mapping of the eight pharmacological classes of ARTF: (a) with eight chemical representatives marked by black circles, and (b) with a diversity sample (A1–A6) and a similarity sample (B1–B3). See Figures 5 and 6 for their chemical structures

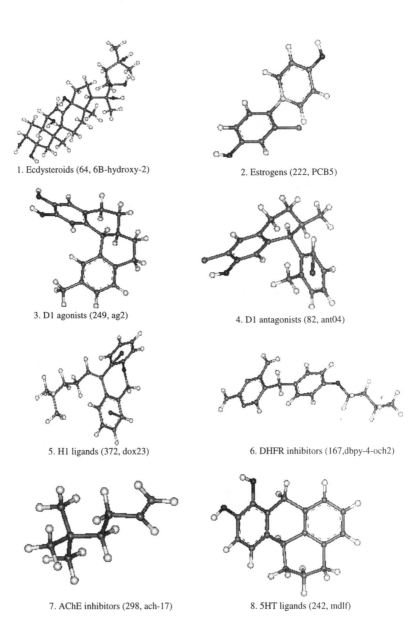

1. Ecdysteroids (64, 6B-hydroxy-2)

2. Estrogens (222, PCB5)

3. D1 agonists (249, ag2)

4. D1 antagonists (82, ant04)

5. H1 ligands (372, dox23)

6. DHFR inhibitors (167,dbpy-4-och2)

7. AChE inhibitors (298, ach-17)

8. 5HT ligands (242, mdlf)

Figure 5. Chemical structure representatives of the eight classes of ARTF
(see Figure 4a)

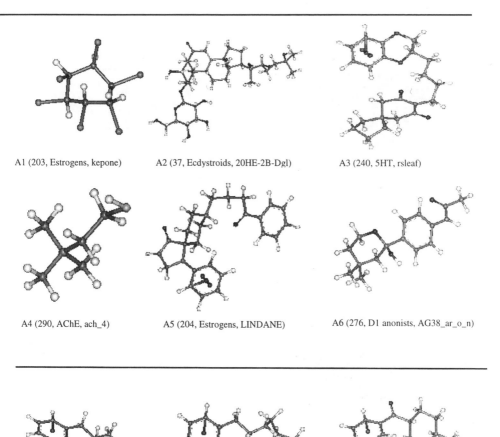

A1 (203, Estrogens, kepone)   A2 (37, Ecdystroids, 20HE-2B-Dgl)   A3 (240, 5HT, rsleaf)

A4 (290, AChE, ach_4)   A5 (204, Estrogens, LINDANE)   A6 (276, D1 anonists, AG38_ar_o_n)

B1 (343, H1 ligands, PPrnorMe5)   B2 (392, H1 ligands, phenbtpat)   B3 (377, H1 ligands,keta8)

Figure 6. Chemical structures for the diversity and similarity applications of the 2D SVD/TNPACK projection for ARTF (see Figure 4b)

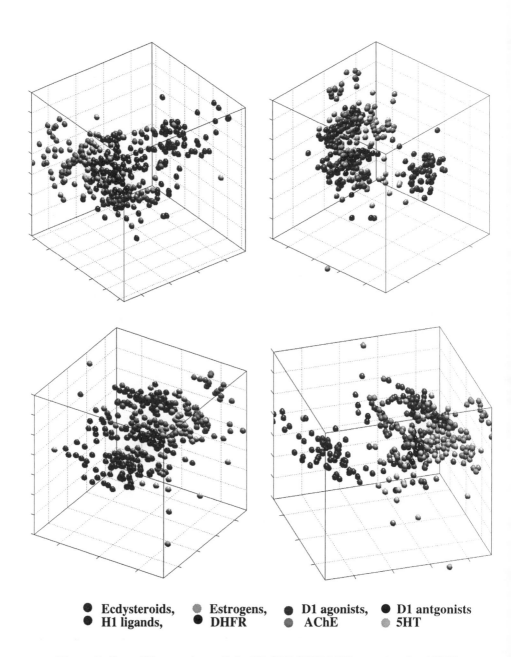

Ecdysteroids,   Estrogens,   D1 agonists,   D1 antgonists
H1 ligands,   DHFR   AChE   5HT

Figure 7: Four different views of the 3D SVD/TNPACK mapping for ARTF

are not easily visualized, the compression of the dataset descriptors can be advantageous in further applications of the compound library (e.g., diversity sampling) as shown here.

When the intercompound distances in 2D/3D approximate the original distance relationships well, the 2D/3D projection offers a simple visualization tool for analyzing the compounds in a large database. We emphasize that these analyses depend on the quality of the original descriptors, an area of research on its own [7]. These clustering analyses can serve as a first step in the study of related combinatorial chemistry questions dealing with large chemical databases, and we hope to examine these possibilities in future work. It will also be important to compare our SVD/TNPACK method to the neural network procedure of Kohonen, both in terms of resulting projection accuracy of clustering and computing performance. Figure 8 shows a mapping of 32 5D-vectors by our SVD/TNPACK vs. Kohonen map, where we used the same data set and Kohonen map figure as given in [18], page 114. For comparison, a reference tree, the so called *minimal spanning tree* (where the most similar pairs of points are linked) [18], is also displayed. The SVD/TNPACK and Kohonen maps have similar clusters with different patterns. However, the SVD/TNPACK map appears more similar to the reference tree.

Further work is also needed on extending our SVD/TNPACK approach to large chemical datasets. The huge database might be subdivided as dictated by computer memory, and the SVD/TNPACK procedure applied to each data subset. To properly assemble these sub-2D-mappings for the purpose of defining a global 2D-mapping, techniques to overlap the database segments will have to be devised. We intend to discuss this extension scheme in detail in our subsequent work. We invite interested readers to contact us about experimenting with our projection software SIEVER (SIngular Values and Error Refinement).

Acknowledgments

We are indebted to Dr. Alexander Tropsha at University of North Carolina for providing the original chemical datasets and for helpful discussions. Support by the National Science Foundation (ASC-9157582 and BIR 94-23827EQ) and the National Institutes of Health (R01 GM55164-01A2) is gratefully acknowledged. T. Schlick is an investigator of the Howard Hughes Medical Institute.

References

[1] Hall Associates Consulting (1995), *Molconn-X version 2.0*, Quincy, Maryland.

[2] NAG Inc. (1995), *NAG Fortran Library, Mark 17*, Opus Place, Suite 200, Downers Grove, Illinois.

[3] Agrafiotis, D.K. (1997), "A new method for analyzing protein sequence relationships based on Sammon maps," *Protein Science, Vol. 6*, 287-293.

[4] Boyd, D.B. (1995), "Rational drug design: Controlling the size of the haystack," *Modern Drug Discovery, Vol. 1, No. 2*, 41-47.

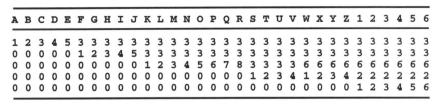

A	B	C	D	E	F	G	H	I	J	K	L	M	N	O	P	Q	R	S	T	U	V	W	X	Y	Z	1	2	3	4	5	6
1	2	3	4	5	3	3	3	3	3	3	3	3	3	3	3	3	3	3	3	3	3	3	3	3	3	3	3	3	3	3	3
0	0	0	0	0	1	2	3	4	5	3	3	3	3	3	3	3	3	3	3	3	3	3	3	3	3	3	3	3	3	3	3
0	0	0	0	0	0	0	0	0	0	1	2	3	4	5	6	7	8	3	3	3	3	6	6	6	6	6	6	6	6	6	6
0	0	0	0	0	0	0	0	0	0	0	0	0	0	0	0	0	0	1	2	3	4	1	2	3	4	2	2	2	2	2	2
0	0	0	0	0	0	0	0	0	0	0	0	0	0	0	0	0	0	0	0	0	0	0	0	0	0	1	2	3	4	5	6

The dataset (32 5D–vectors) used by Kohonen

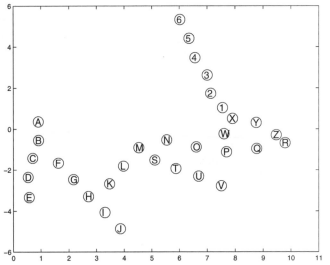

SVD/TNPACK map of the above dataset (ρ=70% with η=0.1)

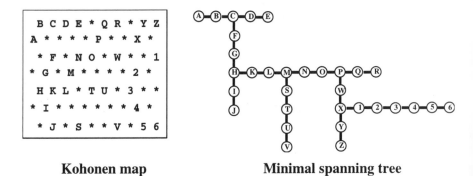

Kohonen map   **Minimal spanning tree**

Figure 8: Comparison of SVD/TNPACK versus the Kohonen methods

[5] Bakonyi, M. and Johnson, C.R. (1995), "The Euclidean distance matrix completion problem," *SIAM J. Matrix Anal. Appl., Vol. 16*, 646-654.

[6] Bienfait, B. (1994), "Applications of high-resolution self-organizing maps to retrosynthetic and QSAR analysis," *J. Chem. Inf. Comput. Sci. Vol. 34*, 890-898.

[7] Brown, R.D. and Martin, Y.C. (1997), "Information content of 2D and 3D structural descriptors relevant to ligand-receptor binding," *J. Chem. Inf. Comput. Sci., Vol. 37*, 1-9.

[8] Cook, W.J., Cunningham, W.H., Pulleyblank, W.R. and Schrijver, A. (1998), *Combinatorial Optimization*, Wiley, New York.

[9] Dembo, R.S. and Steihaug, T. (1983), "Truncated-Newton algorithms for large-scale unconstrained optimization," *Math. Programming, Vol. 26*, 190-212.

[10] Derreumaux, P., Zhang, G., Brooks, B. and Schlick, T. (1994), "A truncated-Newton method adapted for CHARMM and biomolecular applications," *J. Comp. Chem., Vol. 15*, 532-552.

[11] Gasteiger, J.; Zupan, J. (1993), "Neural Networks in Chemistry," *Angew. Chem. Int. Ed. Engl., Vol. 32*, 503-527.

[12] Gill, P.E., Murray, W. and Wright, M.H. (1983), *Practical Optimization*, Academic Press, London.

[13] Crippen, G.M. and Havel, T.F. (1988), *Distance Geometry and Molecular Conformation*, Wiley, New York.

[14] Glunt, W., Hayden, T.L., Hong, S. and Wells, J. (1990), "An alternating projection algorithm for computing the nearest Euclidean distance matrix", *SIAM J. Matrix Anal. Appl., Vol. 11*, 589-600.

[15] Golub, G.H. and Van Loan, C.F. (1996), *Matrix Computations*, John Hopkins University Press, Baltimore, Maryland, third edition.

[16] Gower, J.C. (1985), "Properties of Euclidean and non-Euclidean distance matrices," *Linear Algebra Appl., Vol. 67*, 81-97.

[17] Nemhauser, G.L. and Wolsey, L.A. (1988), *Integer and Combinatorial Optimization*, John Wiley and Sons, New York.

[18] Kohonen, T. (1997), *Self-Organizing Maps, Springer Series in Information Sciences, Vol. 30*, Springer, Berlin, Heidelberg, New York.

[19] Korte, B., Lovász, L. and Schrader, R. (1991), *Greedoids*, Springer-Verlag, New York.

[20] Molecular Design Ltd. *Maccs-II*, 14600 Catalina St., San Leandro, California.

[21] Moré, J.J. and Thuente, D.J. (1994), "Line search algorithms with guaranteed sufficient decrease," *ACM Trans. Math. Softw., Vol. 20*, 286-307.

[22] Moré, J.J. and Wu, Z. (1997), "Distance geometry optimization for protein structures," *Technical Report MCS-P628-1296*, Argonne National Laboratory, Argonne, Illinois.

[23] Oxley, J.G. (1992), *Matroid Theory*, Oxford University Press, New York.

[24] Papadimtriou, C.H. and Steiglitz, K. (1982), *Combinatorial Optimization: Algorithms and Complexity*, Prentice-Hall, Englewood Cliffs, New Jersey.

[25] Pinou, P., Schlick, T., Li, B. and Dowling, H.G. (1996), "Addition of Darwin's third dimension to phyletic trees," *J. Theor. Biol., Vol. 182*, 505-512.

[26] Reeves, C. (1993), *Modern Heuristic Techniques for Combinatorial Problems*, Halsted Press, New York.

[27] Robinson, D.D., Barlow, T.W. and Richard, W.G. (1997), Reduced dimensional representations of molecular structure. *J. Chem. Inf. Comput. Sci., Vol. 37*, 939-942.

[28] Sammon Jr, J. W. (1969), "A nonlinear mapping for data structure analysis," *IEEE Trans. Comp. C-18*, 401-409.

[29] Schlick, T. and Fogelson, A. (1992), "TNPACK — A truncated Newton minimization package for large-scale problems: I. Algorithm and usage," *ACM Trans. Math. Softw., Vol. 14*, 46–70.

[30] Science and Business (1998), "New partnerships for biology and business," *Science, Vol. 282, No. 18*, 2160-2161.

[31] Xie, D. and Schlick, T. (1999), "Efficient implementation of the truncated-Newton algorithm for large-scale chemistry applications," *SIAM J. Optim., Vol. 9*.

[32] Xie, D. and Schlick, T. (1999), "Remark on Algorithm 702 — the updated truncated Newton minimization package," *ACM Trans. Math. Softw., Vol. 25, No. 1*.

[33] Xie, D., Tropsha, A. and Schlick, T. (1999), "An efficient projection protocol for chemical databases: the singular value decomposition combined with truncated Newton minimization," preprint, submitted.

[34] Willett, P. (1998), "Structural similarity measures for database searching." In von Ragué Schleyer, P. (Editor-in Chief), Allinger, N.L., Clark, T., Gasteiger, J., Kollman, P.A. and Schaefer, III, H.F., editors, *Encyclopedia of Computational Chemistry, Vol. 4*, John Wiley & Sons, West Sussex, UK, 2748-2756.

Optimization in Computational Chemistry and Molecular Biology, pp. 287-300
C. A. Floudas and P. M. Pardalos, Editors

Optimization of Carbon and Silicon Cluster Geometry for Tersoff Potential using Differential Evolution

M. M. Ali
Centre for Control Theory and Optimization
Department of Computational and Applied Mathematics
Witwatersrand University, Private Bag-3, Wits-2050, Johannesburg
mali@cam.wits.ac.za

A. Törn
Department of Computer Science, Åbo Akademi University,
SF-20520, Turku, Finland
atorn@abo.fi

Abstract

In this paper we propose a new version of the Differential Evolution (DE) Algorithm for large scale optimization problems. The new algorithm, for exploration and localization of search, periodically uses topographical information on the objective function, in particular the k_g-nearest neighbour graph. The algorithm is tested on hard practical problems from computational chemistry. These are the problems of semi-empirical many-body potential energy functions considered for carbon-carbon and silicon-silicon atomic interactions. The minimum binding energies of both carbon and silicon clusters consisting of upto 15 particles are reported.

Keywords: Many-body, potential function, differential evolution, minimum energy configuration, topographs, graph minima.

1 Introduction

Because of the importance of silicon technology, semiconductor materials and structure-based drug design, empirical many-body potentials are becoming an increasingly important means of various investigations, for instance it has been used to investigate the ion bombardment [1], for dynamic simulation [2, 3] and also used as a sample global optimization problem in search of the stable struture of molecules [4, 5]. Over the last two decades significant advances were made in developing various potentials including the potentials for modelling covalent materials such as carbon and silicon. Many different forms of these potentials can be found in [6, 7] and one of the most successful forms which has been used in a large number of dynamic simulations [3] is due to Tersoff [8, 9]. The determination of the global minimum energy configuration, or the ground state structures, for clusters of carbon and silicon particles, predicted by such potentials, is an important global optimization problem. For instance, the synthesis of highly symmetric spherical carbon molecules, known as fullerences, has stimulated much interests in the geometric structure of small clusters of carbon in their most stable state. Moreover, the increasing trend towards nanoscale devices within the semiconductor industry demands more research into the nature of silicon clusters. The numerical approach to finding the global minimum of empirical potentials poses a very difficult problem as the number of local minima increases rapidly with the number of atoms. To date, several methods have been proposed for this type of optimization problem [5, 10]. These are the spatial smoothing techniques [11, 12] and variants of simulated annealing [13] and the genetic algorithm (GA) [14]. Most of these algorithms, however, were implemented on the Lennard-Jones pair potential. Although, recent optimization involving more realistic many-body potentials is reported in [15, 16] where Tersoff's two different parameterizations for silicon ($Si(B)$ and $Si(C)$) [9] were used, and in [17] where Brenner's, Tersoff-like, Carbon-Hydrogen potential [18] was used, optimization of carbon clusters using Tersoff's carbon potential [19, 20, 21] was not carried out before. In [15] the implementation of eight different recent global optimization algorithms on cluster optimization for a slightly simplified Tersoff potential is reported, but only small problems for silicon clusters of size upto 6 particles were considered for testing different algorithms. The optimization of carbon clusters of upto 60 atoms using Brenner potential is reported in [17] the algorithm used was a binary coded GA driven multistart. The algorithm performs local searches from each child molecule produced by GA in each generation. However, it is not possible to judge their results as the number of function evaluations and cpu times were not given.

The main thrust of this paper is to devise a global optimization algorithm which is robust and efficient in finding the global minimum of silicon (Si) and carbon (C) clusters as well as can handle clusters of large number of particles. In Section 2 and 3 we respectively briefly define the differential evolution and the topographical algorithm. In Section 4 the new algorithm is presented. Details of the potential is given in Section 5 and in Appendix A. The results are discussed and summarised in Section 6, and the conclusion is made in section 7.

2   Differential Evolution (DE)

Storn and Price [22] recently developed a natural evolution based direct search technique, the Differential Evolution (DE), for optimizing functions of continuous variables. The population based DE guides the N points in the set $S = \{x_1, x_2, \cdots, x_N\}$, chosen randomly from the search region $\Omega \subset I\!R^n$, to the vicinity of the global minimum through repeated cycles of mutation, crossover (recombination) and acceptance. In each cycle constituting a generation, N competitions are held to determine the members of S for the next generation. The i-th ($i = 1, 2, \cdots, N$) competition is held to replace x_i in S. Considering x_i as the target point a trial point y_i is found from two points (parents), the point x_i, i.e., the target point and the point $\hat{x}_i$ determined by the mutation operation. In its mutation phase DE randomly selects three distinct points x_{r1}, x_{r2} and x_{r3} from the current set S. None of these points should coincide with the current target point x_i. The point $\hat{x}_i$ is then calculated by

$$\hat{x}_i = x_{r1} + F \times (x_{r2} - x_{r3}) \,. \tag{1}$$

A good value of the scaling factor, $F \leq 1$, is given by

$$F = \begin{cases} max\left(l_{min}, 1 - \left|\frac{f_{max}}{f_{min}}\right|\right) & \text{if } \left|\frac{f_{max}}{f_{min}}\right| < 1 \\ max\left(l_{min}, 1 - \left|\frac{f_{min}}{f_{max}}\right|\right) & \text{otherwise,} \end{cases} \tag{2}$$

where $l_{min} \in [0.4, 0.5]$ and f_{max} and f_{min} respectively are the high and low function values within S [23]. The trial point y_i is found from its parents x_i and $\hat{x}_i$ using the following crossover rule :

$$y_i^j = \begin{cases} \hat{x}_i^j & \text{if } R^j \leq CR \text{ or } j = I_i \\ x_i^j & \text{if } R^j > CR \text{ and } j \neq I_i \,, \end{cases} \tag{3}$$

where I_i is a randomly chosen integer in the set I, i.e., $I_i \in I = \{1, 2, \cdots, n\}$; the superscript j represents the j-th component of respective vectors; $R^j \in (0,1)$, drawn randomly for each j. The entity CR is a constant (eg. 0.5). The acceptance mechanism follows the crossover. This process of targetting x_i and generating the corresponding y_i continues until all members of S have been considered. In the acceptance phase a one to one comparison is made in that the function value at each trial point, $f(y_i)$, is compared to $f(x_i)$, the value at the target point. If $f(y_i) < f(x_i)$ then y_i replaces x_i in S, otherwise, S retains the original x_i. One of the important exploratory features of DE found in a recent study [23] is that it attempts to replace all points in S in each generation. Unlike GA, this replacement is not mandatory in DE.

Although DE is very robust in locating the global minimum, drawbacks remain. Being a direct search method DE does not utilise any properties, for instance, the differentiability properties, of the function being optimized, even if such properties are available. Moreover, its stopping condition depends on the indication that the points in S have formed a dense cluster or that the points have fallen into the region of attraction of the global minimum (or a minimizer). One way to measure this is to see if the absolute difference between the f_{max} and f_{min} falls below some given tolerence. This leads DE to unnecessarily use a large number of function evaluations and its efficiency falls off as the number of dimension increases. Consequently we devise a new DE algorithm which can overcome the above mentioned drawbacks. In particular, we would like to propose a DE algorithm which utilises the complementary strengths of both the existing DE and the Topographical Algorithm (TA) of Törn and Viitanen [24]. How this is done will be described later but first we briefly describe the Topographical Algorithm.

3   Topographical Algorithm (TA)

The TA uses topographical information on the objective function in identifying basins of local minima. For each identified basin a local search is started from its best point (the graph minimum). It is a non-iterative algorithm and therefore enough cover, i.e., a large enough sample size, is needed to identify the basins. A simplified description of the algorithm is the following: For each point in the sample of size N_g the k_g nearest neighbour points are determined ($k_g \ll N_g$). Those points for which all k_g neighbours are inferior points, i.e., the function values are larger, are the graph minima. The number of graph minima is dependent on the sample size N_g and k_g.

TA intends finding out a suitable number of graph minima associated with the number of minima for a given function. However, for a general purpose optimization a fixed value of k_g will be too restrictive to represent the appropriate number of local minima and this becomes even more difficult when the number of dimensions of the function increases. For a discussion on this and the full description of how graph minima are derived, see Törn and Viitanen [24]. Since both TA and DE is incapable of tackling large scale optimization problems we next integrate them in a suitable way, for large scale problems, eg. the many-body potential problems.

4   Topographical Differential Evolution (TDE)

In this section we describe our new algorithm. The new algorithm, TDE, does not simply combine DE and TA rather it combines the modified DE with the localized TA. We first describe the modified DE. In order to use as much information as possible of the points in S we introduce an auxilary set S_a of N points alongside S in DE. Initially, two sets (one set in the original DE) each containing N points are generated in the following way; iteratively sample two points from Ω, the best point x_i going to S and the other x_i' to S_a. The process continues untill each set has N points. The search process then updates both S and S_a simultaniously with generations. The reason for this is to make use of potential trial points which are normally rejected in DE. At each generation, unlike DE which updates one set, S, by the acceptance rule, TDE updates both sets S and S_a. In its acceptance phase, if the trial point y_i, corresponding to the target x_i, does not satisfy the greedy criterion $f(y_i) < f(x_i)$ then the point y_i is not abandoned altogether, rather it competes with its corresponding target x_i' in the set S_a. If $f(y_i) < f(x_i')$ then y_i replaces x_i' in the auxiliary set S_a. The potential points in S_a then can be used for further exploration and exploitation.

Since the DE procedure gradually drives the N points in S towards the global minimizer two measures are introduced in TDE to lessen the chance of missing the global minimizer in the driving process. They are : (a) after each M generations, finding out of the graph minima using the N_g best points from S and then performing a local search from each of the graph minima found, and (b) the replacement of the worst N_g points in S with the best N_g points in S_a immediate after the local searches have been performed. The benefits of (a) are that a local search only starts from a potential point with low function value and these potential points are seperated by higher regions. Since the points in S gradually shift their position these periodically scrutinized local searches will enhance the robustness of the algorithm in locating the global minimum. The benefits of (b) are search diversification and exploitation. We repeatedly find the graph minima locally using N_g best points with k_g nearest neighbours. The best minimum found in the local search phase is recorded and is further updated in the next phase of local search. If a consequitive number, say t, of local search phases does not produce any better minimum value than the previously found best minimum then the algorithm can be terminated. The step by step description of the new algorithm is as follows.

The TDE Algorithm

Step 1 Determine the initial sets $S = \{x_1, x_2, \cdots, x_N\}$ and $S_a = \{x_1', x_2', \cdots, x_N'\}$ with points sampled randomly in Ω. Initialize the generation counter k and the local phase counter t to zero.

Step 2 Determine the points x_{max}, x_{min} and their function values f_{max}, f_{min} such that

$$f_{max} = \max_{x \in S} f(x) \quad \text{and} \quad f_{min} = \min_{x \in S} f(x).$$

Calculate the scaling factor F of the mutation operator using (2). If the stopping condition, say $t \geq 5$, is satisfied then stop.

Step 3 For each $x_i \in S$, determine y_i by the following two operations:

- Mutation : Randomly select three points from S except x_i, the running target and find the second parent $\hat{x}_i$ by the mutation rule (1). If a component $\hat{x}_i^j$ falls outside Ω then it is found randomly in-between the j-th lower and upper limits.

- Crossover : Calculate the trial vector y_i corresponding to the target x_i from x_i and $\hat{x}_i$ using the crossover rule (3).

Step 4 Update both the sets S and S_a for the next generation using the acceptance rule: replace each $x_i \in S$ with y_i if $f(y_i) < f(x_i)$ otherwise replace $x_i' \in S_a$ with y_i if $f(y_i) < f(x_i')$. Set $k := k + 1$. If $k \equiv 0 \,(\text{mod } M)$ then go to Step 5, otherwise go to Step 2.

Step 5 Find the graph minima of the function, $f(x)$, using the best N_g points in S and perform a local search starting from each graph minimum. Keep a record of the very best minimum found so far, replace the worst N_g points in S with the best N_g in S_a. If the current phase of local minimization produces a better minimum than the current best minimum then set $t = 0$ otherwise set $t := t + 1$. Return to Step 2.

5   The Tersoff Potential

The binding energy in the Tersoff formulation [9] is written as a sum over atomic sites in the form

$$E_i = \frac{1}{2} \sum_{j \neq i} f_c(r_{ij})(V_R(r_{ij}) - \beta_{ij} V_A(r_{ij})) \,, \quad \forall i \tag{4}$$

where r_{ij} is the distance between atoms i and j, V_R is a repulsive term, V_A is an attractive term, $f_c(r_{ij})$ is a switching function and β_{ij} is a many-body term that depends on the positions of atoms i and j and the neighbours of atom i. More details of each of these quantities can be found in [9, 18]. The term β_{ij} is given by

$$\beta_{ij} = (1 + \gamma^{n_1} \xi_{ij}^{n_1})^{-1/2n_1} \tag{5}$$

where n_1 and γ are known fitted parameters [9]. The term ξ_{ij} for atoms i and j (i.e., for bond ij) is given by

$$\xi_{ij} = \sum_{k \neq i,j} f_c(r_{ik}) g(\theta_{ijk}) E(r_{ij}, r_{ik}) , \qquad (6)$$

where

$$E(r_{ij}, r_{ik}) = \exp\left(\lambda_3^3 (r_{ij} - r_{ik})^3\right) \qquad (7)$$

and θ_{ijk} is the bond angle between bonds ij and ik and g is given by

$$g(\theta_{ijk}) = 1 + c^2/d^2 - c^2/[d^2 + (h - \cos\theta_{ijk})^2] . \qquad (8)$$

The quantities λ_3, c, d and h which appear in (7) and (8) are also known fitted parameters. The terms $V_R(r_{ij})$ and $V_A(r_{ij})$ are given by

$$
\begin{aligned}
V_R(r_{ij}) &= A e^{-\lambda_1 r_{ij}} & (9a) \\
V_A(r_{ij}) &= B e^{-\lambda_2 r_{ij}} & (9b)
\end{aligned}
$$

where A, B, λ_1 and λ_2 are given fitted parameters. The switching function $f_c(r_{ij})$ is given by

$$f_c(r_{ij}) = \begin{cases} 1 , & r_{ij} \leq R - D \\ \frac{1}{2} - \frac{1}{2}\sin[\pi(r_{ij} - R)/(2D)] , & R - D < r_{ij} < R + D . \\ 0 , & r_{ij} \geq R + D \end{cases} \qquad (10)$$

We consider four different optimization problems, the first two problems are due to two different parameterisations of Tersoff potential for silicon. The two sets of parameter values respectively for $Si(B)$ and $Si(C)$ are taken from [9]. Similarly, the 3rd and the 4th problems are due to two different parameterizations of Tersoff's potential for carbon. The parameter value for the 3rd problem, say C', is taken from [19, 20]. The 4th problem is the Tersoff's potential for carbon, say C'', and the parameter values are taken from [21]. Therefore, each problem has its own set of parameter values that are given in Table 1.

Table 1: Parameters for Si and C.

	$Si(B)$	$Si(C)$	C'	C''
c	4.8381	1.0039E5	38049	19981
d	2.0417	16.216	4.3484	7.0340
h	0.0000	-0.59826	-0.57058	-0.33953
n_1	22.956	0.78734	7.2751E-1	0.99054
γ	0.33675	1.0999E-6	1.5724E-7	4.1612E-6
λ_1 ($\mathring{A}^{-1}$)	3.2394	2.4799	3.4879	3.4653
λ_2 ($\mathring{A}^{-1}$)	1.3258	1.7322	2.2114	2.3064
λ_3 ($\mathring{A}^{-1}$)	1.3258	1.7322	0.0000	0.0000
A (eV)	3.2647E3	1.8308E3	1.3936E3	1544.8
B (eV)	9.5373E1	4.7118E2	3.4674E2	389.63
R ($\mathring{A}$)	3.0	2.85	1.95	2.5
D ($\mathring{A}$)	0.2	0.15	0.15	0.15

5.1   Problem Formulation

In order to calculate the minimum potential energy for, say m atoms we need to calculate the energy for each atom. Each atom, say atom i, has its own potential energy, E_i, given by (4). Therefore, to determine the potential energy of a single particle one has to calculate (4) which involves the calculation of (5) − (10) for each neighbour of that particle. Notice that the energy of a particle depends upon the distances and angles subtended with respect to the other particles and therefore different particles have different energies. To formulate the problem we consider the atomic positions in two and three dimensional space as variables. It is clear that the total energy, say $f(x)$, is a function of atomic coordinates and it is given by

$$f(x) = E_1(x) + E_2(x) + \cdots + E_m(x) \ , \tag{11}$$

for which the global minimum f^* has to be found.

We first fix a particle at the origin and choose our second particle to lie on the positive x-axis. The third particle is chosen to lie in the upper half of the x-axis. Since the position of the first particle is always fixed and the second particle is restricted to the positive x-axis, this gives a minimization problem involving three variables for three particles. For four particles, additionally three variables (the cartesian co-ordinates of the 4-th particle) are required to give a minimisation problem in six independent variables. For each further particle, three variables (cordinates of the position of the particle) are added to determine the energetics of clusters. The first and the third variables are taken to lie in $[0, 4]$ for Si and $[0, 3]$ and $[0, 3.5]$ respectively for C' and C''. The second variable for Si, C' and C'' are taken to lie in $[-4, 4]$, $[-3, 3]$ and $[-3.5, 3.5]$ respectively. The cordinates of the 4-th particle for Si taken to lie in $[-l_i, u_i]$ with $l_i = u_i = 4$ and then for next extra two particles (5-th & 6-th) 6 variables involved are taken to lie in $[-l'_i, u'_i]$, where $l'_i(= u'_i) = l_i(= u_i) + 0.5$. Similarly the next two particles for Si variables are taken to lie in $[-l''_i, u''_i]$ where $l''_i(= u''_i) = l'_i(u'_i) + 0.5$ and then for each extra two particles the same process continues. Therefore, as an example, for Si the six variables for the 5-th & 6-th particles lie in $[-4.5, 4.5]$ and for 15-th & 16-th the associated six variables lie in $[-7, 7]$. In the case of C from the 4-th particle onwards a similar scheme is adopted but with $l_i = u_i = 3$ for C' and $l_i = u_i = 3.5$ for C''.

5.2   The Gradient Calculation

An important feature of the potential function is that it is differentiable. Therefore, the local search incorporated in the new algorithm can be made efficient and reliable by providing the analytical gradient. The gradient of $f(x)$ defined by (11) is the sum of the gradients of the components E_i defined by (4). However, each term of the sum in (4) is the energy attributed to a particular bond and each term in (4) is the function of the same variables. Since the pattern of the gradient of each term in (11) is the same, it suffice to calculate the gradient of E_i. Similarly, it is enough to derive some gradient components of the term associated with the bond ij in (4). For details on the gradient calculation, see Appendix A. Notice that in (4) if a bond length r_{ij}, say for the bond ij, exceeds $R + D$ then the calculation of the corresponding term of the sum (bond's contribution to energy E_i) and hence its derivative can be skipped.

6 Numerical Optimization and Discussion

In this section the numerical results obtained for all four problems are summarized. All computations were carried out on a SGI-Indy Irix-5.3. A limited memory BFGS algorithm (LBFGS) of Lui and Nocedal [25] was used as the local search algorithm. The LBFGS is designed for large scale local optimization problems. The population size in TDE is taken as $N = 10n$ within the suggested range [22]. The number of particles $np = 3, 4$, i.e, for $n = 3$ to $n = 6$ we have used $N_g = 30$ and $k_g = 3$. For $np = 5$ and above N_g is chosen as the nearest integer to $0.4N$, and k_g is increased by one for every four particles, i.e., for $np = 5, 6, 7, 8$ k_g is set to 4 and for $np = 9, 10, 11, 12$ k_g is set to 5 and so on. As an example, for $np = 10$ the parameter values used by the TDE algorithm were $N = 240, N_g = 96, k_g = 5$ and $n = 24$. With the increase of N_g and k_g time required to calculate the graph minima increases rapidly. However, the overhead of the algorithm is less than the time required by a single local minimization even for $np = 10$ and even more so for larger np. However, if needed larger N can be avoided. Since the population size increases with the increase of the number of dimensions (or np), an alternative to using both larger N and N_g is to use fixed values for N and N_g with a smaller k_g and thereby allowing more local searches to be performed. As np grows to a upper limit k_g can be gradually decreased to its lower limit. Indeed the faster and better results were obtained for $np = 13$ to 15 by using N and N_g for $np = 12$, i.e., $N = 300$ and $N_g = 120$ but with $k_g = 5$.

During our numerical experiments using Si for 100 independent runs we noticed that $Si(C)$ was difficult than $Si(B)$, it took on average more function evaluations (FE) and cpu times (CPU) and it was difficult to locate the very best minimum for $Si(C)$, for all particles considered. Similarly for the carbon problems we found that C' was more difficult than C''. This is true for upto 13 particles for C and for more than 13 particles we found it difficult to judge these problem as to which one is easier and which one is most difficult with respect to FE, CPU and locating the best minimum value. However, our numerical experiments suggested that $Si(C)$ was the most difficult of all four problems with C' being the second most difficult problem. We found that all four problems are very rich in local minima and that the number increases with np. Since the general trend in this respect for all four problems are the same, we present the detailed results of $Si(C)$ for upto 10 particles. To give a clear picture we present the detailed results on $Si(C)$ for 100 independent runs in Table 2. The results in Table 2 represent the average result of the runs for which the best minimum was successfully located. We use the following notation : fe is the number of function evaluations required by the algorithm, fl is the number of function evluations required by the local search algorithm, sf is the number of times the best minimum was located by the local search per successful run, lf is the number of different local minima found per successful run, lc is the number of local search performed per successful run, ts is the percentage of successful runs out of 100 runs. Notice that FE is the sum of fe and fl. The data under the columns lc, lf and sf are rounded to the nearest integer. Comparing the second and the third columns of Table 2, we find that the number of function evaluations required by the algorithm is far higher than that of the local minimizations. This gap can be reduced by choosing smaller k_g and thereby allowing more local search to be performed. As was mentioned earlier, for higher number of particles this policy is recommended. The results of the columns under fl and lc tell us that function evaluations per local optimization

Table 2: Summarised Results upto $Si_{10}(C)$

np	fe	fl	ts	lc	lf	sf	CPU
3	1680	435	100	30	5	27	0.99
4	7373	1023	88	47	33	12	5.27
5	19113	2980	96	58	73	13	28.13
6	42707	7388	59	109	98	7	54.72
7	81300	14654	12	162	154	4	87.08
8	111445	20264	7	184	178	4	192.12
9	238563	44864	3	245	233	2	342.44
10	403680	79776	1	482	461	2	1276.39

is not very high. It is simply because a local search starting at a low-lying point of the region of attraction of a local minimizer is conducive to quicker convergence and hence needing less function evaluations. The result for $t \geq 5$ and $M = 3n$ is presented in Table 2. We also studied the effect of M and t on $Si(C)$ for $np = 3$ to 10. When we used $M = 4n$ and $t \geq 5$ the average results of 100 runs showed that the overall FE increased by about 9% although lc decreased by about 6% while ts remained more or less the same. To see the effect of t we took three values, $t \geq 4, 5$ and 6 by fixing $M = 3n$. On average, the best result was obtained for $t \geq 5$. For $t \geq 6$ FE increased by 17% with on average only 3% improvements on ts while for $t \geq 4$ FE decreased but for the expenses of decreasing ts. Further, it appeared that for $t \geq 6$ ts was increased only for higher number of particles, i.e., for $np > 12$. Therefore, for $np \leq 12$ we used $M = 3n$ and $t \geq 5$ throughout our numerical work. With these values for M and t and the values for N_g and k_g as described earlier we next study the effect of the introduction of S_a on the overall results. This time we consider the second most difficult problem, namely the optimization problem associated with C'. We study the effect of S_a for upto 10 particles again using 100 independent runs. For this we ran TDE with and without the consideration of S_a. We denote the TDE algorithm as TDE_S when it only considers the set S. TDE_S, therefore, is TDE but without any replacement of points in S. Total results show that although TDE_S uses 6.7% less FE than TDE the percentage of success ts, in locating the best minimum, decreased by about 13%. Our numerical studies suggested that introduction of S_a increases the exploration of the search region as far as the topographical information is concerned. Therefore, we have used TDE for our optimization purpose. We now present the best minimum values obtained by TDE for all four problems for upto 15 particles. The best result of 100 independent runs are given in Table 3. Since no optimization was carried out of the Tersoff's carbon potential the best minima found for both carbon problems cannot be compared. However, the optimization of Tersoff's silicon potentials using eight different global optimization algorithms and the best known minima for clusters of upto six particles is given in [15]. Among these algorithms, it was found that a Controlled Random Search Algorithm, the CRS4 Algorithm [26], was the best performer in terms of FE, CPU and in locating the best minimum value. However, TDE has proved its superiority over CRS4 not only in terms of FE and CPU but also in locating the very best minimum value for Si clusters of upto six particles. Even for the most difficult problem, $Si(C)$, TDE is superior to CRS4 by about 55% and 82% respectively for FE and CPU. Moreover, within the optimization problems of upto six particles we find

Table 3: The best minimum found

np	$Si(B)$	$Si(C)$	C'	C''
3	-7.87	-5.33	-10.33	-7.42
4	-15.70	-8.63	-15.50	-11.12
5	-20.40	-12.43	-21.52	-18.09
6	-26.52	-15.80	-30.13	-22.19
7	-30.39	-18.66	-36.16	-26.91
8	-36.26	-22.27	-41.32	-32.87
9	-40.67	-26.20	-46.61	-37.57
10	-45.19	-29.26	-51.78	-43.81
11	-49.25	-32.74	-58.27	-48.48
12	-53.07	-36.04	-65.60	-54.29
13	-57.77	-38.42	-70.18	-60.96
14	-62.10	-42.04	-75.80	-65.37
15	-66.02	-44.80	-81.23	-68.75

that the new algorithm was able to produce even better minima for $Si_4(C)$ and $Si_5(C)$. A small number of optimized cluster geometries is shown in Figure 1.

7   Conclusion

We have developed a new global optimization algorithm for large scale problems and the algorithm is implemented on difficult optimization problems involving empirical potentials. The potential function is complex in that the interacting forces are many-body and angle-dependent. The global optimization of carbon and silicon potentials for upto 15 particles consisting of upto 39 variables is carried out. The new algorithm was able to produce better minima for some problems than those previously found. A unique feature of the algorithm is that it carries out local search from potential points scrutinized with the help of topographical information. The robustness of the new algorithm rests with its multiple local search phase in a unique algorithmic framwork. The new algorithm can be applied to large scale optimization problems in other areas of application such as problems in biological chemistry and in plasma physics. Designing a better algorithm for even larger problems will form the basis of our future research.

Acknowledgements : The first Author thanks Professor Roger Smith of Loughborough University of Technology, UK for his support and Professor Leslie Glasser of Witwatersrand University, Johannesburg for his help in plotting the graphs.

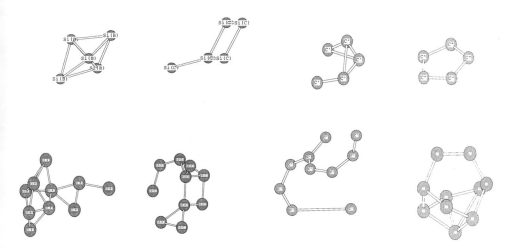

Figure 1: The Structure of Small Clusters of Sizes 5 and 10 for Si and C.

Appendix A

Let the variables involved be (x_i, y_i, z_i) for atom i, (x_j, y_j, z_j) for atom j for the bond ij and (x_k, y_k, z_k) for atom k where k represents the rest of the atoms involved in (6). Since the other partial derivatives are very similar, we only calculate the partial derivative of E_i with respect to one varible, say x_i out of the six variables associated with atoms i and j. This is given by

$$\frac{\partial E_i}{\partial x_i} = \frac{1}{2} \sum_{j \neq i} \left[\frac{df_c(r_{ij})}{dr_{ij}} \frac{\partial r_{ij}}{\partial x_i} \left(V_R(r_{ij}) - \beta_{ij} V_A(r_{ij}) \right) \right.$$
$$\left. + f_c(r_{ij}) \left(\frac{dV_R(r_{ij})}{dr_{ij}} \frac{\partial r_{ij}}{\partial x_i} - \beta_{ij} \frac{dV_A(r_{ij})}{dr_{ij}} \frac{\partial r_{ij}}{\partial x_i} - \frac{\partial \beta_{ij}}{\partial x_i} V_A(r_{ij}) \right) \right]. \tag{12}$$

The partial derivative of β_{ij} is written as

$$\frac{\partial \beta_{ij}}{\partial x_i} = -\frac{1}{2} \gamma^{n_1} \left(1 + \gamma^{n_1} \xi_{ij}^{n_1} \right)^{-\frac{1}{2n_1} - 1} \xi_{ij}^{(n_1 - 1)} \frac{\partial \xi_{ij}}{\partial x_i}. \tag{13}$$

The derivative of ξ_{ij} is given by (14) the components of which in turn are calculated from $(15) - (18)$.

$$\frac{\partial \xi_{ij}}{\partial x_i} = \sum_{k \neq i, j} \left[\frac{df_c(r_{ik})}{dr_{ik}} \frac{\partial r_{ik}}{\partial x_i} g(\theta) E(r_{ij}, r_{ik}) + \frac{\partial g(\theta)}{\partial x_i} f_c(r_{ik}) E(r_{ij}, r_{ik}) + f_c(r_{ik}) g(\theta) \frac{\partial E(r_{ij}, r_{ik})}{\partial x_i} \right].$$
$$\tag{14}$$

$$\frac{\partial E(r_{ij}, r_{ik})}{\partial x_i} = \frac{\partial E(r_{ij}, r_{ik})}{\partial r_{ij}} \frac{\partial r_{ij}}{\partial x_i} + \frac{\partial E(r_{ij}, r_{ik})}{\partial r_{ik}} \frac{\partial r_{ik}}{\partial x_i} . \tag{15}$$

$$\frac{\partial g(\theta)}{\partial x_i} = \frac{dg(\theta)}{d\cos\theta} \frac{\partial\cos\theta}{\partial x_i} . \tag{16}$$

$$\frac{\partial\cos\theta}{\partial x_i} = \frac{\partial\cos\theta}{\partial r_{ij}} \frac{\partial r_{ij}}{\partial x_i} + \frac{\partial\cos\theta}{\partial r_{ik}} \frac{\partial r_{ik}}{\partial x_i} . \tag{17}$$

The $\cos\theta$ between the bonds ij and ik is given by

$$\cos\theta = \frac{r_{ij}^2 + r_{ik}^2 - r_{jk}^2}{2r_{ij}r_{ik}} . \tag{18}$$

The derivative with respect to remaining 5 variables involved within the bond ij are easy to calculate but the slightly different partial derivative of E_i at x_k of (x_k, y_k, z_k) is given by (19) followed by the its relevent terms defined by (20) $-$ (24).

$$\frac{\partial E_i(x)}{\partial x_k} = -\frac{1}{2} \sum_{j\neq i} \frac{\partial\beta_{ij}}{\partial x_k} V_A(r_{ij}) f_c(r_{ij}) . \tag{19}$$

$$\frac{\partial\beta_{ij}}{\partial x_k} = -\frac{1}{2}\gamma^{n_1} \left(1 + \gamma^{n_1}\xi_{ij}^{n_1}\right)^{-\frac{1}{2n_1}-1} \xi_{ij}^{(n_1-1)} \frac{\partial\xi_{ij}}{\partial x_k} . \tag{20}$$

$$\frac{\partial\xi_{ij}}{\partial x_k} = \sum_{k\neq i,j} \left[\frac{df_c(r_{ik})}{dr_{ik}} \frac{\partial r_{ik}}{\partial x_k} g(\theta) E(r_{ij}, r_{ik}) + \frac{\partial g(\theta)}{\partial x_k} f_c(r_{ik}) E(r_{ij}, r_{ik}) + f_c(r_{ik}) g(\theta) \frac{\partial E(r_{ij}, r_{ik})}{\partial x_k} \right] . \tag{21}$$

$$\frac{\partial E(r_{ij}, r_{ik})}{\partial x_k} = \frac{\partial E(r_{ij}, r_{ik})}{\partial r_{ik}} \frac{\partial r_{ik}}{\partial x_k} . \tag{22}$$

$$\frac{\partial g(\theta)}{\partial x_k} = \frac{dg(\theta)}{d\cos\theta} \frac{\partial\cos\theta}{\partial x_k} . \tag{23}$$

$$\frac{\partial\cos\theta}{\partial x_k} = \frac{\partial\cos\theta}{\partial r_{ik}} \frac{\partial r_{ik}}{\partial x_k} + \frac{\partial\cos\theta}{\partial r_{jk}} \frac{\partial r_{jk}}{\partial x_k} . \tag{24}$$

The partial derivatives with respect to y_k and z_k can be calculated similarly.

References

[1] Garrison, B. J., Winograd, N., Deavon, D. M., Reimann, C. T., Lo, D. Y., Tombrello, T. A., Harrison D. E. Jr. and Shapiro, M. H. (1988), "Many-body Embedded-atom Potential for describing the Energy and Angular distributions of Rh atoms distorted from Ion-bombarded Rh{111}", *Physics Review B, Vol. 37,* 7197-7204.

[2] Smith, R. (1992), "A Semi-empirical Many-body interatomic Potential for Modelling Dynamical Processes in Gallium Arsenide", *Nuclear Instruments and Methods in Physics Research, Section B, Vol. 67,* 335-339.

[3] Smith, R., Harrison, D. E. Jr, and Garrison, B. J. (1989), "keV Particle Bombardment of Semiconductors : A Molecular Dynamic Simulation", *Physics Review B, Vol. 40,* 93-107.

[4] Ali, M. M. and Smith, R. (1993), "The Structure of Small Clusters Ejected by Ion Bombardment of Solids", *Vacuum, Vol. 44*, 377-379.

[5] Scheraga, H. A. (1996), "Recent Developments in the Theory of Protein Folding : Searching for the Global Minimum", *Biophysical Chemistry, Vol. 59*, 329-339.

[6] Carlsson, A. E. (1990), "Beyond Pair Potentials in Elemental Transition Metals and Semiconductors", *Solid State Physics, Vol. 43*, 1-90.

[7] Beardmore, K. and Smith, R. (1996), "Empirical Potentials for $C - Si - H$ Systems with Application to C_{60} interactions with Si Crystal Surface", *Philosophical Magazine A, Vol. 74*, 1439-1466.

[8] Tersoff, J. (1988), "New Empirical approach for the Structure and Energy of Covalent Systems", *Physics Review B, Vol. 37*, 6991-7000.

[9] Tersoff, J. (1988), "Empirical Interatomic Potential for Silicon with improved Elastic Properties", *Physics Review B, Vol. 38*, 9902-9905.

[10] Pardalos, P. M., Shalloway, D. and Xue, G. L. (1994), "Optimization Methods for Computing Global Minima of Non-convex Potential Energy Function", *Journal of Global Optimization, Vol. 4*, 117-133.

[11] Moré, J. J. and Wu, Z. (1995), "Global Smoothing and Continuation for Large Scale Molecular Optimization", *MCS-P539-1095*.

[12] Kostrowicki, J., Piela, L., Cherayil, B. J. and Scheraga, A. (1991), "Performance of the Diffusion Equation Method in Searches for Optimum Structures of Clusters of Lennard-Jones Atoms", *Journal of Physical Chemistry, Vol. 95*, 4113-4119.

[13] Xue, G. L. (1994), "Molecular Conformation on the CM-5 by Parallel Two-level Simulated Annealing", *Journal of Global Optimization, Vol. 4*, 187-208.

[14] Deaven, D. M., Tit, N., Morris, J. R. and Ho, K. M. (1996), "Structural Optimization of Lennard-Jones Clusters by a Genetic Algorithm", *Chemical Physics Letters, Vol. 256*, 195-198.

[15] Ali, M. M., Storey, C. and Törn, A. (1997), "Application of Stochastic Global Optimization Algorithms to Practical Problems", *Journal of Optimization Theory and Applications, Vol. 95*, 545-563.

[16] Hobday, S. (1998), "Artificial Intelligence and Simulations Applied to Interatomic Potentials", *PhD Thesis*, Department of Mathematical Sciences, Loughborough University of Technology, UK.

[17] Hobday, S. and Smith, R. (1997), "Optimization of Carbon Cluster Geometry Using a Genetic Algorithm", *Journal of Chemical Society, Faraday Transactions, Vol. 93*, 3919-3926.

[18] Brenner, D. W. (1990), "Empirical Interatomic Potential for Hydrocarbon for Use in Simulating the Chemical Vapor Deposition of Diamond Films", *Physics Review B, Vol. 42*, 9458-9471.

[19] Tersoff, J. (1988), "Empirical Interatomic Potential for Carbon with Applications to Amorphous Carbon", *Physical Review Letters, Vol. 21*, 2879-2882.

[20] Tersoff, J. (1989), "Modelling Solid-State Chemistry : Interatomic Potential for Multicomponent Systems", *Physical Review B, Vol. 39*, 5566-5568.

[21] Tersoff, J. (1990), "Carbon Defects and Defect Reactions in Silicon", *Physical Review Letters, Vol. 64*, 1757-1760.

[22] Storn, R. and Price, K. (1997), "Differential Evolution – A Simple and Efficient Heuristic for Global Optimization over Continuous Spaces," *Journal of Global Optimization, Vol. 11*, 341-359.

[23] Ali, M. M. and Törn, A. (1999), "Evolution based Global Optimization Techniques and the Controlled Random Search Algorithm : Proposed Modifications and Numerical Studies," submitted to the *Journal of Global Optimization*.

[24] Törn, A. and Viitanen, S. (1992), "Topographical Global Optimization," in *Recent Advances in Global Optimization*, Edited by A. Floudas and M. Pardalos, Princeton University Press, USA.

[25] Lui, D. C. and Nocedal, J. (1989), "On the Limited Memory BFGS Method for Large Scale Optimization", *Mathematical Programming, Vol. 45*, 503-528.

[26] Ali, M. M. and Storey, C. (1995), "Modified Controlled Random Search Algorithms", *International Journal of Computer Mathematics, Vol. 54*, 229-235.

Optimization in Computational Chemistry and Molecular Biology, pp. 301-339
C. A. Floudas and P. M. Pardalos, Editors
©2000 Kluwer Academic Publishers

D.C. Programming Approach for Large-Scale Molecular Optimization via the General Distance Geometry Problem

Le Thi Hoai An and Pham Dinh Tao
Mathematical Modelling and Applied Optimization Group
Laboratory of Mathematics, CNRS UPRES A 60 85
National Institute for Applied Sciences-Rouen
BP 8, F 76 131 Mont Saint Aignan Cedex, France
lethi@insa-rouen.fr   pham@insa-rouen.fr

Abstract

In this paper we are concerned with a new d.c.(difference of convex functions) approach to the general distance geometry problem and the two phase solution algorithm DCA. We present a thorough study of this d.c. program in its elegant matrix formulation including substantial subdifferential calculus for related convex functions. It makes it possible to express DCA in its simplest form and to exploit sparsity. In Phase 1 we extrapolate all pairwise dissimilarities from given bound constraints and then apply DCA to the resulting Euclidean Multidimensional Scaling (EMDS) problem. In Phase 2 we solve the original problem by applying DCA from the point obtained by Phase 1. Requiring only matrix-vector products and one Cholesky factorization, DCA seems to be robust and efficient in the large scale setting as proved by numerical simulations which furthermore indicated that DCA always converges to global solutions.

Keywords: d.c. programming, d.c. duality, d.c. algorithm (DCA), multidimensional scaling problem, distance geometry problems, molecular optimization, subdifferential calculations.

1 Introduction

Distance geometry problems, which play a key role in the molecular optimization, have earned active researches in recent years ([2], [9] - [14], [16], [23] - [25], [36]). These problems for the determination of protein structures are specified by a subset $\mathcal{S}$ of all atoms pairs and by the Euclidean distances δ_{ij} between atoms i and j for $(i, j) \in \mathcal{S}$. They initially consist in finding a set of $x^1, \ldots, x^n$ in $\mathbb{R}^3$ such that

$$\|x^i - x^j\| = \delta_{ij}, (i, j) \in \mathcal{S}. \tag{1}$$

It is called the exact distance geometry problem and has the following simple formulations as global optimization problems([2], [9] - [14], [16], [23] - [25], [36]):

$$\text{(EDP)} \quad 0 = \inf \left\{ \sum_{(i,j)\in S} p_{ij}\theta_{ij}(x^i - x^j) : x^1, ..., x^n \in \mathbb{R}^3 \right\} \tag{2}$$

where $p_{ij} > 0$ for $i \neq j$ and the pairwise potential $\theta_{ij} : \mathbb{R}^n \longrightarrow \mathbb{R}$ is defined for problem (1) by either

$$\theta_{ij}(x) = \left(\delta_{i,j}^2 - \|x\|^2 \right)^2 \tag{3}$$

or

$$\theta_{ij}(x) = (\delta_{ij} - \|x\|)^2. \tag{4}$$

The exact geometry problem is a special case of the Euclidean Multidimensional Scaling (EMDS) problem where the quantities δ_{ij} represent pairwise dissimilarities (which are not necessarily distances) between objects i and j ([2], [10], [36] and references therein).

In practice, lower and upper bounds on the distances are specified instead of their exact values. We then are faced with the so called general distance geometry problem

$$l_{ij} \leq \|x^i - x^j\| \leq u_{ij}, \ (i, j) \in \mathcal{S}. \tag{5}$$

The standard formulation of (5), due to Crippen and Havel [9], is in terms of globally solving the nonconvex program

$$0 = \inf \left\{ f(x^1, \ldots, x^n) = \sum_{(i,j)\in S} p_{ij}\theta_{ij}(x^i - x^j) : x^1, ..., x^n \in \mathbb{R}^3 \right\}, \tag{6}$$

where the pairwise function $\theta_{ij} : \mathbb{R}^n \mapsto \mathbb{R}$ is defined by

$$\theta_{ij}(x) = \min^2 \left\{ \frac{\|x\|^2 - l_{ij}^2}{l_{ij}^2}, 0 \right\} + \max^2 \left\{ \frac{\|x\|^2 - u_{ij}^2}{u_{ij}^2}, 0 \right\}.$$

An important case of the general distance geometry problem is to obtain an ε-optimal solution of (1), namely a configuration $x^1, \ldots, x^n$ in $\mathbb{R}^3$ satisfying

$$\mid \|x^i - x^j\| - \delta_{ij} \mid \leq \varepsilon, (i, j) \in \mathcal{S}, \tag{7}$$

for some $\varepsilon > 0$. An ε-optimal solution is useful when the exact solution to the distance geometry problem (1) does not exist because of small errors in the data. Such a situation can happen, for example, when the triangle inequality

$$\delta_{ij} \leq \delta_{ik} + \delta_{kj}$$

is violated for atoms $\{i, j, k\}$ because of possible inconsistencies of the experimental data.

In this paper we are interested in the large-scale molecular conformation from the general distance geometry problem (5) within the d.c. optimization framework. It is based on the new formulation of the general distance geometry problem as a d.c. program (i.e. a problem of minimizing a difference of convex functions):

$$0 = \inf \left\{ \sum_{(i,j)\in S} p_{ij}(\|x^i - x^j\| - t_{ij})^2 : x^1, ..., x^n \in \mathbb{R}^3, l_{ij} \le t_{ij} \le u_{ij}, (i,j) \in S \right\}, \quad (8)$$

where $p_{ij} = p_{ji} > 0$ for $i \ne j, (i,j) \in S$. It is a *linearly constrained nonsmooth nonconvex optimization problem* (see Section 2). Different formulations of (5) as d.c. programs are possible, for instance problem (6) is also a d.c. program. However the formulation (8) seems to be advantageous to d.c. algorithms (DCA) for solving the general distance geometry problem because DCA then requires only matrix-vector products and one Cholesky factorization. It is worth noting that the exact distance geometry problem (EDP) with θ_{ij} defined by (3), (resp. (EDP) with θ_{ij} defined by (4)), is a special case of the general distance geometry problem (6), (resp. (8)), with $l_{ij} = u_{ij} = \delta_{ij}$, for every $(i,j) \in S$.

When all pairwise distances are available and a solution exists, the exact distance geometry problem (1) can be solved by a polynomial time algorithm (Blumenthal [7], Crippen and Havel [9]). However, in practice, one knows only a subset of the distances, and it is well known (Saxe [39]) that $p-$ dimensional distance geometry problems are strongly NP-complete with $p = 1$ and strongly NP-hard for all $p > 1$. The visible sources of difficulties of these problems are

- the question of existence of a solution,

- the nonuniqueness of solutions,

- the presence of a large number of local minimizers,

- the large dimension of problems that arise in practice.

Several methods have been proposed for solving the distance geometry problems (1) and/or (7). De Leeuw ([10], [11]) proposed the well-known majorization method for solving the Euclidean MDS (EMDS) problem which includes (EDP) with θ_{ij} given by (4). Cripen and Havel [9] used the formulation (6) for solving the general distance geometry problem (5) by the *embed* algorithm. Their method consists of solving a sequence of exact distance geometry problems where all pairwise distances are included. More precisely the *embed* algorithm breaks down into three distinct steps. The first step, called *bound smoothing*, determines unknown bounds l_{ij} and u_{ij} from the given bounds by using the relationships

$$u_{ij} = \min_{k=1,...,n} \{u_{ij}, u_{ik} + u_{kj}\}, \quad l_{ij} = \max_{k=1,...,n} \{l_{ij}, l_{ik} - u_{kj}, l_{jk} - u_{ki}\}$$

which can be deduced from the triangle inequality. Given a full set of bounds, distances $\delta_{ij} \in [l_{ij}, u_{ij}]$ are chosen, and an attempt (which is the second step called *embedding*) is made to compute coordinates $x^1, \ldots, x^n$ in $\mathbb{R}^3$ such that

$$\|x^i - x^j\| = \delta_{ij}, \forall (i,j) \quad (9)$$

by solving the special complete distance geometry problem with, for example, the majorization algorithm applied to the problem defined by (2) and (4) as suggested in [13].

This attempt usually fails because the bounds δ_{ij} tend to be inconsistent, but it can be used to generate an approximate solution. This approximation can be refined (in the third step called *optimization*) by minimizing a function of the form (2)-(3). The *embed* algorithm may require many trial choices of δ_{ij} in $[l_{ij}, u_{ij}]$ before a solution to problem (9) is found. Current implementations of the *embed* algorithm use a local minimizer of the problem defined by (2) and (3). Glunt, Hayden and Raydan [14] studied a special gradient method for determining a local minimizer of (EDP) with θ_{ij} defined in (4). Using a graph-theoretic viewpoint Hendrickson [16] developed an algorithm to solve (EDP) where θ_{ij} is given by (3). His method works well for his test problems where a protein contains at most 124 amino acids (at most 777 atoms). In a smoothing technique and a continuation approach based on the Gaussian transform of the objective function and on the trust region method, Moré and Wu [23] proposed an algorithm for solving problem (EDP) with θ_{ij} defined by (3). By Gaussian transform, the original function becomes a smoother function with fewer local minimizer. Computational experiments with up to 648 variables ($n = 216$) in [23] proved that the continuation method is more reliable and efficient than the multistart approach, a standard procedure for finding the global minimizer to this problem. Also by Gaussian transform, Moré and Wu [25] solved the general distance geometry problem using the formulation (6). Their dgsol algorithm ([24], [25]) can be used to obtain solutions to the general distance geometry problem (6). In [24], the dgsol with multistarting points globally solved problem (6) with $n = 100$ and $n = 200$, and its reliabiltity varies between 40% and 100%. A stochastic/perturbation algorithm was proposed by Zou, Bird and Schnabel [45], using the standard formulations of (6) and (EDP) with θ_{ij} defined by (3), for both general and exact distance geometry problems. This is a combination of a stochastic phase that identifies an initial set of local minimizers and a more deterministic phase that moves from low to even lower local minimizer. The numerical experiments presented there (with the same data as in Moré and Wu [23] and Hendrickson [16]) showed that this approach is promising. It is worth noting that the distance geometry problem is intimately related to the Euclidean distance matrix completion problem ([1], [22]). This problem has been formulated as a semidefinite programming problem and solved by . A.Y. Alfakih, A. Khandami and H. Wolkowicz with an adapted interior-point method.

In convex analysis approach to nondifferentiable nonconvex programming, the d.c. optimization and its solution algorithms (DCA) developed by Pham and Le Thi ([2] - [6], [33] - [36] and references therein) constitute a natural and logical extension of Pham's earlier works concerning convex maximization and its subgradient algorithms ([27] - [32] and references therein). The majorization algorithm ([10]) is a suitable adaptation of the just mentioned subgradient methods for maximizing a seminorm over the unit ball of another seminorm, given that the latter is shown by de Leew to be equivalent to the Euclidean MDS problem. But the passage is not straightforward because the stepsize is computed by taking into account the original problem. De Leeuw's algorithm is actually a special case of DCA applied to EMDS problem ([2], [35], [36] and references therein). Remark that the argument used by de Leew ([10]) is no longer valid in the general distance geometry problem (8). Our method in this work, based on the d.c. optimization approach, serves to solve the general distance geometry problem (5) via the new d.c. program (8).

The aim of this paper is to demonstrate that the DCA can be used to develop efficient algorithms for solving large-scale general distance geometry problems via the new formulation (8). The DCA is a primal-dual subgradient method for solving a general d.c. program that consists in the minimization of difference of convex functions. It is at the present time one of a few algorithms in the local approach which has been successfully applied to many large-scale d.c. optimization problems and proved to be more robust and efficient than related standard methods. Using local optimality conditions and duality in d.c. programming, it cannot guarantee the globality of computed solutions for general d.c. programs. However, we observe that with a suitable starting point it converges quite often to a global one (see e.g. [2] - [6], [34] - [36]). This property motivates us to investigate a technique for computing a "good" starting point for the DCA in the solution of (8).

For initializing DCA applied to Problem (8) we used procedures C1 and C2 presented in Subsection 4.4. Procedure C1 (deducing dissimilarities) is a particular case of the triangle bound smoothing due to Crippen while Procedure C2 (imposing dissimilarities) is quite different because the quantities l_{ij} do not necessarily represent lower bounds for unknown distances. In any case it does not cause trouble to DCA applied to the resulting EMDS problem ([36]) in which the dissimilarities are not necessarily distances. Algorithm 2 (of our two phase DCA for solving the general distance geometry problem (8)) then consists of applying DCA to the resulting EMDS problem with a *full dissimilarity matrix*. Algorithm 1 is exactly DCA applied to the *sparse* general distance geometry problem (8) with initial point computed by Algorithm 2.

The advantages of the present method are:

Firstly, we need to work only once with both *dense* and *sparse* sets of constraints. In contrast, the existing methods for the general distance geometry problem work many times with full and sparse sets of constraints (see e.g. the *embed* algorithm) or a sparse set of constraints ([25], [45]).

Secondly, we can exploit sparsity of the given bound matrices. This is important because only a small subset of constraints is known in practice.

Our algorithms are quite simple and easy to implement. They only require matrix-vector products and one Cholesky factorization. We have tested our codes on the artificial general distance geometry problems (Moré & Wu [23], [24], [25]) with up to 10125 variables (the molecule contains 3375 atoms).

Section 2 is devoted to the new formulation of the general distance geometry problem (5) in the suitable matrix working space. The main tools of our work are the d.c. optimization approach and the DCA. The background indispensable for understanding the d.c. programming and DCA is described in Section 3 where we emphasize the crucial role played by d.c. decompositions and initial points for DCA in the global solution of a d.c. program. Section 4 is the core of the paper where is presented a thorough study of problem (8) in its elegant matrix formulation (including substantial subdifferential calculus for related convex functions) which allows to express the DCA (for solving the general distance geometry problem (8)) in explicit form and to exploit sparsity. Finally extensive numerical simulations and comments are reported in Section 5.

2   D.C. Formulation

In Euclidean distance geometry problems, we must take into consideration the symmetry of both the subset $\mathcal{S}$ (i.e. $(i,j) \in \mathcal{S}$ implies $(j,i) \in \mathcal{S}$) and the weight matrix $P = (p_{ij})$.

For the sake of simplifying calculations in DCA, the general distance geometry problem (8) will be reformulated as:

$$(\text{GEDP}) \quad 0 = \inf \left\{ \frac{1}{4} \sum_{(i,j) \in S, i<j} p_{ij} (\|x^i - x^j\| - t_{ij})^2 : x^1, ..., x^n \in \mathbb{R}^3, \, l_{ij} \le t_{ij} \le u_{ij}, (i,j) \in S \right\}$$

In this section we first prove that problem (GEDP) is a d.c. program and point out simple convex functions (on the convex constraint set) whose difference is the objective function of this problem. Since

$$-2t_{ij}\|x^i - x^j\| = -(\|x^i - x^j\| + t_{ij})^2 + \|x^i - x^j\|^2 + t_{ij}{}^2,$$

the objective function of (GEDP) can be expressed as

$$\frac{1}{2} \sum_{(i,j) \in S, i<j} p_{ij}\|x^i - x^j\|^2 + \frac{1}{2} \sum_{(i,j) \in S, i<j} p_{ij}\, t_{ij}^2 - \frac{1}{4} \sum_{(i,j) \in S, i<j} p_{ij}(\|x^i - x^j\| + t_{ij})^2. \quad (10)$$

Remark that by setting $p_{ij} = 0$ for $(i,j) \notin S$ the constraint $(i,j) \in S$ can be omitted in (10). Since the functions

$$\frac{1}{2} \sum_{(i,j) \in S, i<j} p_{ij}\|x^i - x^j\|^2 + \frac{1}{2} \sum_{(i,j) \in S, i<j} p_{ij}\, t_{ij}^2 \quad \text{and} \quad \frac{1}{4} \left\{ \sum_{(i,j) \in S, i<j} p_{ij}(\|x^i - x^j\| + t_{ij})^2 \right\}$$

are convex with respect to the variables $(x^1, ..., x^n), T$) with $x^1, ..., x^n \in \mathbb{R}^3$ and $\mathrm{T} = (t_{ij})$ on the convex constraint set, it is clear that this expression is difference of convex functions (d.c. function in short). The matrix spaces that we shall present below are useful for various calculations of subgradients in DCA.

Let $\mathcal{M}_{n,p}(\mathbb{R})$ and $\mathcal{M}_{n,n}(\mathbb{R})$ denote the spaces of real matrices of order $n \times p$ and $n \times n$ respectively. For $X \in \mathcal{M}_{n,p}(\mathbb{R})$, X_i (resp. X^i) is its ith row (resp. column), while X^T is the transpose of X. By identifying a set of positions $x^1, ..., x^n$ with the matrix X (i.e. $(X^T)^j = (X_j)^T = x^j$ for j = 1,..., n), and let $T = (t_{ij})$ be a symmetric matrix in $\mathcal{M}_{n,n}(\mathbb{R})$, we shall advantageously express the general distance geometry problem in the product of matrix spaces $\mathcal{M}_{n,p}(\mathbb{R}) \times \mathcal{M}_{n,n}(\mathbb{R})$. First, note that we can identify a matrix $X \in \mathcal{M}_{n,p}(\mathbb{R})$ with a row-vector (resp. column-vector) in $(\mathbb{R}^p)^n$ (resp. $(\mathbb{R}^n)^p$) by writing, respectively,

$$X \longleftrightarrow \mathcal{X} = (\, X_1 ... X_n), \quad X_i^T \in \mathbb{R}^p, \, \mathcal{X}^T \in (\mathbb{R}^p)^n, \quad (11)$$

and

$$X \longleftrightarrow \overline{\mathcal{X}} = \begin{pmatrix} X^1 \\ . \\ . \\ . \\ X^p \end{pmatrix}, \quad X^i \in \mathbb{R}^n, \overline{\mathcal{X}} \in (\mathbb{R}^n)^p. \quad (12)$$

The inner product in $\mathcal{M}_{n,p}(\mathbb{R})$ is then defined as the inner product in $(\mathbb{R}^p)^n$ or $(\mathbb{R}^n)^p$. That is

$$\langle X, Y \rangle_{\mathcal{M}_{n,p}(\mathbb{R})} = \langle \mathcal{X}^T, \mathcal{Y}^T \rangle_{(\mathbb{R}^p)^n} = \sum_{i=1}^n \langle X_i^T, Y_i^T \rangle_{\mathbb{R}^p} = \sum_{i=1}^n X_i Y_i^T$$

$$= \langle \overline{\mathcal{X}}, \overline{\mathcal{Y}} \rangle_{(\mathbb{R}^n)^p} = \sum_{k=1}^{p} \langle X^j, Y^j \rangle_{\mathbb{R}^n} = \sum_{k=1}^{p} (X^j)^T Y^j = Tr(X^T Y).$$

Likewise, a matrix $T \in \mathcal{M}_{n,n}(\mathbb{R})$ can be identified with a vector $\mathcal{T} \in (\mathbb{R}^n)^n$, and the inner product in $\mathcal{M}_{n,n}(\mathbb{R})$ is then defined as the inner product in $(\mathbb{R}^n)^n$. In the sequel, for simplicity, we shall suppress, if no possible ambiguity, the indices for the inner product and denote by $\|.\|$ the corresponding Euclidean norm. Evidently we must choose either representation in a convient way.

Let $d_{ij} : \mathcal{M}_{n,p}(\mathbb{R}) \longmapsto \mathbb{R}$ and $\varphi_{ij} : \mathcal{M}_{n,n}(\mathbb{R}) \longmapsto \mathbb{R}$ be the pairwise functions defined by $d_{ij}(X) = \|x^i - x^j\|, \varphi_{ij}(T) = t_{ij}$.

Problem (GEDP) can be now written in the matrix form

$$\begin{cases} 0 = \inf & \{F(X,T) := L(X,T) - H(X,T)\} \\ s.t. & (X,T) \in \Omega := \mathcal{M}_{n,p}(\mathbb{R}) \times \mathcal{C}, \end{cases} \tag{13}$$

with

$$L(X,T) := \eta(X) + \zeta(T),$$

$$\eta(X) := \frac{1}{2} \sum_{i<j} p_{ij} \, d_{ij}^2(X), \; \zeta(T) := \frac{1}{2} \sum_{i<j} p_{ij} \, \varphi_{ij}^2(T),$$

$$H(X,T) := \frac{1}{4} \sum_{i<j} p_{ij} [d_{ij}(X) + \varphi_{ij}(T)]^2,$$

$$\mathcal{C} := \{T \in \mathcal{M}_{n,n}(\mathbb{R}) : l_{ij} \le t_{ij} \le u_{ij}, (i,j) \in \mathcal{S}\}. \tag{14}$$

Clearly, the function L is finite and convex on $\mathcal{M}_{n,p}(\mathbb{R}) \times \mathcal{M}_{n,n}(\mathbb{R})$. Since for every $(i,j) \in \mathcal{S}$, the function: $(X,T) \to d_{ij}(X) + \varphi_{ij}(T)$ is finite and convex on $\mathcal{M}_{n,p}(\mathbb{R}) \times \mathcal{M}_{n,n}(\mathbb{R})$ and nonnegative on Ω, the function H then is convex on Ω too . Let χ_Ω be the indicator function of Ω defined by $\chi_\Omega(X,T) = 0$ if $(X,T) \in \Omega, +\infty$ otherwise, then problem (GEDP) can be expressed in the standard form of d.c. programs:

$$\begin{cases} 0 = \inf & \{F(X,T) := G(X,T) - H(X,T)\} \\ s.t. & (X,T) \in \mathcal{M}_{n,p}(\mathbb{R}) \times \mathcal{M}_{n,n}(\mathbb{R}), \end{cases} \tag{15}$$

where $G(X,T) := L(X,T) + \chi_\Omega(X,T)$ is the separable function in its variables X and T. Before going further let us specify the obvious relation between problems (5) and (15).

Proposition 1 *(i) If a set of positions $(x^1, ..., x^n)$ is a solution to problem (5), then the couple of matrices (X,T), with $X = (x^1, ..., x^n)^T$ and $t_{ij} = \|x^i - x^j\|$ for $(i,j) \in \mathcal{S}$, is a solution to (15).*

(ii) If a couple of matrices (X,T) is a solution to (15), then the set of positions $(x^1, ..., x^n) = X^T$ is a solution to (5) and $t_{ij} = \|X_i^T - X_j^T\|$ for $(i,j) \in \mathcal{S}$.

Throughout this paper we assume that the weight matrix $P = (p_{ij})$ is irreducible, which can be viewed as the associated graph $G(N, S)$ with $N = \{1, ..., n\}$ is connected. This assumption is not restrictive for problem (5) since it can be decomposed into a number of smaller problems otherwise. Then we work under the next assumptions:

(a1) for $i \neq j$, $l_{ij} > 0$ when $(i, j) \in S$ (i.e., two different atoms are not in the same position),

(a2) for $i \neq j$, $p_{ij} = 0$ if and only if $(i, j) \notin S$,

(a3) the weight matrix P is irreducible.

Under assumption (a2) and (a3), we can restrict the working space to an appropriate set which is, as will be seen in the next, favourable to our calculations. Indeed, let $\mathcal{A}$ denote the subspace composed of matrices in $\mathcal{M}_{n,p}(\mathbb{R})$ whose rows are identical, i.e.,

$$\mathcal{A} := \{X \in \mathcal{M}_{n,p}(\mathbb{R}) : X_1 = \cdots = X_n\}$$

and let $P_{\mathcal{A}}$ (resp. $\mathcal{A}^{\perp}$) be the orthogonal projection on $\mathcal{A}$ (resp. the orthogonal complement of $\mathcal{A}$), we have:

Lemma 1 *(i)* $\mathcal{A} = \{ev^T : v \in \mathbb{R}^p\}$ *is a* $p-$*dimensional subspace of* $\mathcal{M}_{n,p}(\mathbb{R})$ *and* $\mathcal{A}^{\perp} = \{Y \in \mathcal{M}_{n,p}(\mathbb{R}) : \sum_{i=1}^n Y_i = 0\}$.

(ii) $\mathcal{A} \subset \eta^{-1}(0)$.

(iii) $P_{\mathcal{A}} = (1/n)ee^T$; $P_{\mathcal{A}^{\perp}} = I - (1/n)ee^T$ *(e is the vector of ones in* $\mathbb{R}^n$*).*

(iv) If the weight matrix P is irreducible, then $\mathcal{A} = \eta^{-1}(0)$. *In this case Problem (15) is equivalent to*

$$\begin{cases} 0 = \inf & \{F(X,T) := \ G(X,T) - H(X,T)\} \\ s.t. & (X,T) \in \mathcal{A}^{\perp} \times \mathcal{M}_{n,n}(\mathbb{R}), \end{cases} \tag{16}$$

in the sense that if (X^*, T^*) *is an optimal solution to (16), then* $(X^* + X, T^*)$ *is an optimal solution to (15) for all* $X \in \mathcal{A}$.

Proof. (i) and (ii) are straightforward from the definition of $\mathcal{A}$. The proof of (iii) is easy.
Let $X \in \mathcal{M}_{n,p}(\mathbb{R})$ such that $\eta(X) = 0$ and $(i, j) \in \{1, \ldots, n\}^2$ with $i \neq j$. Since the matrix P is irreducible, there is a finite sequence $\{i_1, \ldots, i_r\} \subset \{1, \ldots, n\}$ verifying $p_{ii_1} > 0, p_{i_k i_{k+1}} > 0$ for $k = 1, \ldots, r-1$, and $p_{i_r j} > 0$. It follows that $X_i = X_{i_1} = \cdots = X_{i_r} = X_j$, and then $\eta^{-1}(0) = \mathcal{A}$.
Finally, the rest of property (iv) follows from the fact that the kernel of the seminorms $\eta^{\frac{1}{2}}$ and d_{ij} contain the subspace $\mathcal{A}$. □

3   D.C. Programming and DCA

We present here the material needed for an easy understanding of d.c. programming and DCA which will be used to solve the general distance geometry problem (16). Our working space is $E = \mathbb{R}^n$ equipped with the canonical inner product $\langle \cdot, \cdot \rangle$ and the corresponding Euclidean norm $\| \cdot \|$, thus the dual space E^* of E can be identified with E itself. We follow [37] for definitions of usual tools of modern convex analysis where functions could take infinite values $\pm\infty$. A function $\theta : E \to \mathbb{R} \cup \{\pm\infty\}$ is said to be proper if it takes nowhere the value $-\infty$ and is not identically equal to $+\infty$. The effective domain of θ, denoted by dom θ, is

$$\text{dom } \theta = \{x \in E : \theta(x) < \infty\}.$$

The set of all lower semicontinuous proper convex functions on E is denoted $\Gamma_0(E)$. For $g \in \Gamma_0(E)$, the conjugate function g^* of g is a function belonging to $\Gamma_0(E)$ and defined by

$$g^*(y) = \sup\{\langle x, y \rangle - g(x) : x \in E\}.$$

and we have $g^{**} = g$.

Let $g \in \Gamma_0(E)$ and let $x^0 \in$ dom g and $\epsilon > 0$, then $\partial_\epsilon g(x^0)$ stands for the $\epsilon-$ *subdifferential* of g at x^0 and is given by

$$\partial_\epsilon g(x^0) = \{y^0 \in E^* : g(x) \geq g(x^0) + \langle x - x^0, y^0 \rangle - \epsilon, \forall x \in E\}$$

while $\partial g(x^0)$ corresponding to $\epsilon = 0$, stands for the usual (or exact) subdifferential of g at x^0. Recall that

$$y^0 \in \partial g(x^0) \iff x^0 \in \partial g^*(y^0) \iff \langle x^0, y^0 \rangle = g(x^0) + g^*(y^0).$$

One says that g is subdifferentiable at x^0 if $\partial g(x^0)$ is nonempty. It has been proved that [37]

$$ri(\text{dom } g) \subset \text{dom } \partial g \subset \text{dom } g$$

where ri(dom g) stands for the relative interior of dom g and dom $\partial g := \{x \in E : \partial g(x) \neq \emptyset\}$.

Also, the indicator function χ_C of a closed convex set is defined by $\chi_C(x) = 0$ if $x \in C$, $+\infty$ otherwise.

A function $\theta \in \Gamma_0(E)$ is said to be polyheral convex if ([37])

$$\theta(x) = \max\{\langle a^i, x \rangle - \alpha_i : i = 1, \ldots, m\} + \chi_S(x), \quad \forall x \in E,$$

where $a^i \in E^*$, $\alpha_i \in \mathbb{R}$ for $i = 1, \ldots, m$ and S is a nonempty polyhedral convex set in E.

Let $\rho \geq 0$ and C be a convex subset of E. One says that a function $\theta : C \longrightarrow \mathbb{R} \cup \{+\infty\}$ is $\rho - convex$ if

$$\theta[\lambda x + (1-\lambda)x'] \leq \lambda\theta(x) + (1-\lambda)\theta(x') - \frac{\lambda(1-\lambda)}{2}\rho\|x - x'\|^2, \forall \lambda \in]0,1[, \forall x, x' \in C.$$

It amounts to saying that $\theta - (\rho/2)\| \cdot \|^2$ is convex on C. The modulus of strong convexity of θ on C, denoted by $\rho(\theta, C)$ or $\rho(\theta)$ if $C = E$, is given by:

$$\rho(\theta, C) = \sup\{\rho \geq 0 : \theta - (\rho/2)\| \cdot \|^2 \text{ is convex on } C\}. \tag{17}$$

Clearly, θ is convex on C if and only if $\rho(\theta, C) = 0$. One says that θ is *strongly convex* on C if $\rho(\theta, C) > 0$.

For f_1 and f_2 belonging to $\Gamma_0(E)$, the infimal convolution of f_1 and f_2, denoted $f_1 \triangledown f_2$, is a convex function on E, defined by [18]

$$f_1 \triangledown f_2(x) = \inf\{f_1(x_1) + f_2(x_2) : x_1 + x_2 = x\}, \ \forall x \in E.$$

In convex analysis, this functional operation aims, as the convolution in functional analysis, at regularizing convex functions [18]. The proximal regularization corresponding to $\theta = \frac{\lambda}{2}\| \cdot \|^2$.

For $f \in \Gamma_0(E)$ and $\lambda > 0$ the Moreau-Yosida regularization of f with parameter λ, denoted by f_λ, is the inf-convolution of f and $\frac{1}{2\lambda}\| \cdot \|^2$. The function f_λ is continuously differentiable, underapproximates f without changing the set of minimizers and $(f_\lambda)_\mu = f_{\lambda+\mu}$. More precisely, $\nabla f_\lambda = \frac{1}{\lambda}[I - (I + \lambda \partial f)^{-1}]$ is Lipschitzian with ratio $\frac{1}{\lambda}$. The operator $(I + \lambda \partial f)^{-1}$ is called the proximal mapping associated with λf ([20], [21], [38]).

A general d.c. program is of the following form with $g, h \in \Gamma_0(E)$

$$(P_{dc}) \qquad \begin{cases} \alpha = \inf & f(x) := g(x) - h(x) \\ s.t. & x \in E, \end{cases}$$

where we adopt the convention $+\infty - (+\infty) = +\infty$ to avoid ambiguity. One says that $g - h$ is a d.c. decomposition (or d.c. representation) of f, and g, h are its convex d.c. components. If g and h are finite on E, then $f = g - h$ is said to be finite d.c. function on E. The set of d.c. functions (resp. finite d.c. functions) on E is denoted by $\mathcal{DC}(E)$ (resp. $\mathcal{DC}_f(E)$).

Note that the finiteness of α merely implies that

$$\text{dom } g \subset \text{dom } h \quad \text{and} \quad \text{dom } h^* \subset \text{dom } g^*. \tag{18}$$

Such inclusions will be assumed throughout the paper.

A point x^* is said to be *a local minimizer* of $g - h$ if $g(x^*) - h(x^*)$ is finite (i.e., $x^* \in \text{dom } g \cap \text{dom } h$) and there exists a neighbourhood U of x^* such that

$$g(x^*) - h(x^*) \leq g(x) - h(x), \quad \forall x \in U. \tag{19}$$

Under the convention $+\infty - (+\infty) = +\infty$, the property (19) is equivalent to $g(x^*) - h(x^*) \leq g(x) - h(x), \quad \forall x \in U \cap \text{dom } g$.

x^* is said to be *a critical point* of $g - h$ if $\partial g(x^*) \cap \partial h(x^*) \neq \emptyset$.

It is worth noting the richness of $\mathcal{DC}(E)$ and $\mathcal{DC}_f(E)$ ([2], [35] and references therein):

(i) $\mathcal{DC}_f(E)$ is a subspace containing the class of lower-$\mathcal{C}^2$ functions (f is said to be lower-$\mathcal{C}^2$ if f is locally a supremum of a family of $\mathcal{C}^2$ functions). In particular, $\mathcal{DC}_f(E)$ contains the space $\mathcal{C}^{1,1}(E)$ of functions whose gradient is locally Lipschitzian on E.

(ii) Under some caution we can say that $\mathcal{DC}(E)$ is the subspace generated by the convex cone $\Gamma_0(E)$: $\mathcal{DC}(E) = \Gamma_0(E) - \Gamma_0(E)$. This relation marks the passage from convex optimization to nonconvex optimization and also indicates that $\mathcal{DC}(E)$ constitutes a minimal realistic extension of $\Gamma_0(E)$.

(iii) $\mathcal{DC}_f(E)$ is closed under all the operations usually considered in optimization. In particular, a linear combination of $f_i \in \mathcal{DC}_f(E)$ belongs to $\mathcal{DC}_f(E)$, a finite supremum of d.c. functions is d.c. This result has been extended to $\mathcal{DC}(E)$ under some restrictions.

D.c. programming has been a natural extension of convex maximization in which the function g is the indicator function χ_C of a nonempty closed convex set C, (i.e. $\chi_C(x) = 0$ if $x \in C, +\infty$ otherwise). In convex approach to nonsmooth nonconvex optimization (not to be confused with the global combinatorial approach due to Hoang Tuy[41], [42], [43]), convex maximization has been extensively studied since 1974 by Pham ([27]- [28]-[29] - [32] and references therein) who has introduced subgradient algorithms for solving convex maximization problems.

The d.c. duality, (due to Toland [40], who generalized in a very elegant and natural way the early works, just mentioned above, of Pham on convex maximization programming), associates the d.c. program (P_{dc}) with the following one called its dual d.c. program

$$(D_{dc}) \qquad \begin{cases} \alpha = \inf \; h^*(y) - g^*(y) \\ s.t. \qquad y \in E^* \end{cases}$$

with the help of the functional conjugate notion and states relationships between them. More precisely, as a sort of getting to the root of convex functions (namely a convex function $\theta \in \Gamma_0(E)$ is characterized as the supremum of a collection of affine minorizations, in particular there holds the following expression

$$\theta(x) = \sup\{\langle x, y \rangle - \theta^*(y) : y \in E^*\}, \; \forall x \in E \qquad (20)$$

that will appear in the concept of our DCA again, the d.c. duality is built by replacing, in problem (P_{dc}), the function h with its corresponding expression of (20).

Thanks to a symmetry in the d.c. duality (the bidual d.c. program) is exactly the primal one) and the d.c. duality transportation of global minimizers, solving a d.c. program implies solving the dual one and *vice versa*. The equality of the optimal value in the primal and dual programs can be easily translated (with the help of *ϵ-subdifferential* of the d.c. components) in global optimality conditions. These nice conditions mark the passage from convex optimization to nonconvex optimization but are rather difficult to use for devising solution methods to d.c. programs.

Local d.c. optimality conditions constitute (with the d.c. duality) the basis of the DCA. In general, it is not easy to state them as in global d.c. optimality and there have been found very few properties which are useful in practice.

We will briefly present the main results on local and global optimality conditions in d.c. programming in the Appendix for the convenience of the reader.

The DCA was introduced as an extension of the aforementioned subgradient algorithms (for convex maximization programming) to d.c. programming by Pham in 1986. But this field has been really developed from 1994 only with joint works by Le Thi and Pham ([2] - [6], [33] - [36] and references therein) in which d.c. programming approach has been successfully applied to solving nonsmooth nonconvex optimization problems. To our knowledge, DCA is actually one of a few algorithms (in the convex analysis approach to d.c. programming) which allows to solve large-scale d.c. programs.

3.1 Description of DCA for a general d.c. program

Based on local optimality conditions and duality in d.c. programming, the DCA consists in the construction of two sequences $\{x^k\}$ and $\{y^k\}$ such that x^{k+1} (resp. y^k) is a solution to the convex program (P_k) (resp. (D_k)) defined by

$$(P_k) \qquad \begin{cases} \inf & \{g(x) - [h(x^k) + \langle x - x^k, y^k \rangle]\} \\ s.t. & x \in E \end{cases}$$

$$(D_k) \qquad \begin{cases} \inf & \{h^*(y) - [g^*(y^{k-1}) + \langle x^k, y - y^{k-1} \rangle]\} \\ s.t. & y \in E^*. \end{cases}$$

In view of the relation: (P_k) (resp. (D_k)) is obtained from (P_{dc}) (resp. (D_{dc})) by replacing h (resp. g^*) with its affine minorization defined by $y^k \in \partial h(x^k)$ (resp. $x^k \in \partial g^*(y^{k-1})$), the DCA yields the next scheme:

$$y^k \in \partial h(x^k); \quad x^{k+1} \in \partial g^*(y^k). \tag{21}$$

It is proved in Pham and Le Thi [34], [35] that

(i) The sequences $\{g(x^k) - h(x^k)\}$ and $\{h^*(y^k) - g^*(y^k)\}$ are decreasing and

- $g(x^{k+1}) - h(x^{k+1}) = g(x^k) - h(x^k)$ if and only if $y^k \in \partial g(x^k) \cap \partial h(x^k)$, $y^k \in \partial g(x^{k+1}) \cap \partial h(x^{k+1})$ and $[\rho(g) + \rho(h)]\|x^{k+1} - x^k\| = 0$.

- $h^*(y^{k+1}) - g^*(y^{k+1}) = h^*(y^k) - g^*(y^k)$ if and only if $x^{k+1} \in \partial g^*(y^k) \cap \partial h^*(y^k)$, $x^{k+1} \in \partial g^*(y^{k+1}) \cap \partial h^*(y^{k+1})$ and $[\rho(g^*) + \rho(h^*)]\|y^{k+1} - y^k\| = 0$.

  In such a case DCA terminates at the kth iteration.

(ii) If $\rho(g) + \rho(h) > 0$ (resp. $\rho(g^*) + \rho(h^*) > 0$), then the series $\{\|x^{k+1} - x^k\|^2\}$ (resp. $\{\|y^{k+1} - y^k\|^2\}$) converges.

(iii) If the optimal value α of problem (P_{dc}) is finite and the sequences $\{x^k\}$ and $\{y^k\}$ are bounded then every limit point x^∞ (resp. y^∞) of the sequence $\{x^k\}$ (resp. $\{y^k\}$) is a critical point of $g - h$ (resp. $h^* - g^*$).

(iv) DCA has a linear convergence for general d.c. programs.

To make the reading easier, we present in the Appendix the principal convergence theorem of DCA for general d.c. program (for more details, see [34], [35] and references therein).

Remark 1 *The d.c. objective function (of a d.c. program) has infinitely many d.c. decompositions which may have an important influence on the qualities (robustness, stability, rate of convergence and globality of sought solutions). For example if g and h are convex* *d.c. components of the d.c. function then so are the functions $g + \theta$ and $h + \theta$ with θ being a finite convex function on E. In particular, if θ is the kernel $\frac{\lambda}{2}\|.\|^2$, $(\lambda > 0)$, we make the d.c. components (of the primal objective function $f = g - h$) strongly convex and the convex d.c. components (of the dual objective function) continuously differentiable. This operation amounts to applying the proximal regularization to the dual d.c. program (D_{dc}).*

So there are as many DCA as there are d.c. decompositions and it is of particular interest to study various equivalent d.c. forms for the primal and dual d.c. problems. It is worth mentioning, for instance, that by using conjointly suitable d.c. decompositions of convex functions and proximal regularization techniques we can obtain the proximal point algorithm and the Goldstein-Levitin-Polyak subgradient method (in convex programming) as special cases of DCA.

The choice of the d.c. decomposition of the objective function and the initial point for DCA are open questions to be studied for the specific structure of the problem being considered. In practice, for solving a given d.c. program, we try to choose g and h such that sequences of iterates can be easily calculated, i.e. either they are in explicit form or their computations are inexpensive.

A deeper insight into DCA.

The above description of DCA does not really reveal the main features of this approach which could partly explain the qualities (robustness, stability, rate of convergence and globality of sought solutions) of DCA from the computational viewpoint. We will introduce below a deeper interpretation of DCA that has the merit of offering a valuable insight into DCA.

Denote by h_k (resp. h^k) the following affine (resp. polyhedral convex) minorization of the convex function h defined by:

$$h_k(x) := h(x^k) + \langle x - x^k, y^k \rangle = \langle x, y^k \rangle - h^*(y^k), \forall x \in E$$

$$h^k(x) := \sup\{h_i(x) : i = 0, ..., k\} = \sup\{\langle x, y^i \rangle - h^*(y^i) : i = 0, ..., k\}, \forall x \in E,$$

where the sequences $\{x^k\}$ and $\{y^k\}$ are generated as above, i.e. as solutions respectively of the relaxed convex program (P_k) (resp. (D_k)) which is obtained from the original program (P_{dc}) (resp. (D_{dc})) by replacing the convex function h (resp. g^*) with h_k (resp. the affine minorizationn of g^* given by : $y \longrightarrow g^*(y^k) + \langle x^{k+1}, y - y^k \rangle$).

Since a proper lower semicontinuous convex function $\theta(x)$ is characterized as the supremum of a collection of its affine minorizations, in particular

$$\theta(x) = \sup\{\langle x, y \rangle - \theta^*(y) : y \in E^*\}, \forall x \in E,$$

it seems better, instead of the affine minorization h_k, using the polyhedral convex function h^k to under approximate the convex function h. In other words we are dealing with the following relaxed programs (which still are (nonconvex) d.c. programs as opposed to the relaxed convex programs (P_k))

$$\inf\{g(x) - h^k(x) : x \in E\}. \tag{22}$$

Problem (22), called *polyhedral d.c. program*, is a d.c. program in which at least one of convex d.c. components is polyhedral convex. The special class of polyhedral d.c. programs, which plays a key role in nonconvex optimization, possesses worthy properties, from both theoretical and computational viewpoints, *as necessary and sufficient local optimality conditions, and finite convergence for DCA* (see e.g. [2], [34], [35]).

It naturally leads us to the following crucial questions:

- how to solve the polyhedral d.c. program (22)?

- what exactly is the relationship between the sequence $\{x^k\}$ and solutions to (22)?

Answers to these questions are presented below. First, by writing problem(22) in the form

$$\inf\{g(x) - \sup\{h_i(x) : i = 0, ..., k\} : x \in E\},$$

i.e.,

$$\inf\{g(x) + \inf\{-h_i(x) : i = 0, ..., k\} : x \in E\},$$

or equivalently

$$\inf_{x \in E} \inf_{i=0,...,k}\{g(x) - h_i(x)\},$$

and finally

$$\inf_{i=0,...,k} \inf_{x \in E}\{g(x) - h_i(x)\},$$

and since x^{i+1} is a solution to (P_i), we easily deduce that x^l, with $l \in \arg\min\{g(x^{i+1}) - h_i(x^{i+1}) : i = 0, ..., k\}$, is a solution to problem (22).

On the other hand, we have

$$g(x^{i+1}) - h(x^{i+1}) \le g(x^{i+1}) - h_i(x^{i+1}), \text{ for } i = 0, ..., k.$$

So

$$g(x^{k+1}) - h(x^{k+1}) \le \inf\{g(x^{(i+1)}) - h_i(x^{(i+1)}) : i = 0, ..., k\} \tag{23}$$

since the sequence $\{ g(x^i) - h(x^i)\}$ is decreasing.

It is clear that if equality holds in (23), then x^{k+1} is a solution to the polyhedral d.c. program (22).

We shall now find conditions ensuring such an equality. For this we will distinguish two cases:

(A) There is some k such that $h_k(x^{k+1}) = h(x^{k+1})$. In this case the above equality holds since we have the double inequality

$$\begin{aligned} g(x^{k+1}) - h(x^{k+1}) &\le \inf\{g(x^{i+1}) - h_i(x^{i+1}) : i = 0, ..., k\} \\ &\le g(x^{k+1}) - h(x^{k+1}) \le g(x^{k+1}) - h(x^{k+1}). \end{aligned}$$

According to the main properties above mentioned of the sequences $\{x^k\}$ and $\{y^k\}$ generated by DCA, it is easy to prove that the equality $g(x^{k+1}) - h(x^{k+1}) = g(x^k) - h(x^k)$, *(which occurs when DCA has a finite convergence, especially in polyhedral d.c. programming)*, implies that $h_k(x^{k+1}) = h(x^{k+1})$.

(B) $h_k(x^{k+1}) < h(x^{k+1})$ for every k. In this case we can state the following asymptotic behaviour: Assume the optimal value of problem (P_{dc}) be finite and the sequences $\{x^k\}$ and $\{y^k\}$ generated by DCA be bounded. Then

$$g(x^\infty) - h(x^\infty) = \inf\{g(x^{i+1}) - h_i(x^{i+1}) : i = 0, ..., \infty\}$$

for every limit point x^∞ of the sequence $\{x^k\}$, where h_∞ denotes the affine minorization of h given by

$$h_\infty(x) := h(x^\infty) + \langle x - x^\infty, y^\infty \rangle = \langle x, y^\infty \rangle - h^*(y^\infty), \ \forall x \in E$$

with $y^\infty \in \partial h(x^\infty)$. Consequently the point x^∞ is a solution to the following d.c. program

$$\inf\{g(x) - h^\infty(x) : x \in E\}, \tag{24}$$

where the convex function h^∞ is defined by

$$h^\infty(x) := \sup\{\langle x, y^i \rangle - h^*(y^i) : i = 0, ..., \infty\}, \ \forall x \in E.$$

Similarly, for the dual d.c. program (D_{dc}), the related minorizations of g^* are defined by

$$(g^*)_k(y) := g^*(y^k) + \langle y - y^k, x^{k+1} \rangle = \langle y, x^{k+1} \rangle - g(x^{k+1}), \ \forall y \in E^*,$$

$$(g^*)_\infty(y) := g^*(y^\infty) + \langle y - y^\infty, x^\infty \rangle = \langle y, x^\infty \rangle - g(x^\infty), \ \forall y \in E^*,$$

and

$$(g^*)^\infty(y) := \sup\{(g^*)_i(y) : i = 0, ..., \infty\} = \sup\{\langle y, x^{i+1} \rangle - g(x^{i+1}) : i = 0, ..., \infty\}, \ \forall y \in E^*.$$

These affine functions satisfy, simultaneously with the functions h_i, conditions similar to (A) and (B). To sum up, DCA generates two sequences $\{x^k\}$ and $\{ y^k\}$, (candidates to primal and dual solutions respectively), which are improved at each iteration (the sequences $\{g(x^k) - h(x^k)\}$ and $\{h^*(y^k) - g^*(y^k)\}$ are decreasing) and serve, in turn, to construct affine minorizations of h, (the affine functions h_i), and affine minorizations of g^* (the affine functions $(g^*)_i$) such that:

Theorem 1 *If the optimal value α of the d.c. program (P_{dc}) is finite and the sequences $\{x^k\}, \{y^k\}$ are bounded, then every limit point x^∞ (resp. y^∞) of $\{x^k\}$ (resp. $\{y^k\}$) is a solution to the approximated d.c. program*

$$\inf\{g(x) - h^\infty(x) : x \in E\}$$

and

$$\inf\{h^*(y) - (g^*)^\infty(y) : y \in E^*\}$$

respectively. Consequently, the nearer the function h^∞ (resp. $(g^*)^\infty$) comes to the function h (resp. g^*), the better x^∞ (resp. y^∞) approximates a solution to the d.c. program (P_{dc}) (resp. (D_{dc})). Furthermore if either of the following conditions holds

(i) The functions h^∞ and h coincide at some solution to (P_{dc}),

(ii) The functions $(g^*)^\infty$ and g^* coincide at some solution to (D_{dc}),

then x^∞ and y^∞ are also solutions to (P_{dc}) and (D_{dc}) respectively.

Proof. It is a consequence of the results just mentioned above and Theorem 3 (Appendix). □

In practice, in order to globally solve (P_{dc}), it is particularly interesting to get the function h^∞ tight underapproximate the function h. To this aim, we have to find suitable d.c. decompositions and initial points for DCA according to the specific structure of the d.c. program being considered.

4   Solution of the General Distance Geometry Problem (15) by the DCA

Our interest in the DCA has increased recently motivated by its success to a great deal of various large-scale d.c. programs ([2] - [6], [34] - [36]). The positive aspects of the DCA that come out of numerical solutions of these problems are

- It often converges to a global solution;

- The number of concave variables do not affect the complexity for the algorithm;

- It can be used for large-scale problems at little cost.

- It is particularly suitable to nonsmoothness.

As indicated in the introduction, the general distance geometry problem (5) can be formulated as a global optimization problem, for example, the unconstrained nonconvex optimization problem (6) (with piecewise twice differentiable objective function) due to Crippen and Havel [9], or the linearly constrained nondifferentiable nonconvex optimization problem (15) proposed by ourselves. Both are d.c. programs, however we have chosen to apply DCA to the latter because the construction of the sequences$\{x^k\}$ and $\{y^k\}$ then is quite simple. Indeed, as will be seen in the next, DCA requires only matrix-vector products and one Cholesky factorization and hence is suitable to the large-scale setting.

On the other hand we have studied the problem of computing initial points for DCA applied to (15). Since DCA is very efficient in the solution of the Euclidean EMDS problem with a complete set of dissimilarities([2], [36]), we proposed a method (to compute intial points for DCA applied to (15)) based on DCA and a procedure of completing a given set of dissimilarities. We now use the general scheme (21) to solve Problem (15). Performing this scheme thus is reduced to calculating subdifferentials of the functions H and G^*:

$$(Y^{(k)}, W^{(k)}) \in \partial H(X^{(k)}, T^{(k)}), (X^{(k+1)}, T^{(k+1)}) \in \partial G^*(Y^{(k)}, W^{(k)}). \qquad (25)$$

According to Lemma 1 the sequence $\{X^{(k)}\}$ can be taken in the subspace $\mathcal{A}^\perp$. Likewise, we shall show that (see Proposition 4, Appendix) the sequence $\{Y^{(k)}\}$ is contained in $\mathcal{A}^\perp$ too.

We shall first present the crucial results on the computation of ∂H and ∂G^* which have given rise to our two phase DCA for solving the general distance geometry problem (15).

4.1   Computing ∂H.

We will see in the next that a subgradient of H, say (Y, W), can be determined in explicit form. Remark that Problem (13) only involves the restriction of H to Ω where this function is convex. Actually we shall compute the conditional subdifferential of H with respect to the convex set Ω ([12]) that we again denote by ∂H for simplicity, i.e., (Y^0, W^0) is a conditional subgradient of H at $(X^0, T^0) \in \Omega$ with respect to Ω if

$$H(X,T) \geq H(X(^0,T^0) + \langle (X,T) - (X^0,T^0), (Y^0, W^0) \rangle \quad \text{for } (X,T) \in \Omega.$$

By the very definition, it is easy to state the next inclusion for $(X,T) \in \Omega$

$$\partial H(X,T) \supset \frac{1}{2} \sum_{i<j} p_{ij}[d_{ij}(X) + \varphi_{ij}(T)][\partial d_{ij}(X) \times \{0\} + \{0\} \times \partial \varphi_{ij}(T)].$$

It follows that

Proposition 2 *The subdifferential $\partial H(X,T)$ contains the couples (Y,W) such that $W = (w_{ij})$ is defined by*

$$w_{ij} = \begin{cases} \frac{1}{2} p_{ij}(d_{ij}(X) + t_{ij}) & \text{if } (i,j) \in \mathcal{S}, \\ 0 & \text{if } (i,j) \notin \mathcal{S}, \end{cases}$$

and

$$Y = \frac{1}{2}(B + C(X,T))X, \tag{26}$$

where $B = (b_{ij})$ is the $n \times n$ matrix given by

$$b_{ij} = \begin{cases} -p_{ij} & \text{if } i \neq j \\ -\sum_{k=1,k\neq i}^{n} b_{ik} & \text{if } i = j, \end{cases} \tag{27}$$

and $C(X,T) = (c_{ij}(X,T))$ is the $n \times n$ matrix valued function given by

$$c_{ij}(X,T) = \begin{cases} -p_{ij}t_{ij}s_{ij}(X) & \text{if } i \neq j, \\ -\sum_{k=1,k\neq i}^{n} c_{ik} & \text{if } i = j, \end{cases} \tag{28}$$

with

$$s_{ij}(X) = \begin{cases} 1/(\|X_i^T - X_j^T\|) & \text{if } X_i \neq X_j \text{ and } (i,j) \in \mathcal{S}, \\ 0 & \text{otherwise.} \end{cases} \tag{29}$$

Proof. See Subsection 6.5 (Appendix).

Remark 2 *According to (26) the sequence $\{Y^{(k)}\}$ defined by (25) is contained in the subspace $\mathcal{A}^\perp$ if the ranges of B and $C(X,T)$, for a given couple (X,T), are contained in $\mathcal{A}^\perp$. This condition holds in virtue of Proposition 4 (Appendix).*

4.2 Computing ∂G^*.

As aforementioned, the calculation of $\partial G^*(Y,W)$ consists of solving the next problem of the form (P_k):

$$\min\{G(X,T) - \langle (Y,W),(X,T)\rangle : (X,T) \in \mathcal{M}_{n,p}(\mathbb{R}) \times \mathcal{M}_{n,n}(\mathbb{R}) \}.$$

Since the functions G is separable in its variables, the last problem can be decomposed into the following two problems:

$$\min\left\{\eta(X) - \langle Y,X\rangle := \left[\frac{1}{2}\sum_{i<j}p_{ij}\ d^2_{ij}(X)\right] - \langle Y,X\rangle : X \in \mathcal{M}_{n,p}(\mathbb{R})\right\}, \qquad (30)$$

$$\min\left\{\zeta(T) - \langle W,T\rangle := \left[\frac{1}{2}\sum_{i<j}p_{ij}t^2_{ij}\right] - \langle W,T\rangle : T \in \mathcal{C}\right\}. \qquad (31)$$

Proposition 3 *The subdifferential of G^* at (Y,W) is the set of couples (X,T) such that $T = (t_{ij})$ is the $n \times n$ matrix defined by*

$$t_{ij} = \begin{cases} \frac{1}{2}(d_{ij}(X) + t_{ij}) & \text{if} \quad l_{ij} \le \frac{1}{2}(d_{ij}(X) + t_{ij}) \le u_{ij},\ (i,j) \in \mathcal{S}, \\ l_{ij} & \text{if} \quad \frac{1}{2}(d_{ij}(X) + t_{ij}) < l_{ij},\ (i,j) \in \mathcal{S}, \\ u_{ij} & \text{if} \quad \frac{1}{2}(d_{ij}(X) + t_{ij}) > u_{ij},\ (i,j) \in \mathcal{S}, \end{cases} \qquad (32)$$

and

$$\left(B + \frac{1}{n}ee^T\right)X = Y, \qquad (33)$$

where $B = (b_{ij})$ is the $n \times n$ matrix given by (27).

Proof. See Subsection 6.6 (Appendix).

According to Propositions 2 and 3 and (58) we can now provide the description of the DCA applied to (15) or more exactly, the DCA for solving (16).

4.3 Description of the DCA for solving (16)

Algorithm 1 *(DCA applied to (16)).*

The primal sequence $\{(X^{(k)},T^{(k)})\}$ with $X^{(k)} \in \mathcal{A}^\perp$ is generated as follows:
Let $\varepsilon > 0$, and $X^{(0)} \in \mathcal{A}^\perp \setminus \{0\}, T^{(0)} \in \mathcal{C}$ be given.
For $k = 0,1,\ldots$ until

$$l_{ij} \le \|X_i^{(k)} - X_j^{(k)}\| \le u_{ij}, \quad \text{for all } (i,j) \in \mathcal{S} \qquad (34)$$

set $T^{(k+1)} = (t_{ij}^{(k+1)})$ as follows, according to (32)

$$t_{ij}^{(k+1)} = \begin{cases} \frac{1}{2}(d_{ij}(X^{(k)}) + t_{ij}^{(k)}) & \text{if} \quad l_{ij} \le \frac{1}{2}(d_{ij}(X^{(k)}) + t_{ij}^{(k)}) \le u_{ij},\ (i,j) \in \mathcal{S}, \\ l_{ij} & \text{if} \quad \frac{1}{2}(d_{ij}(X^{(k)}) + t_{ij}^{(k)}) < l_{ij},\ (i,j) \in \mathcal{S}, \\ u_{ij} & \text{if} \quad \frac{1}{2}(d_{ij}(X^{(k)}) + t_{ij}^{(k)}) > u_{ij},\ (i,j) \in \mathcal{S}. \end{cases} \qquad (35)$$

$$t_{ij}^{(k+1)} = 0 \text{ for } (i,j) \notin \mathcal{S} \tag{36}$$

and solve the following nonsingular linear system to obtain $X^{(k+1)}$

$$(B + \frac{1}{n}ee^T)X = \frac{1}{2}(B + C(X^{(k)}, T^{(k)}))X^{(k)} \tag{37}$$

where $B = (b_{ij})$ is the $n \times n$ matrix defined by

$$b_{ij} = \begin{cases} -p_{ij} & \text{if } i \neq j \\ -\sum_{k=1, k\neq i}^{n} b_{ik} & \text{if } i = j, \end{cases}$$

and $C(X,T) = (c_{ij}(X,T))$ is the $n \times n$ matrix valued function given by

$$c_{ij}(X,T) = \begin{cases} -p_{ij}t_{ij}s_{ij}(X) & \text{if } i \neq j, \\ -\sum_{k=1, k\neq i}^{n} c_{ik} & \text{if } i = j, \end{cases}$$

with

$$s_{ij}(X) = \begin{cases} 1/(\|X_i^T - X_j^T\|) & \text{if } X_i \neq X_j \text{ and } (i,j) \in \mathcal{S}, \\ 0 & \text{otherwise.} \end{cases}$$

Remark 3 *(i) According to Proposition 1, if (34) occurs then $F(X^{(k)}, \overline{T} = (\overline{t}_{ij})) = 0$, with $\overline{t}_{ij} = \|X_i^{(k)T} - X_j^{(k)T}\|$.*
(ii) To solve the linear system with positive definite constant matrix (37) one can use only one Cholesky factorization. So Algorithm 1 requires only matrix-vector products and one Cholesky factorizattion.

As indicated above, one of interesting features of the DCA is the nice effect of a *good* starting point. This property motivates us to investigate a technique for computing a "good" starting point $X^{(0)}$ in Algorithm 1, taking into account the specific structure of the general distance geometry problem (16).

4.4   Finding a good starting point for Algorithm 1.

It has been pointed out, in the introduction, that we will use a method based on DCA and the standard procedure of completing the set of constraints due to Crippen ([9], [13]).

First, with the help of the triangle inequality, we complete the set of dissimilarities by either deducing dissimilarities from the given bounds or imposing additional dissimilarities (see procedures C1 and C2 below). We then solve the Euclidean MDS problem (EMDS) given below where all pairwise dissimilarities δ_{ij} are known, and take its solution as $X^{(0)}$.

$$(\text{EMDS}) \quad \min\left\{ \sigma(X) := \sum_{i<j} \left(\|X_i^T - X_j^T\| - \delta_{ij} \right)^2 : \quad X \in \mathcal{M}_{n,p}(\mathbb{R}) \right\}.$$

The idea of this technique comes from two facts:

- There exits an efficient algorithm in d.c. optimization approach [2], [36] for solving the EMDS problem when all pairwise disimilarities are known. This algorithm is very simple, it is explicit form which requires only matrix-vector products, and works very well in practice.

- In the general case where only a small subset of bound constraints are known one can *approximate* a solution to (GEDP) by considering the *resulting* EMDS problem in which all pairwise disimilarities are extrapolated from given constraints.

The algorithm that we use for solving (EMDS) has been studied in [2], [36]. Many versions of DCA (with and without regularization techniques, and/or via the lagrangian duality in d.c. programming) to solve (EMDS) were presented in there. We give below the description of one of these algorithms. For all details about these methods, we refer the reader to [2], [36].

Algorithm 2: *(DCA applied to (EMDS)).*
Let $\varepsilon > 0$, and $0 \neq X^{(0)} \in \mathcal{A}^{\perp} \setminus \{0\}$ be given.
For $k = 0, 1, \dots$ until either $\|X^{(k+1)} - X^{(k)}\| \leq \varepsilon$ or $|\sigma(X^{(k)}) - \sigma(X^{(k+1)})| \leq \varepsilon$ take

$$X^{(k+1)} = \frac{1}{n}\widetilde{C}(X^{(k)})X^{(k)},$$

where $\widetilde{C}(X) = (\widetilde{c}_{ij}(X))$ is the $n \times n$ matrix given as

$$\widetilde{c}_{ij}(X) = \begin{cases} -\delta_{ij}s_{ij}(X) & \text{if } i \neq j, \\ -\sum_{k=1, k \neq i}^{n} \widetilde{c}_{ik} & \text{if } i = j, \end{cases} \tag{38}$$

with $s_{ij}(X)$ defined by

$$s_{ij}(X) = \begin{cases} 1/(\|X_i^T - X_j^T\|) & \text{if } X_i \neq X_j \text{ and } (i,j) \in \mathcal{S}, \\ 0 & \text{otherwise.} \end{cases}$$

This algorithm is not expensive because it requires only matrix-vector products. Moreover it provides often a global solution to (EMDS) ([2], [36]).

To complete the matrix of dissimilarities (δ_{ij}), as indicated in the introduction, we first consider that $u_{ij} = +\infty$, $l_{ij} = -\infty$ for $(i,j) \notin \mathcal{S}$, and then use one of the following procedures:

Procedure C1 *(deducing dissimilarities)* If $(i,j) \notin \mathcal{S}$, then

$$u_{ij} = \min_{k=1,\dots,n} \{u_{ij}, u_{ik} + u_{kj}\}, \qquad l_{ij} = \max_{k=1,\dots,n} \{l_{ij}, l_{ik} - u_{kj}, l_{jk} - u_{ki}\}$$

Knowing two complete matrices $(l_{ij}), (u_{ij})$, the dissimilarity matrix (δ_{ij}) is taken as

$$\delta_{ij} = \frac{l_{ij} + u_{ij}}{2} \quad \text{for all } i, j \in \{1,\dots,n\}.$$

Procedure C2 *(imposing dissimilarities)* If $(i,j) \notin \mathcal{S}$, then

$$u_{ij} = \min_{k=1,\dots,n} \{u_{ij}, u_{ik} + u_{kj}\}, \qquad l_{ij} = \max_{k=1,\dots,n} \{l_{ij}, |l_{ik} - l_{kj}|\}.$$

Knowing two complete matrices $(l_{ij}), (u_{ij})$, the dissimilarity matrix (δ_{ij}) is taken as

$$\delta_{ij} = \frac{l_{ij} + u_{ij}}{2} \quad \text{for all } i, j \in \{1,\dots,n\}.$$

Finally, our two phase DCA for solving *(16)* can be summarized as follows:

4.5   The main algorithm GDCA for solving (16)

Phase 1. Finding an initial point.
  Step 1. *Determining* the dissimilarity matrix (δ_{ij}) by using either Procedure C1 or Procedure C2.
  Step 2. *Solving* Problem (EMDS) by Algorithm 2 to obtain the point denoted $\tilde{X}$.
Phase 2. Solving the original problem (16) by Algorithm 1 from the point $X^{(0)} := \tilde{X}$.

5   Implementation and Computational Experiments

We have implemented our algorithm in FORTRAN and run on a SGI Origin 2000 multiprocessor with IRIX system. We have tested our algorithm on the second model problems from Moré-Wu [23] where the molecule has $n = s^3$ atoms located in the three-dimensional lattice

$$\{(i_1, i_2, i_3) : 0 \le i_1 < s, 0 \le i_2 < s, 0 \le i_3 < s\}$$

for some integer $s \ge 1$. The set $\mathcal{S}$ is specified by $(X_i^T = (i_1, i_2, i_3))$

$$\mathcal{S} = \{(i,j) : \|X_i^T - X_j^T\| \le \sqrt{r}\} \tag{39}$$

with some integer $r \ge 1$.
  The aim of our computational experiments is to show that GDCA is an efficient approach to solve (16) with large dimension. We consider molecules containing up to 3375 atoms (then (16) has 10125 variables).
  We study the efficiency of GDCA on various bounds l_{ij} and u_{ij}. For this reason, we used the same procedure from [25] to generate the given bounds:

$$l_{ij} = (1 - \|X_i^T - X_j^T\|)\epsilon, \quad u_{ij} = (1 + \|X_i^T - X_j^T\|)\epsilon \tag{40}$$

for some $\epsilon \in (0,1)$. We are then able to examine the behavior of the algorithm as ϵ varies over $(0,1)$. As in [25], we varied ϵ over $[0.04, 0.16]$.
  On the other hand, the computational experiments allow studying the effect of the number of given bounds on the performance of our algorithm. We have tested the algorithm on different values of r that vary the cardinality of $\mathcal{S}$.
  Finally, we are also interested in the influence of the starting point on the rate of convergence of Algorithm 1 (Experiment 2).
  We considered $p_{ij} = 1$ for all $i \ne j$ in (GEDP).
  For starting Algorithm 2, we first took a random point X in $(0, s - 1)$ and then set $X^{(0)} = P_{A^\perp}(X)$. We terminated this algorithm when

$$|\sigma(X^{(k+1)}) - \sigma(X^{(k)})| \le 10^{-6}|\sigma(X^{(k+1)})|.$$

In our computational results a matrix $X^* \in \mathcal{M}_{n,3}(\mathbb{R})$ solves the general distance geometry problem (GEDP) if

$$(1 - \tau)l_{ij} \le \|X_i^{*T} - X_j^{*T}\| \le u_{ij}(1 + \tau), \ (i,j) \in \mathcal{S} \tag{41}$$

with some tolerance τ. When $n < 3375$ we used the same tolerance considered in [25], that is $\tau = 0.01$, and when $n = 3375$ we took $\tau = 0.02$. Then the stopping criterion of Algorithm 1 is

$$(1 - \tau)l_{ij} \leq \|X_i^{(k)T} - X_j^{(k)T}\| \leq (1 + \tau)u_{ij}, \ (i, j) \in \mathcal{S}. \tag{42}$$

For solving the linear system (37) in Phase 2 we first decomposed the matrix $2(B + \frac{1}{n}ee^T) = R^T R$ by the Cholesky factorization, and then at each iteration we solved two triangular linear systems $R^T U = (B + C(X^{(k)}, T^{(k)}))X^{(k)}$ and $RX = U$.

In the tables presented below we indicate the following results:

- t0: CPU time for determining δ_{ij}.

- it1 and time1: respectively, the number of iterations and CPU time, of Algorithm 2,

- it2 and time2: respectively, the number of iterations and CPU time, of Phase 2 (Algorithm 1)

- ttotal: the total CPU time of the main algorithm GDCA,

- data: the number of given distances, i.e. $(1/2)|\mathcal{S}|$, where $|\mathcal{S}|$ is the cardinality of $\mathcal{S}$.

  Note that in the complete bound matrices, i.e., $\mathcal{S} = N \times N$, we have data $= n(n-1)$.

All CPU times were computed in seconds.

5.1 Experiment 1

In the first experiment we study the influence of the number of given constraints (when r varies) and of the length of bounds (when ϵ varies) on the performance of Algorithm GDCA. For completing the dissimilarity matrix in Phase 1, we used Procedure C1.

Table 1 gives the computing results (in detail) of the two phase algorithm GDCA to solve 36 problems (s varies from 3 to 15 while $r = 1$, $r = 2$, and $r = s$). To generate the bound constraints we took $\epsilon = 0.04$ in (40).

The curves in Figure 1 show the behavior of GDCA (the total time in seconds) when the length of bounds varies. We consider three cases where $\epsilon = 0.04$, $\epsilon = 0.08$, and $\epsilon = 0.16$. To generate $\mathcal{S}$ we took $r = 1$ in (39).

In Figure 2 we present the behavior of GDCA (the total time in seconds) when the number of given constraints varies. We consider three cases where $r = 1$, $r = 2$, and $r = s$. There the bound constraints are defined with $\epsilon = 0.08$ in (40).

In these figures, the size n of the test problems varies on $[216, 3375]$ (s varies from 6 to 15).

5.2 Experiment 2

In this experiment we study the efficiency of Algorithm GDCA with two different procedures for completing the dissimilarity matrix in Phase 1. In other words, we study the effect of the starting point on the performance of Algorithm 1 applied to (16). For generating the bound constraints we took $\epsilon = 0.16$.

n	r	data	t0	iter1	time1	iter2	time2	ttotal
27	1	54	0.000	54	0.025	5	0.001	0.026
	2	126	0.000	307	0.120	10	0.003	0.124
	s	347	0.000	75	0.029	0	0.000	0.030
64	1	144	0.006	60	0.021	7	0.007	0.146
	2	360	0.006	125	0.274	10	0.012	0.293
	s	1880	0.001	89	0.197	97	0.250	0.448
125	1	300	0.050	91	0.773	14	0.042	0.866
	2	780	0.496	123	1.045	7	0.029	1.123
	s	7192	0.007	102	0.865	170	1.707	2.580
216	1	540	0.265	67	1.782	19	0.167	2.214
	2	1440	0.261	78	2.081	7	0.087	2.429
	s	21672	0.033	84	2.217	243	7.763	10.012
343	1	882	1.364	98	7.060	19	0.590	9.014
	2	2394	1.369	83	5.965	9	0.343	7.677
	s	53799	0.254	108	7.727	406	37.127	45.108
512	1	1344	6.288	77	14.560	18	1.696	22.545
	2	3696	6.321	77	14.542	11	1.213	22.076
	s	119692	1.062	92	17.112	550	133.238	151.413
729	1	1944	21.568	77	34.721	23	6.180	62.469
	2	5400	21.520	88	40.089	13	4.020	65.628
	s	243858	3.182	97	43.506	777	462.851	509.539
1000	1	2700	61.444	70	65.703	33	19.939	147.086
	2	7560	61.653	87	82.269	14	10.559	154.481
	s	456872	8.975	90	83.950	750	928.538	1021.464
1331	1	3630	200.538	77	172.612	41	70.598	443.749
	2	10230	198.051	83	185.684	15	29.840	413.575
	s	809763	25.768	109	213.182	846	2227.612	2466.562
1728	1	4752	445.023	73	279.023	51	149.920	873.966
	2	13464	446.334	81	308.554	17	59.954	814.843
	s	1359216	65.929	95	361.217	1157	5900.815	6327.962
2197	1	6084	1135.899	79	559.327	62	363.544	2058.771
	2	17316	1106.973	100	707.987	17	122.642	1937.603
2744	1	7644	2478.885	83	1071.762	74	890.792	4441.439
	2	21840	2430.332	99	1268.619	18	274.561	3973.511
3375	1	9450	5305.042	68	1440.196	16	666.760	7412.002
	2	27090	5306.743	83	1759.097	22	674.270	7740.112

Table 1: The performance of Algorithm GDCA with $\epsilon = 0.04$.

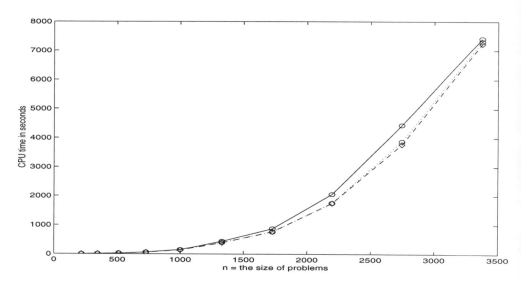

Figure 1: The behavior of GDCA with different lengths of bounds: $\epsilon = 0.04$ (solid), $\epsilon = 0.08$ (dotted), $\epsilon = 0.16$ (dashed)

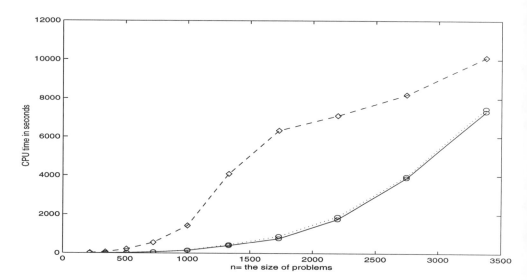

Figure 2: The behavior of GDCA with different numbers of given contraints: $r = 1$ (solid), $r = 2$ (dotted), $r = s$ (dashed)

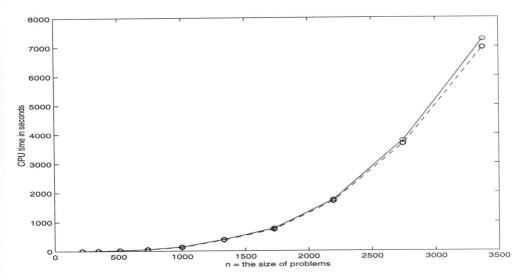

Figure 3: CPU time of GDCA-C1 (solid) and GDCA-C2 (dashed) in case $r = 1$

We denote by GDCA-C1 (resp. GDCA-C2) Algorithm GDCA that uses Procedure C1 (resp. C2) in Phase 1. Table 2 gives the computing results of GDCA-C1 and GDCA-C2 to solve 38 problems (s varies from 3 to 15).

Figure 3 and Figure 4 show, respectively, the total time of GDCA-C1 and GDCA-C2 in the cases where $r = 1$ and $r = s$ (s varies from 6 to 15)).

5.3   Comments

The most important fact is that in all experiments Algorithm GDCA gives a global solution to (GEDP). Moreover, since the basic DCA is efficient, GDCA can solve large-scale problems in a reasonable time (the maximum running time of GDCA-C2 is about 2 hours for a problem with 10125 variables).

About the influence of the length of bounds (when ϵ varies), from Experiment 1 (computational results are not all presented here) in which Procedure C1 is used for determining the dissimilarity matrix in Phase 1 we note that

- In the cases $r = 1$ and/or $r = 2$, the more ϵ increases (i.e., the more the length of bounds increases), the faster GDCA is (except for the cases where $n = 1331$, $n = 1728$ with $r = 2$ for which GDCA is the most expensive when $\epsilon = 0.08$).

- In the case $r = s$, GDCA is the most efficient when $\epsilon = 0.16$, and it is the most expensive when $\epsilon = 0.08$. The ratio of CPU time between these cases goes up to 2.

The influence of the number of given bounds (when r varies) on the behavior of Algorithm GDCA depends upon the procedure for constructing the dissimilarity matrix in Phase 1. If we use Procedure C1 for determining, then

n	r	GDCA-C1				GDCA-C2			
		iter1	iter2	t0	ttotal	iter1	iter2	t0	ttotal
27	1	53	0	0.000	0.021	62	1	0.000	0.025
	2	133	2	0.000	0.053	116	1	0.000	0.046
	s	75	0	0.000	0.029	74	0	0.000	0.029
64	1	59	1	0.007	0.138	72	0	0.006	0.164
	2	95	11	0.006	0.202	104	15	0.005	0.205
	s	94	0	0.001	0.207	86	0	0.001	0.193
125	1	101	1	0.051	0.919	70	0	0.043	0.639
	2	100	5	0.050	0.907	86	17	0.042	0.836
	s	97	1	0.007	0.849	99	0	0.006	0.854
216	1	68	1	0.265	2.122	64	1	0.221	1.978
	2	70	7	0.261	2.228	91	3	0.218	2.720
	s	88	1	0.033	2.426	84	1	0.026	2.356
343	1	99	1	1.382	8.663	67	1	1.138	6.126
	2	80	10	1.376	7.621	80	15	1.148	7.438
	s	76	2	0.256	6.015	113	1	0.213	8.653
512	1	76	1	6.252	21.061	69	1	5.329	18.985
	2	80	11	6.321	22.852	93	8	5.362	24.175
	s	87	229	1.060	72.868	81	1	0.862	17.094
729	1	75	1	21.861	57.510	68	1	19.247	51.863
	2	85	13	22.005	65.592	87	11	19.290	62.848
	s	95	287	3.184	216.226	97	1	2.705	49.044
1000	1	75	1	62.373	136.584	69	1	55.388	124.417
	2	86	14	62.121	153.497	87	16	55.652	150.106
	s	100	734	8.966	1007.941	96	1	7.990	103.298
1331	1	79	7	201.153	394.176	79	8	185.121	378.658
	2	80	16	199.261	407.705	81	19	185.157	400.933
	s	106	1189	26.066	3349.828	107	2	25.744	278.941
1728	1	75	9	450.351	772.104	72	12	418.289	738.542
	2	79	18	446.618	808.542	81	17	415.923	781.178
	s				> 7000	90	2	60.292	424.232
2197	1	84	6	1090.061	1750.941	87	4	1040.691	1702.609
	2	90	19	1099.075	1860.344	94	19	1038.127	1830.544
	s				> 7000	108	3	147.398	961.255
2744	1	81	21	2432.432	3777.915	86	16	2312.027	3667.723
	2	92	26	2472.011	4032.231	96	20	2384.731	3974.735
	s				> 7000	109	134	330.175	4223.827
3375	1	79	6	5319.071	7253.235	74	11	5056.181	6972.817
	2	81	11	5305.273	7362.221	84	13	5132.266	7298.032

Table 2: Compare two procedures for completing the dissimilarity matrix, $\epsilon = 0.16$.

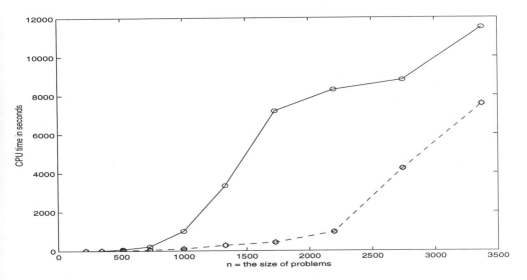

Figure 4: CPU time of GDCA-C1 (solid) and GDCA-C2 (dashed) in case $r = s$

- When $\epsilon = 0.08$ and/or $\epsilon = 0.16$ the more r increases (i.e., the more the number of given distances increases), the more expensive GDCA is (except for the cases where $n = 27$ and $n = 343$).

- When $\epsilon = 0.04$ (Table 1) the last property remains true in the case $n \leq 1000$ (except for the case $n = 343$). When $n > 1000$ and $n = 343$ GDCA is the most efficient with $r = 2$, and it is always the most expensive with $r = s$.

In contrast with the results in Experiment 1, if we use Procedure C2 for determining the dissimilarity matrix (Experiment 2), then GDCA is the most efficient when $r = s$ (it solves the problem of 1728 variables in 424 seconds), and it is the most expensive when $r = 2$ (except the case $n = 343$).

We also observe that the rate of convergence of the DCA in Phase 1 (Algorihm 2) does not seem to depend on data. Algorithm 2 converges after at most 125 iterations (except the case $r = 2$, $n = 27$). In contrast, the DCA in Phase 2 (Algorithm 1) is quite sensitive to data. On the other hand, although the sequence $\{X^{(k)}\}$ in Algorithm 1 is not in explicit form, the cost of one iteration of this algorithm (that contains the cost of the computation of matrices $C(X^{(k)}, T^{(k)})$, $T^{(k+1)}$ and the cost of the solution of two triangular linear systems) is not more expensive than the cost of one iteration of Algorithm 2. This shows that Algorithm 1 exploits well sparsity of S (in the determination of matrices $C(X^{(k)}, T^{(k)})$, $T^{(k+1)}$, and the product $C(X^{(k)}, T^{(k)})X^{(k)}$).

Moreover, from Experiment 1, we observe that when the number of given bounds is small ($r = 1$ and $r = 2$) Phase 1 is much more expensive than Phase 2 (the cost of Phase 1 occupies up to 96.4% of the total cost when $n \geq 1728$). However this phase is indispensable for GDCA to obtain a global solution of (GEDP) in these cases. On the contrary, when the number of given bounds is large ($r = s$), the cost of Phase 2 is the most important. But the last property is no more true in Experiment 2 where Phase 2 is inexpensive.

Finally, from the second experiment we see that Algorithm 1 *is very sensitive to the choice of the starting point*, or in other words, it *depends on the procedure for determining the dissimilarity matrix* in Phase 1. In fact, when $r = s$, the ratio of CPU time of GDCA with two different procedures C1 and C2 goes up to 17. However, when $r = 1$ and $r = 2$ the difference is much less significant: in general, GDCA with Procedure C2 is faster than GDCA with Procedure C1.

5.4   Concluding Remarks

We have proposed a two phase DCA (called GDCA) for solving the general distance geometry problem under a new formulation as a d.c. program. The first phase consists of applying DCA to a complete EMDS problem, the full dissimilarity matrix of which is generated by the triangle bound estimating technique. The second one is the application of DCA to the general distance geometry problem (under the new form of a d.c. program) with an initial point given by Phase 1. Our preliminary computational results suggest that our method is successful in locating the large configurations satisfying given bound-constraints. The DCA actually requires matrix-vector products and only one Cholesky factorization, and allows exploiting sparsity in the large-scale setting. Algorithm GDCA can then be used in the large-scale molecular optimization from the general distance geometry problem.

We wish to find less expensive procedures for computing an acceptable initial point (Phase 1) to improve the efficiency of GDCA and to expand our testing to distance data generated from NMR experiments. We plan to address these issues in future work.

Acknowledgments The authors are grateful to the referees for their helpful comments and suggestions which have improved the presentation of the revised paper.

6   Appendix

We summarize in the Appendix the main results on local and global optimality conditions in d.c. programming and convergence theorem of DCA for general d.c. programs. For more details, see [34], [35] and references therein). We also report herein proofs of results concerning subdifferential calculations of the functions H and G^*.

6.1   Optimality conditions in d.c. programming

Let $\mathcal{P}$ and $\mathcal{D}$ denote the solution sets of problems (P_{dc}) and (D_{dc}), respectively, and let
$$\mathcal{P}_l = \{x^* \in E : \partial h(x^*) \subset \partial g(x^*)\}, \quad \mathcal{D}_l = \{y^* \in F : \partial g^*(y^*) \subset \partial h^*(y^*)\}.$$
We state first the following fundamental result on d.c. programming which constitute the basis of DCA presented in Subsection 3.1.

Theorem 2 *([2], [34]) (i)* $\cup\{\partial h(x) : x \in \mathcal{P}\} \subset \mathcal{D} \subset dom\, h^*$.
*The first inclusion becomes equality if g^* is subdifferentiable in $\mathcal{D}$ (in particular if $\mathcal{D} \subset ri(dom\, g^*)$ or if g^* is subdifferentiable in $dom\, h^*$). In this case $\mathcal{D} \subset (dom\, \partial g^* \cap dom\, \partial h^*)$.
(ii) If x^* is a local minimum of $g - h$, then $x^* \in \mathcal{P}_l$.*

(iii) Let x^* be a critical point of $g - h$ and $y^* \in \partial g(x^*) \cap \partial h(x^*)$. Let U be a neighbourhood of x^* such that $(U \cap \text{dom } g) \subset \text{dom } \partial h$. If for any $x \in U \cap \text{dom } g$ there is $y \in \partial h(x)$ such that $h^*(y) - g^*(y) \geq h^*(y^*) - g^*(y^*)$ then x^* is a local minimum of $g - h$. More precisely,

$$g(x) - h(x) \geq g(x^*) - h(x^*), \quad \forall x \in U \cap \text{dom } g.$$

(iv) Let $x^* \in \text{dom } \partial h$ be a local minimum of $g - h$ and let $y^* \in \partial h(x^*)$ (i.e., $\partial h(x^*)$ is nonempty and x^* admits a neighbourhood U such that $g(x) - h(x) \geq g(x^*) - h(x^*), \quad \forall x \in U \cap \text{dom } g$.) If

$$y^* \in \text{int } (\text{dom } g^*) : \quad \text{and} \quad \partial g^*(y^*) \subset U, \tag{43}$$

then y^* is a local minimum of $h^* - g^*$ ((43) holds if g^* is differentiable at y^*).

6.2   The complete DCA for general d.c.   programs

In the description of DCA given in Section 3, x^{k+1} (resp. y^k) is arbitrarily chosen in $\partial g^*(y^k)$ (resp. $\partial h(x^k)$). Since they are candidates to primal and dual solutions respectively, it is natural to impose the following choice

$$x^{k+1} \in \arg\min\{g(x) - h(x) : x \in \partial g^*(y^k)\} \tag{44}$$

and

$$y^k \in \arg\min\{h^*(y) - g^*(y) : y \in \partial h(x^k)\} \tag{45}$$

Problems (44) and (45) are equivalent to convex maximization problems (46) and (47) respectively

$$x^{k+1} \in \arg\min\{\langle x, y^k \rangle - h(x) : x \in \partial g^*(y^k)\} \tag{46}$$

$$y^k \in \arg\min\{\langle x^k, y \rangle - g^*(y) : y \in \partial h(x^k)\} \tag{47}$$

The resulting DCA is called complete DCA. It allows approximating a couple (x^*, y^*). It can be viewed as a sort of decomposition approach of the primal and dual problems (P_{dc}), (D_{dc}). From a practical point of view, although problems (44) and (45) are simpler than (P_{dc}), (D_{dc}) (we work in $\partial h(x^{k+1})$ and $\partial g^*(y^k)$ with convex maximization problems), they remain nonconvex programs and thus still hard to solve. In practice, except the cases where the convex maximization problems (46) and (47) are easy to solve, one generally uses the former DCA (presented in Subsection 3.1), which is also called the simplified DCA.

6.3   Convergence of the DCA for general d.c.   programs

Let ρ_i and ρ_i^*, $(i = 1, 2)$ be real nonnegative numbers such that $0 \leq \rho_i < \rho(f_i)$ (resp. $0 \leq \rho_i^* < \rho(f_i^*)$) where $\rho_i = 0$ (resp. $\rho_i^* = 0$) if $\rho(f_i) = 0$ (resp. $\rho(f_i^*) = 0$) and ρ_i (resp. ρ_i^*) may take the value $\rho(f_i)$ (resp. $\rho(f_i^*)$) if it is attained in (17). We next set $f_1 = g$ and $f_2 = h$.
Also let $dx^k := x^{k+1} - x^k$ and $dy^k := y^{k+1} - y^k$.
    The basic convergence theorem of DCA for general d.c. programming will be stated below. Its proof is very technical and long ([2], [35]) and is omitted here. Main outlines of a like proof can be found in [35], and [34] which deals with the DCA for solving the trust region problem.

Theorem 3 *Suppose that the sequences $\{x^k\}$ and $\{y^k\}$ are defined by the DCA. Then we have*

(i) $$(g-h)(x^{k+1}) \le (h^*-g^*)(y^k) - \max\{\frac{\rho_2}{2}\|dx^k\|^2, \frac{\rho_2^*}{2}\|dy^k\|^2\} \le (g-h)(x^k)$$

$$-\max\{\frac{\rho_1+\rho_2}{2}\|dx^k\|^2, \frac{\rho_1^*}{2}\|dy^{k-1}\|^2 + \frac{\rho_2}{2}\|dx^k\|^2, \frac{\rho_1^*}{2}\|dy^{k-1}\|^2 + \frac{\rho_2^*}{2}\|dy^k\|^2\}.$$

The equality $(g-h)(x^{k+1}) = (g-h)(x^k)$ holds if and only if $x^k \in \partial g^*(y^k), y^k \in \partial h(x^{k+1})$ and $(\rho_1+\rho_2)dx^k = \rho_1^* dy^{k-1} = \rho_2^* dy^k = 0$. In this case
* $(g-h)(x^{k+1}) = (h^*-g^*)(y^k)$ and x^k, x^{k+1} are the critical points of $g-h$ satisfying $y^k \in (\partial g(x^k) \cap \partial h(x^k))$ and $y^k \in (\partial g(x^{k+1}) \cap \partial h(x^{k+1}))$,
* y^k is a critical point of $h^* - g^*$ satisfying $[x^k, x^{k+1}] \subset ((\partial g^*(y^k) \cap \partial h^*(y^k))$,
* $x^{k+1} = x^k$ if $\rho(g) + \rho(h) > 0, y^k = y^{k-1}$ if $\rho(g^*) > 0$ and $y^k = y^{k+1}$ if $\rho(h^*) > 0$.

(ii) *Similarly, for the dual problem we have*

$$(h^*-g^*)(y^{k+1}) \le (g-h)(x^{k+1}) - \max\{\frac{\rho_1}{2}\|dx^{k+1}\|^2, \frac{\rho_1^*}{2}\|dy^k\|^2\} \le (h^*-g^*)(y^k)$$

$$-\max\{\frac{\rho_1}{2}\|dx^{k+1}\|^2 + \frac{\rho_2}{2}\|dx^k\|^2, \frac{\rho_1^*}{2}\|dy^k\|^2 + \frac{\rho_2}{2}\|dx^k\|^2, \frac{\rho_1^*+\rho_2^*}{2}\|dy^k\|^2\}.$$

The equality $(h^*-g^*)(y^{k+1}) = (h^*-g^*)(y^k)$ holds if and only if $x^{k+1} \in \partial g^*(y^{k+1}), y^k \in \partial h(x^{k+1})$ and $(\rho_1^*+\rho_2^*)dy^k = \rho_2 dx^k = \rho_1 dx^{k+1} = 0$. In this case
* $(h^*-g^*)(y^{k+1}) = (g-h)(x^{k+1})$ and y^k, y^{k+1} are the critical points of $h^* - g^*$ satisfying $x^{k+1} \in (\partial g^*(y^k) \cap \partial h^*(y^k))$ and $x^{k+1} \in (\partial g^*(y^{k+1}) \cap \partial h^*(y^{k+1}))$,
* x^{k+1} is a critical point of $g - h$ satisfying $[y^k, y^{k+1}] \subset ((\partial g(x^{k+1}) \cap \partial h(x^{k+1}))$,
* $y^{k+1} = y^k$ if $\rho(g^*) + \rho(h^*) > 0, x^{k+1} = x^k$ if $\rho(h) > 0$ and $x^{k+1} = x^{k+2}$ if $\rho(g) > 0$.

(iii) *If α is finite, then*

* *the decreasing sequences $\{(g-h)(x^k)\}$ and $\{(h^*-g^*)(y^k)\}$ converge to the same limit $\beta \ge \alpha$, i.e., $\lim_{k\to+\infty}(g-h)(x^k) = \lim_{k\to+\infty}(h^*-g^*)(y^k) = \beta$.*

* *If $\rho(g) + \rho(h) > 0$ (resp. $\rho(g^*) + \rho(h^*) > 0$), then the series $\{\|x^{i+1} - x^i\|^2\}$ (resp. $\{\|y^{i+1} - y^i\|^2\}$) converges. More precisely we have in this case*

$$\frac{\rho_1+\rho_2}{2}\sum_{i=0}^{+\infty}\|x^{i+1} - x^i\| \le (g-h)(x^0) - \alpha$$

$$(resp. \frac{\rho_1^*+\rho_2^*}{2}\sum_{i=0}^{+\infty}\|y^{i+1} - y^i\| \le (h^*-g^*)(y^0) - \alpha).$$

(iv) *If α is finite and the sequences $\{x^k\}$ and $\{y^k\}$ are bounded, then for every limit point x^∞ of $\{x^k\}$ (resp. y^∞ of $\{y^k\}$) there exists a limit point y^∞ of $\{y^k\}$ (resp. x^∞ of $\{x^k\}$) such that*

* $(x^\infty, y^\infty) \in [\partial g^*(y^\infty) \cap \partial h^\infty(y^\infty)] \times [\partial g(x^*) \cap \partial h(x^*)]$ and $(g-h)(x^\infty) = (h^*-g^*)(y^\infty) = \beta$,
* $\lim_{k\to+\infty}\{g(x^k) + g^*(y^k)\} = \lim_{k\to+\infty}\langle x^k, y^k \rangle$.

Comments on Theorem 3.

(i) Properties (i) and (ii) prove that the DCA is a descent method for both primal and dual programs. DCA provides critical points for (P_{dc}) and (D_{dc}) after finitely many operations if there is no strict decrease of the primal (or dual) objective function.

(ii) If C and D are convex sets such that $\{x^k\} \subset C$ and $\{y^k\} \subset D$ then Theorem 3 remains valid if we replace $\rho(f_i)$ by $\rho(f_i, C)$ and $\rho(f_i^*)$ by $\rho(f_i^*, D)$ for $i = 1, 2$. By this way we may improve the results in the theorem.

(iii) In (ii) of Theorem 3, the convergence of the whole sequence $\{x^k\}$ (resp. $\{y^k\}$) can be ensured under the following conditions ([26], [29]):

- $\{x^k\}$ is bounded;
- The set of limit points of $\{x^k\}$ is finite;
- $\lim_{k\to+\infty} \|x^{k+1} - x^k\| = 0$.

(iv) In general, the qualities (robustness, stability, rate of convergence and globality of sought solutions) of the DCA depend upon the d.c. decomposition of the function f. Theorem 3 shows how strong convexity of d.c. components in primal and dual problems can influence on the DCA. To make the d.c. components (of the primal objective function $f = g - h$) strongly convex we usually apply the following process

$$f = g - h = (g + \frac{\lambda}{2}\| \cdot \|^2) - (h + \frac{\lambda}{2}\| \cdot \|^2).$$

In this case the d.c. components in the dual problem will be differentiable.
In the same way, inf-convolution of g and h with $\frac{\lambda}{2}\| \cdot \|^2$ will make the d.c. components (in dual problem) strongly convex and the d.c. components of the primal objective function differentiable. For a detailed study of regularization techniques in d.c. programming, see [2], [33], [35].

(v) The only difference between the simplified DCA and the complete DCA lies on the choice of y^k in $\partial h(x^k)$ and x^{k+1} in $\partial g^*(y^{(k)})$. The convergence result of the complete DCA is thus improved: in Theorem 3, the nonemptiness of a subdifferential intersection is replaced by a subdifferential inclusion ([2], [33], [35]). In other words, the complete DCA permits to obtain a couple of elements $(x^*, y^*) \in \mathcal{P}_l \times \mathcal{D}_l$. In general, DCA converges to a local solution, however we observed from our numerous experiments that DCA converges quite often to a glogal one (see e.g. [2] - [6], [34] - [36]).

(vi) It has been proved in [2], [3], [35] that the DCA is finite for polyhedral d.c. programming which is a d.c. program where either f or g is a polyhedral convex function.

6.4   Well-definiteness of the DCA Applied to General D.C. Programs.

The DCA is well defined if one can construct two sequences $\{x^k\}$ and $\{y^k\}$ as above from an arbitrary initial point $x^0 \in \text{dom } g$. We have $x^{k+1} \in \partial g^*(y^k)$ and $y^k \in \partial h(x^k) : \forall k \geq 0$. So $\{x^k\} \subset \text{range } \partial g^* = \text{dom } \partial g$ and $\{y^k\} \subset \text{range } \partial h = \text{dom } \partial h^*$. Then it is clear that

Lemma 2 *Sequences* $\{x^k\}, \{y^k\}$ *in the DCA are well defined if and only if*

$$dom \, \partial g \subset dom \, \partial h, \quad dom \, \partial h^* \subset dom \, \partial g^*.$$

Since for $\theta \in \Gamma_0(E)$, we have ([37])

$$ri(\text{dom } g) \subset \text{dom } \partial g \subset \text{dom } g$$

we can say, under the essential assumption (18), that DCA is in general well defined.

6.5   The proof of Proposition 2

Since

$$\varphi_{ij}(T) = t_{ij} = \langle T, E_{ij} \rangle_{\mathcal{M}_{n,n}(\mathbb{R})},$$

with $E_{ij} = e_i e_j^T$ ($e_i \in \mathbb{R}^n$ is the unit vector with value one in the ith component and zero otherwise), φ_{ij} is differentiable on $\mathcal{M}_{n,n}(\mathbb{R})$, and $\nabla \varphi_{ij}(T) = E_{ij}$. Hence $\partial H(X, T)$ contains the couples (Y, W) defined by

$$(Y, W) = \frac{1}{2} \sum_{i<j} p_{ij} d_{ij}(X) + \varphi_{ij}(T))(Y(i, j), E_{ij}),$$

with $Y(i, j) \in \partial d_{ij}(X)$, i.e.

$$Y = \frac{1}{2} \sum_{i<j} p_{ij} d_{ij}(X) Y(i, j) + \frac{1}{2} \sum_{i<j} p_{ij} \varphi_{ij}(T) Y(i, j),$$

and

$$W = \frac{1}{2} \sum_{i<j} p_{ij} (d_{ij}(X) + \varphi_{ij}(T)) E_{ij}.$$

Hence $W = (w_{ij})$ is defined by

$$w_{ij} = \begin{cases} \frac{1}{2} p_{ij}(d_{ij}(X) + t_{ij}) & \text{if } (i, j) \in \mathcal{S}, \\ 0 & \text{if } (i, j) \notin \mathcal{S}. \end{cases}$$

For determining Y, we first compute $Z := \sum_{i<j} p_{ij} d_{ij}(X) Y(i, j)$. By using the row-representation of $\mathcal{M}_{n,p}(\mathbb{R})$, d_{ij} can be expressed as :

$$d_{ij} = \|.\| \circ \phi_{ij} : (\mathbb{R}^p)^n \longrightarrow \mathbb{R}^p \longrightarrow \mathbb{R}, \ X \longmapsto \phi_{ij}(X) = X_i^T - X_j^T \longmapsto \|X_i^T - X_j^T\|,$$

we have ([17], [37])

$$\partial d_{ij}(X) = \phi_{ij}^T \partial(\|.\|)(\phi_{ij}(X)).$$

Hence

$$Y(i, j) \in \partial d_{ij}(X) \Leftrightarrow Y(i, j) = \phi_{ij}^T y, \quad y \in \partial(\|.\|)(X_i^T - X_j^T),$$

which implies

$$Y(i, j)_k^T = 0 \text{ if } k \notin \{i, j\} \text{ and } Y(i, j)_i^T = -Y(i, j)_j^T \in \partial(\|.\|)(X_i^T - X_j^T). \quad (48)$$

Then $Y(i, j)$ can be chosen as

$$Y(i, j)_i = -Y(i, j)_j = \begin{cases} \frac{X_i - X_j}{\|X_i^T - X_j^T\|} & \text{if } X_i \neq X_j, \\ 0 & \text{otherwise.} \end{cases} \quad (49)$$

And so

$$Z_k := \sum_{i<j} p_{ij}d_{ij}(X)Y(i,j)_k = \sum_{i<k} p_{ik}d_{ik}(X)Y(i,k)_k + \sum_{j>k} p_{kj}d_{kj}(X)Y(k,j)_k$$

$$= \sum_{i<k} p_{ik}(X_k - X_i) + \sum_{j>k} p_{jk}(X_k - X_j) = \left[\sum_{i=1}^n p_{ik}\right]X_k - \sum_{i=1}^n p_{ik}X_i.$$

It follows that

$$Z = \sum_{i<j} p_{ij}d_{ij}(X)Y(i,j) = BX, \tag{50}$$

where $B = (b_{ij})$ is the $n \times n$ matrix given by

$$b_{ij} = \begin{cases} -p_{ij} & \text{if } i \neq j \\ -\sum_{k=1,k\neq i}^n b_{ik} & \text{if } i = j. \end{cases}$$

By the same way, we get

$$\sum_{i<j} p_{ij}\varphi_{ij}(T)Y(i,j) = C(X,T)X, \tag{51}$$

where $C(X,T) = (c_{ij}(X,T))$ is the $n \times n$ matrix valued function given by

$$c_{ij}(X,T) = \begin{cases} -p_{ij}t_{ij}s_{ij}(X) & \text{if } i \neq j, \\ -\sum_{k=1,k\neq i}^n c_{ik} & \text{if } i = j, \end{cases}$$

with

$$s_{ij}(X) = \begin{cases} 1/(\|X_i^T - X_j^T\|) & \text{if } X_i \neq X_j \text{ and } (i,j) \in \mathcal{S}, \\ 0 & \text{otherwise.} \end{cases}$$

Finally, we have

$$Y = \frac{1}{2}(B + C(X,T))X. \quad \Box \tag{52}$$

We are now proving that the sequence $\{Y^{(k)}\}$ defined by (25) is contained in the subspace $\mathcal{A}^\perp$. According to (52) it suffices to prove that the ranges of B and C(X,T), for a given couple (X,T), are contained in $\mathcal{A}^\perp$.

Proposition 4 *(1) Let B be the matrix defined by (27). Then*

(i) B is positive semidefinite, $\nabla\eta(X) = BX$ and $\eta(X) = \frac{1}{2}\langle X, BX \rangle$.

(ii) Furthermore if the weight matrix P is irreducible, then

$$\mathcal{A} = \eta^{-1}(0) = \{X \in \mathcal{M}_{n,p}(\mathbb{R}) : BX = 0\}, \tag{53}$$

rank $B = n - 1$ and $\mathcal{A}^\perp = \{Y = BX : X \in \mathcal{M}_{n,p}(\mathbb{R})\} = \{Y = B^+X : X \in \mathcal{M}_{n,p}(\mathbb{R})\}$.

(2) Let $T = (t_{ij})$ be a given matrix in the set $\mathcal{C}$ given by (14). Then the function ξ_T defined on $\mathcal{M}_{n,p}(\mathbb{R})$ by

$$\xi_T(X) := \sum_{i<j} p_{ij} t_{ij} d_{ij}(X) \tag{54}$$

is a seminorm such that $C(X,T)X \in \partial \xi_T(X)$ whose kernel contains the subspace $\mathcal{A}$ and

$$\xi_T(X) = \langle X, C(X,T)X \rangle, \forall X \in \mathcal{M}_{n,p}(\mathbb{R}). \tag{55}$$

Furthermore, for every $X \in \mathcal{M}_{n,p}(\mathbb{R})$, the symmetric matrix $C(X,T)$ is positive semidefinite and its range is contained in the subspace $\mathcal{A}^{\perp}$.

Proof. (i) The positive semidefiniteness of B comes from [44]. According to the preceding computation of Z and the relations (48), (49) and (50), the function η is differentiable, and

$$\nabla \eta(X) = \sum_{i<j} p_{ij} \; d_{ij}(X) \partial d_{ij}(X) = BX.$$

Hence from Euler's relation follows $\eta(X) = \frac{1}{2}\langle X, BX \rangle$.
(ii) The first equality of (53) is immediate from Lemma 1. If $BX = 0$, then $\eta(X) = \frac{1}{2}\langle X, BX \rangle = 0$, so $\{X \in \mathcal{M}_{n,p}(\mathbb{R}) : BX = 0\} \subset \eta^{-1}(0)$. Conversely, if $\eta(X) = 0$, then $\langle X, BX \rangle = 0$, i.e., $BX = 0$ because B is semidefinite positive matrix. Hence the second equality of (53) holds.
Denote by $\mathcal{B}$ the symmetric matrix having p diagonal blocks and each block is B. Obviously, $\mathcal{N}(\mathcal{B})^{\perp} = \text{Im}(\mathcal{B})$, and for $X, Y \in \mathcal{M}_{n,p}(\mathbb{R})$ one has, using the column-identification (12)

$$\mathcal{B}\overline{\mathcal{X}} = 0 \Leftrightarrow BX^k = 0, \quad k = 1, \ldots, p \quad \Leftrightarrow BX = 0, \tag{56}$$

$$\overline{\mathcal{Y}} = \mathcal{B}\overline{\mathcal{X}} \Leftrightarrow Y^k = BX^k, k = 1, \ldots, p \quad \Leftrightarrow Y = BX. \tag{57}$$

It follows that

$$\{Y = BX : X \in \mathcal{M}_{n,p}(\mathbb{R})\} = \{X \in \mathcal{M}_{n,p}(\mathbb{R}) : BX = 0\}^{\perp}.$$

As for rank B, we have rank $B \leq n - 1$ because $Be = 0$ with $e \in \mathbb{R}^n$ being the vector of ones. On the other hand if rank $B < n - 1$, then there exists $u \notin \mathbb{R}e$ such that $Bu = 0$. Let $X = uv^T$ with $v \in \mathbb{R}^p \setminus \{0\}$. Clearly $BX = 0$, i.e., $X \in \mathcal{A}$ in virtue of (53). It implies $u \in \mathbb{R}e$ since the rows of X are all identical. This contradiction proves that rank $B = n-1$.
  (2) It is clear that the function ξ_T is a seminorm on $\mathcal{M}_{n,p}(\mathbb{R})$ whose kernel contains the subspace $\mathcal{A}$. Like the matrix B, the positive semidefiniteness of the matrix C(X, T), for every $X \in \mathcal{M}_{n,p}(\mathbb{R})$, comes from [44]. The preceding computations in Subsection 4.1 leading to formulations (51) and (28) show that, for a given matrix $T = (t_{ij}) \in \mathcal{C}$, $C(X,T)X \in \partial \xi_T(X)$ for every $X \in \mathcal{M}_{n,p}(\mathbb{R})$. The extended Euler's relation (54) for positively homogeneous convex functions is then immediate. Finally since $C(X,T)e = 0$, according to Lemma 1, the range of $C(X,T)$ is contained in the subspace $\mathcal{A}^{\perp}$. $\square$

Remark 4 Under the assumptions (a1),(a2) and (a3) of Section 2, one can prove, as in Lemma 1, that $\mathcal{A}^{\perp}$ is exactly the range of $C(X,T)$.

6.6   The Proof of Proposition 3

It is easy to see that the solutions of (31) can be explicitly determined. Indeed, by definition of $W = (w_{ij})$ we have

$$\zeta(T) - \langle W, T \rangle = \frac{1}{2} \sum_{i<j,(i,j)\in\mathcal{S}} p_{ij}t_{ij}^2 - \sum_{i,j=1}^{n} w_{ij}t_{ij}$$

$$= \frac{1}{2} \sum_{i<j,(i,j)\in\mathcal{S}} p_{ij}t_{ij}^2 - \frac{1}{2} \sum_{i<j,(i,j)\in\mathcal{S}} p_{ij}(d_{ij}(X) + t_{ij})t_{ij}.$$

The following result allows computing explicit solutions to (31).

Lemma 3 *Let M be a subset of $\{1, ..., m\}$. Let a_i, b_i, d_i and f_i be real numbers such that $a_i \le b_i$ and $d_i > 0$ for $i \in M$. Then the solution set of the following convex program in $\mathbb{R}^m$*

$$\min\{\frac{1}{2} \sum_{i\in M} d_i u_i^2 - \sum_{i\in M} f_i u_i : a_i \le u_i \le b_i, : i \in M\}$$

is $\{u : u_i = \frac{1}{d_i} f_i$ if $d_i a_i \le f_i \le d_i b_i$, a_i if $f_i \le d_i a_i$, and b_i if $f_i \ge d_i b_i\}$.

Proof. Since the objective function is separable in its variables $u_i, i \in M$ and the constraint set is a box, a vector $v = (v_1, ..., v_m)$ is a solution to the preceding problem if and ony if, for each $i \in M, v_i$ is a solution to the following one-dimensional convex quadratic optimization problem

$$\min\{\frac{1}{2} d_i u_i^2 - f_i u_i : a_i \le u_i \le b_i\}$$

i.e.

$$v_i = \frac{1}{d_i} f_i \text{ if } d_i a_i \le f_i \le d_i b_i, \; a_i \text{ if } f_i \le d_i a_i, \text{ and } b_i \text{ if } f_i \ge d_i b_i.$$

The proof then is complete.   □

According to Lemma 3 the solution set of Problem (31) is determined by: $T^* = (t_{ij}^*)$ is a solution to this problem if and only if t_{ij} is arbitrary for $(i, j) \notin S$ and

$$t_{ij}^* = \begin{cases} \frac{1}{2}(d_{ij}(X) + t_{ij}) & \text{if} & l_{ij} \le \frac{1}{2}(d_{ij}(X) + t_{ij}) \le u_{ij}, \; (i,j) \in \mathcal{S}, \\ l_{ij} & \text{if} & \frac{1}{2}(d_{ij}(X) + t_{ij}) < l_{ij}, \; (i,j) \in \mathcal{S}, \\ u_{ij} & \text{if} & \frac{1}{2}(d_{ij}(X) + t_{ij}) > u_{ij}, \; (i,j) \in \mathcal{S}. \end{cases}$$

Only the components $t_{ij}^{(k+1)}$ with $(i, j) \in \mathcal{S}$ actually intervene in Problem (15), so we set $T^{(k+1)} = (t_{ij}^{(k+1)})$ as follows:

$$t_{ij}^{(k+1)} = 0 \text{ for } (i,j) \notin \mathcal{S} \quad \text{and} \quad t_{ij}^{(k+1)} = t_{ij}^* \text{ for } (i,j) \in \mathcal{S}. \tag{58}$$

It remains to solve (30). Clearly X^* is a solution of (30) if and only if $Y \in \partial\eta(X^*)$. So, according to Proposition 4,

$$Y = BX^*.$$

Hence, solving (30) amounts to computing the pseudo-inverse of B denoted B^+. The next result permits to compute B^+:

Lemma 4 *Let Λ be an $n \times n$ symmetrix matrix and $a \in \mathbb{R}^n$, $a \neq 0$ such that $\Lambda a = 0$. Then $\Lambda + \frac{1}{\|u\|^2} aa^T$ is nonsingular if and only if rank $\Lambda = n - 1$.*
In this case for every $y \in \mathbb{R}^n$ there exists $x \in Im\,\Lambda$ satisfying $\Lambda x = P_{Im\,\Lambda} y$ and

$$(\Lambda + \frac{1}{\|a\|^2} aa^T)(x + \frac{a^T y}{\|a\|^2} a) = y \quad i.e., \Lambda^+ = (\Lambda + \frac{1}{\|a\|^2} aa^T)^{-1} - \frac{1}{\|a\|^2} aa^T.$$

Proof. If $(\Lambda + \frac{1}{\|a\|^2} aa^T)$ is nonsingular, then $Im\left(\Lambda + \frac{1}{\|a\|^2} aa^T\right) = \mathbb{R}^n$. Thus

$$Im\,\Lambda + Im\,\left(\frac{1}{\|a\|^2} aa^T\right) = \mathbb{R}^n = Im\,\Lambda + \mathcal{N}(\Lambda).$$

This implies $\mathcal{N}(\Lambda) = Im\,(\frac{1}{\|a\|^2} aa^T) = \mathbb{R}a$. Thus rank $\Lambda = $ n-1.
Conversely, if rank $\Lambda = n - 1$, then $\mathcal{N}(\Lambda) = \mathbb{R}a$. Let $1x \in \mathbb{R}^n$ such that

$$(\Lambda + \frac{1}{\|a\|^2} aa^T)x = 0, \text{i.e.,} \Lambda x = -\frac{1}{\|a\|^2} aa^T x.$$

This implies $\Lambda x = 0$ and $a^T x = 0$. Hence $x = 0$ and we can deduce that $\mathcal{N}(\Lambda + \frac{1}{\|a\|^2} aa^T) = \{0\}$, i.e., $\Lambda + \frac{1}{\|a\|^2} aa^T$ is nonsingular. In this case the projection on $\mathcal{N}(\Lambda) = \mathbb{R}a$ is given by $P_{\mathcal{N}(\Lambda)} = (1/\|a\|^2)aa^T$ and

$$P_{Im\,\Lambda} = I - \frac{1}{\|a\|^2} aa^T. \tag{59}$$

Let y be an arbitrary vector in $\mathbb{R}^n$. Since $P_{Im\,\Lambda}(y)$ belongs to $Im\,\Lambda$, there exists $\bar{x} \in \mathbb{R}^n$ such that $\Lambda \bar{x} = P_{Im\,\Lambda}(y)$. The decomposition $\mathbb{R}^n = Im\,\Lambda + \mathcal{N}(\Lambda)$ insures the existence of $x \in Im\,\Lambda, x_1 \in \mathcal{N}(\Lambda)$ such that $\bar{x} = x + x_1$ and $\Lambda \bar{x} = \Lambda x = P_{Im\,\Lambda}(y)$. Observing $\Lambda a = 0, a^T x = 0$ (since $x \in Im\,\Lambda$) we have from (59):

$$(\Lambda + \frac{1}{\|a\|^2} aa^T)(x + \frac{u^T y}{\|a\|^2} a) = \Lambda x + \frac{1}{\|a\|^2} aa^T \frac{a^T y}{\|a\|^2} a = P_{Im\,\Lambda} y + \frac{a^T y}{\|u\|^2} a = y. \tag{60}$$

This implies

$$x = (\Lambda + \frac{1}{\|a\|^2} aa^T)^{-1} y - \frac{a^T y}{\|a\|^2} a.$$

Therefore

$$\Lambda^+ = (\Lambda + \frac{1}{\|a\|^2} aa^T)^{-1} - \frac{1}{\|a\|^2} aa^T. \quad \square \tag{61}$$

The matrix B satisfies the assumptions of Lemma 4, then using (61) we have, for $Y \in \mathcal{M}_{n,p}(\mathbb{R})$,

$$B^+ Y = \left(B + \frac{1}{n} ee^T\right)^{-1} Y - \frac{1}{n} ee^T Y.$$

That implies, for $Y \in \mathcal{A}^\perp$,

$$X = B^+ Y = \left(B + \frac{1}{n} ee^T\right)^{-1} Y, \tag{62}$$

i.e.,

$$\left(B + \frac{1}{n} ee^T\right) X = Y. \quad \square$$

References

[1] ABDO Y. ALFAKIH, A. KHANDANI & H. WOLKOWICZ, *An interior-point method for the Euclidean distance matrix completion problem*, Research Report CORR 97-9, University of Waterloo, Waterloo, Ontario N2L 3G1, Canada.

[2] LE THI HOAI AN, *Contribution à l'optimisation non convexe et l'optimisation globale: Théorie, Algorithmes et Applications* , Habilitation à Diriger des Recherches, Université de Rouen, Juin 1997.

[3] LE THI HOAI AN and PHAM DINH TAO (1997), *Solving a class of linearly constrained indefinite quadratic problems by D.c. algorithms*, Journal of Global Optimization, **11** pp. 253-285.

[4] LE THI HOAI AN and PHAM DINH TAO (1998) *A Branch-and-Bound method via D.C. Optimization Algorithm and Ellipsoidal technique for Box Constrained Nonconvex Quadratic Programming Problems*, Journal of Global Optimization **13**(1998), pp. 171-206.

[5] LE THI HOAI AN and PHAM DINH TAO (1999) *A continuous approach for large-scale linearly constrained quadratic zero-one programming*, To appear in Optimization.

[6] LE THI HOAI AN, PHAM DINH TAO and L.D. MUU (1996), *Numerical solution for Optimization over the efficient set by D.c. Optimization Algorithm*, Operations Research Letters, 19, pp. 117-128.

[7] L.M. BLUMENTHAL, *Theory and Applications of Distance Geometry*, Oxford University Press, 1953.

[8] G.M. CRIPPEN, *Rapid calculation of coordinates from distance measures*, J. Computational Physics, Vol 26 (1978), pp. 449-452.

[9] G.M. CRIPPEN & T.F. HAVEL, *Distance Geometry and Molecular Conformation*, John Wiley & Sons, 1988

[10] J. DE LEEUW, *Applications of convex analysis to multidimensional scaling*, Recent developments in Statistics, J.R. Barra *et al.*, editors, North-Holland Publishing Company, pp. 133-145, 1977.

[11] J. DE LEEUW, *Convergence of the majorization method for multidimansional scaling*, Journal of Classification, Vol 5 (1988), pp. 163-180.

[12] V.F. DEMYANOV & L.V. VASILEV, *Nondifferentiable optimization*, Optimization Software, Inc. Publications Division, New York, (1985)

[13] T.F. HAVEL, *An evaluation of computational strategies for use in the determination of protein structure from distance geometry constraints obtained by nuclear magnetic resonance*, Prog. Biophys. Mol. Biol., 56 (1991), pp. 43-78.

[14] W. GLUNT, T.L. HAYDEN & M. RAYDAN, *Molecular Conformation from distance matrices*, J. Comp. Chem., 14 (1993), pp. 114-120.

[15] L. GUTTMAN, *A general nonmetric technique for finding the smallest coordinate space for a configuration of point*, Psychometrika, Vol 33 (1968), pp. 469-506.

[16] B.A. HENDRICKSON, *The molecule problem: Exploiting structure in global optimization*, SIAM J. Optimization, 5(1995), pp. 835-857.

[17] J.B. HIRIART URRUTY and C. LEMARECHAL, *Convex Analysis and Minimization Algorithms*, Springer-Verlag Berlin Heidelberg, 1993.

[18] P.J. LAURENT, *Approximation et Optimisation*, Hermann, Paris, 1972.

[19] P. MAHEY, NGUYEN VAN HIEN & PHAM DINH TAO, *Proximal techniques for convex programming*, Technical Report, Lab ARTEMIS, IMAG, CNRS-Grenoble, 1992.

[20] P. MAHEY and PHAM DINH TAO, *Partial regularization of the sum of two maximal monotone operators*, Math. Modell. Numer. Anal. M^2 *AN*, Vol. 27 (1993), pp. 375-395.

[21] P. MAHEY and PHAM DINH TAO, *Proximal decomposition of the graph of maximal monotone operator*, SIAM J. Optim. 5 (1995), pp. 454-468.

[22] M. LAURENT, *Cuts, matrix completions and a graph rigidity*, Mathematical Programming, Vol. 79 (1997), Nos 1-3, pp. 255-283.

[23] J.J. MORE & Z. WU, *Global continuation for distance geometry problems*, SIAM J. Optimization, 8(1997), pp. 814-836.

[24] J.J. MORE & Z. WU, *Issues in large-scale Molecular Optimization*, preprint MCS-P539-1095, Argonne National Laboratory, Argonne, Illinois 60439, Mars 1996.

[25] J.J. MORE & Z. WU, *Distance geometry optimization for protein structures*, preprint MCS-P628-1296, Argonne National Laboratory, Argonne, Illinois 60439, December 1996.

[26] A.M. OSTROWSKI, *Solutions of equations and systems of equations*, Academic Press, New York, 1966.

[27] PHAM DINH TAO, Eléments homoduaux d'une matrice A relatifs à un couple de normes (ϕ,ψ). Applications au calcul de $S_{\phi\psi}(A)$, Séminaire d'Analyse Numérique, Grenoble, n°236,1975

[28] PHAM DINH TAO, Calcul du maximum d'une forme quadratique définie positive sur la boule unité de la norme du maximum, Séminaire d'Analyse Numérique, Grenoble, n°247,1976

[29] PHAM DINH TAO, *Contribution à la théorie de normes et ses applications à l'analyse numérique*, Thèse de Doctorat d'Etat Es Science, Université Joseph Fourier- Grenoble, 1981.

[30] PHAM DINH TAO, Convergence of subgradient method for computing the bound-norm of matrices, Linear Alg. and Its Appl., Vol 62 (1984), pp. 163-182.

[31] PHAM DINH TAO, *Algorithmes de calcul d'une forme quadratique sur la boule unité de la norme maximum*, Numer. Math., Vol 45 (1985), pp. 377-440.

[32] PHAM DINH TAO, *Algorithms for solving a class of non convex optimization problems. Methods of subgradients*, Fermat days 85. Mathematics for Optimization, Elsevier Science Publishers B.V. North-Holland, (1986), pp. 249-271.

[33] PHAM DINH TAO, *Duality in d.c. (difference of convex functions) optimization. Subgradient methods*, Trends in Mathematical Optimization, International Series of Numer Math. Vol 84 (1988), Birkhauser, pp. 277-293.

[34] PHAM DINH TAO and LE THI HOAI AN, *D.c. optimization algorithms for the trust region problem.* SIAM J. Optimization, Vol. 8, No 2, (1998), pp. 476-505.

[35] PHAM DINH TAO and LE THI HOAI AN, Convex analysis approach to d.c. programming: Theory, Algorithms and Applications (dedicated to Professor Hoang Tuy on the occasion of his 70th birthday), *Acta Mathematica Vietnamica*, Vol. 22, No 1, 1997, pp 289-355.

[36] PHAM DINH TAO and LE THI HOAI AN, D.C. Programming approach and solution algorithm to the Multidimensional Scaling Problem, *To appear.*

[37] R.T. ROCKAFELLAR, *Convex Analysis*, Princeton University, Princeton, 1970.

[38] R.T. ROCKAFELLAR, *Monotone operators and the proximal point algorithm*, SIAM J. Control and Optimization, Vol.14, N^o5 (1976)

[39] J. B. SAXE, *Embeddability of Weighted Graphs in k-space is Strongly NP-hard*, Proc. 17 Allerton Conference in Communications, Control and Computing, 1979, pp. 480-489.

[40] J.F. TOLAND, *On subdifferential calculus and duality in nonconvex optimization*, Bull. Soc. Math. France, Mémoire 60 (1979), pp. 177-183.

[41] H. TUY, *A general deterministic approach to global optimization via d.c. programming*, Fermat Days 1985: Mathematics for Optimization, North-Holland, Amsterdam, (1986), pp. 137-162.

[42] H. TUY, *Introduction to Global Optimization*, Les Cahiers du GERAD, Groupe d'Etudes et de Recherche en Analyse des Décision, Montréal, Québec, 1994.

[43] H. TUY, *Convex Analysis and Global Optimization*, Kluwer Academic Publishers, 1998.

[44] R. VARGA *Matrix iterative analysis*, Prentice Hall, 1962.

[45] Z. ZOU, RICHARD.H. BIRD, & ROBERT B. SCHNABEL, *A Stochastic/Pertubation Global Optimization Algorithm for Distance Geometry Problems*, J. of Global Optimization, 11(1997), pp. 91-105.

Nonconvex Optimization and Its Applications

1. D.-Z. Du and J. Sun (eds.): *Advances in Optimization and Approximation*. 1994
    ISBN 0-7923-2785-3
2. R. Horst and P.M. Pardalos (eds.): *Handbook of Global Optimization*. 1995
    ISBN 0-7923-3120-6
3. R. Horst, P.M. Pardalos and N.V. Thoai: *Introduction to Global Optimization* 1995
    ISBN 0-7923-3556-2; Pb 0-7923-3557-0
4. D.-Z. Du and P.M. Pardalos (eds.): *Minimax and Applications*. 1995
    ISBN 0-7923-3615-1
5. P.M. Pardalos, Y. Siskos and C. Zopounidis (eds.): *Advances in Multicriteria Analysis*.
    1995   ISBN 0-7923-3671-2
6. J.D. Pintér: *Global Optimization in Action*. Continuous and Lipschitz Optimization:
    Algorithms, Implementations and Applications. 1996   ISBN 0-7923-3757-3
7. C.A. Floudas and P.M. Pardalos (eds.): *State of the Art in Global Optimization*.
    Computational Methods and Applications. 1996   ISBN 0-7923-3838-3
8. J.L. Higle and S. Sen: *Stochastic Decomposition*. A Statistical Method for Large
    Scale Stochastic Linear Programming. 1996   ISBN 0-7923-3840-5
9. I.E. Grossmann (ed.): *Global Optimization in Engineering Design*. 1996
    ISBN 0-7923-3881-2
10. V.F. Dem'yanov, G.E. Stavroulakis, L.N. Polyakova and P.D. Panagiotopoulos: *Quasi-differentiability and Nonsmooth Modelling in Mechanics, Engineering and Economics*. 1996   ISBN 0-7923-4093-0
11. B. Mirkin: *Mathematical Classification and Clustering*. 1996 ISBN 0-7923-4159-7
12. B. Roy: *Multicriteria Methodology for Decision Aiding*. 1996 ISBN 0-7923-4166-X
13. R.B. Kearfott: *Rigorous Global Search: Continuous Problems*. 1996
    ISBN 0-7923-4238-0
14. P. Kouvelis and G. Yu: *Robust Discrete Optimization and Its Applications*. 1997
    ISBN 0-7923-4291-7
15. H. Konno, P.T. Thach and H. Tuy: *Optimization on Low Rank Nonconvex Structures*.
    1997   ISBN 0-7923-4308-5
16. M. Hajdu: *Network Scheduling Techniques for Construction Project Management*.
    1997   ISBN 0-7923-4309-3
17. J. Mockus, W. Eddy, A. Mockus, L. Mockus and G. Reklaitis: *Bayesian Heuristic Approach to Discrete and Global Optimization*. Algorithms, Visualization, Software, and Applications. 1997   ISBN 0-7923-4327-1
18. I.M. Bomze, T. Csendes, R. Horst and P.M. Pardalos (eds.): *Developments in Global Optimization*. 1997   ISBN 0-7923-4351-4
19. T. Rapcsák: Smooth Nonlinear Optimization in R^n. 1997   ISBN 0-7923-4680-7
20. A. Migdalas, P.M. Pardalos and P. Värbrand (eds.): *Multilevel Optimization: Algorithms and Applications*. 1998   ISBN 0-7923-4693-9
21. E.S. Mistakidis and G.E. Stavroulakis: *Nonconvex Optimization in Mechanics*.
    Algorithms, Heuristics and Engineering Applications by the F.E.M. 1998
    ISBN 0-7923-4812-5

Nonconvex Optimization and Its Applications

22. H. Tuy: *Convex Analysis and Global Optimization*. 1998   ISBN 0-7923-4818-4
23. D. Cieslik: *Steiner Minimal Trees*. 1998   ISBN 0-7923-4983-0
24. N.Z. Shor: *Nondifferentiable Optimization and Polynomial Problems*. 1998
     ISBN 0-7923-4997-0
25. R. Reemtsen and J.-J. Rückmann (eds.): *Semi-Infinite Programming*. 1998
     ISBN 0-7923-5054-5
26. B. Ricceri and S. Simons (eds.): *Minimax Theory and Applications*. 1998
     ISBN 0-7923-5064-2
27. J.-P. Crouzeix, J.-E. Martinez-Legaz and M. Volle (eds.): *Generalized Convexitiy,
     Generalized Monotonicity: Recent Results*. 1998   ISBN 0-7923-5088-X
28. J. Outrata, M. Kočvara and J. Zowe: *Nonsmooth Approach to Optimization Problems
     with Equilibrium Constraints*. 1998   ISBN 0-7923-5170-3
29. D. Motreanu and P.D. Panagiotopoulos: *Minimax Theorems and Qualitative Proper-
     ties of the Solutions of Hemivariational Inequalities*. 1999   ISBN 0-7923-5456-7
30. J.F. Bard: *Practical Bilevel Optimization*. Algorithms and Applications. 1999
     ISBN 0-7923-5458-3
31. H.D. Sherali and W.P. Adams: *A Reformulation-Linearization Technique for Solving
     Discrete and Continuous Nonconvex Problems*. 1999   ISBN 0-7923-5487-7
32. F. Forgó, J. Szép and F. Szidarovszky: *Introduction to the Theory of Games*. Concepts,
     Methods, Applications. 1999   ISBN 0-7923-5775-2
33. C.A. Floudas and P.M. Pardalos (eds.): *Handbook of Test Problems in Local and
     Global Optimization*. 1999   ISBN 0-7923-5801-5
34. T. Stoilov and K. Stoilova: *Noniterative Coordination in Multilevel Systems*. 1999
     ISBN 0-7923-5879-1
35. J. Haslinger, M. Miettinen and P.D. Panagiotopoulos: *Finite Element Method for
     Hemivariational Inequalities*. Theory, Methods and Applications. 1999
     ISBN 0-7923-5951-8
36. V. Korotkich: *A Mathematical Structure of Emergent Computation*. 1999
     ISBN 0-7923-6010-9
37. C.A. Floudas: *Deterministic Global Optimization: Theory, Methods and Applications*.
     2000   ISBN 0-7923-6014-1
38. F. Giannessi (ed.): *Vector Variational Inequalities and Vector Equilibria*. Mathemat-
     ical Theories. 1999   ISBN 0-7923-6026-5
39. D.Y. Gao: *Duality Principles in Nonconvex Systems*. Theory, Methods and Applica-
     tions. 2000   ISBN 0-7923-6145-3
40. C.A. Floudas and P.M. Pardalos (eds.): *Optimization in Computational Chemistry
     and Molecular Biology*. Local and Global Approaches. 2000   ISBN 0-7923-6155-5

KLUWER ACADEMIC PUBLISHERS – DORDRECHT / BOSTON / LONDON

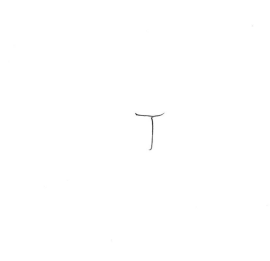